ELECTRONIC COMPOSITES

T0192559

Electronic composites, whose properties can be controlled by thermal or electro-magnetic means, play an important role in micro- and nano-electromechanical systems (MEMS/NEMS) such as sensors, actuators, filters and switches. This book describes the processing, simulation, and applications of electronic composites, beginning with a review of their mechanical, thermal, electromagnetic, and coupling behavior; their major applications are then discussed. All the key simulation models are described in detail and illustrated by reference to real examples. The book closes with a discussion of electronic-composite processing, including MEMS design and packaging. It contains a comprehensive list of references and is aimed at graduate students of electrical engineering, mechanical engineering and materials science. It will also be a useful reference for researchers and engineers in the electronic packaging and MEMS industry.

MINORU TAYA has been a professor of mechanical engineering, and Adjunct Professor of Materials Science and Engineering and Electrical Engineering at the University of Washington (UW) since 1986, and he is currently Boeing–Pennell Professor of Engineering at UW. He is also currently running a Center for Intelligent Materials and Systems (CIMS) as Director, where he is involved in supervising a number of projects. He is a fellow of the American Society of Mechanical Engineers (ASME) and also of the American Academy of Mechanics, serving as an associate editor and editorial board member for several journals. He has also served as the Chair of the Materials Division of ASME in the past. Professor Taya has published over 235 journal and conference papers, co-authored a monograph, and edited five books.

ELECTRONIC COMPOSITES

MINORU TAYA

University of Washington, USA

CAMBRIDGE
UNIVERSITY PRESS

CAMBRIDGE UNIVERSITY PRESS
Cambridge, New York, Melbourne, Madrid, Cape Town, Singapore, São Paulo

Cambridge University Press
The Edinburgh Building, Cambridge CB2 8RU, UK

Published in the United States of America by Cambridge University Press, New York

www.cambridge.org
Information on this title: www.cambridge.org/9780521841740

First published 2005
This digitally printed version 2008

A catalogue record for this publication is available from the British Library

ISBN 978-0-521-84174-0 hardback
ISBN 978-0-521-05731-8 paperback

To My Family, Makiko, Ken, Satoko and Michio

Contents

Preface

The subject of "electronic composites" is both old and new. Electromagnetic properties (particularly dielectric constant and electric conductivity) of electronic composites have been studied extensively since the time of James Maxwell in the nineteenth century, while electronic composites are key materials for microelectronics that today include computer packages, actuators, sensors and micro-electromechanical systems (MEMS), nano-electromechanical systems (NEMS) and BioMEMS.

The aim of this book is to provide readers with an introductory knowledge of various models that can relate the parameters of nanostructure and/or microstructure of the constituent materials to the overall properties of the electronic composites. The readers that the author wishes to reach are graduate students and engineers who are interested in and/or involved in designing microelectronic packages, actuators, sensors, and MEMS/NEMS/BioMEMS. To determine the optimum micro- and nanostructure of electronic composites, a knowledge of the modeling of electronic composites necessarily precedes processing of the composites. This book provides a summary of such modeling. The contents of this book are introductory in early chapters (1–3) and more comprehensive in later chapters (4–8). To help readers who want in-depth knowledge, the book contains detailed appendices and a long list of references.

The author wrote a paper on "Micromechanics modeling of electronic composites" (Taya, 1995) and has been teaching "Electronic composites" as a graduate course at the University of Washington since 1998; the contents of the book originate from his lecture notes. There are no textbooks on electronic composites, presumably because electronic composites are strongly interdisciplinary, covering a wide variety of subjects. The author was motivated to edit his lecture notes into this book by including more recent subjects related to electronic composites.

Chapter 1 introduces a definition of electronic composites and their early study during the nineteenth and twentieth centuries. Chapter 2 discusses various types of electronic composites that are used in current applications: electronic packaging and MEMS, BioMEMS, sensors and actuators, and also the control of electromagnetic waves. In Chapter 3, the foundation is given of the basic equations that govern the physical behavior of electronic composites, ranging from thermomechanical to electromagnetic behavior. Chapter 4 is a key chapter of the book, discussing in detail modeling of electronic composites based on effective medium theory, which ranges from the rudimentary theory of the law of mixtures to the more rigorous Eshelby model of coupled behavior such as piezoelectricity. Chapters 5 and 6 discuss the resistor network model and the percolation model, respectively, which are effective for cases that cannot be modeled by the effective medium theory, providing a nanostructure–macro-behavior relation for electronic composites. Chapter 7 discusses the lamination model, which is simple but effective in estimating the overall behavior of electronic composites with laminated microstructure associated often with a number of modern microelectronic devices. This chapter includes design of (i) piezoelectric actuators for bending mode, (ii) thermal stress analysis in a thin film on a substrate, and (iii) electromagnetic wave propagation in laminated composites with two examples: switchable window and surface plasmon resonance. The final chapter (Chapter 8) opens with a discussion on selected engineering problems associated with the processing of electronic composites, to acquaint readers with current processing routes, ranging from lithography to deposition of organic films. This is followed by an account of standard measurements of key thermophysical properties and electromigration.

In all chapters, both index formulations with subscripts and symbolic notations are used. The former is sometimes needed for readers to grasp the exact relations in the governing equations while the latter provides readers with a simple expression, yet leading to matrix formulation, the most convenient form for numerical calculations.

Acknowledgements

Colleagues and collaborators

The author acknowledges past and present graduate students and research associates who contributed to the major part of the subject of the book. Among past graduate students, Dr. Marty Dunn, currently at the University of Colorado, deserves a special acknowledgement as he worked extensively on the framing of the Eshelby model as applied to inclusion problems in a coupled field in an electroelastic medium. Dr. Yozo Mikata (Lockheed–Martin) also deserves a similar acknowledgement as he provided the details of the Eshelby tensors for piezoelectric composites, which are detailed in Appendix A. The author's thanks go also to Dr. Woo-Jin Kim (General Electric Infrastructure Sensing) and Dr. N. Ueda (Kureha Chemical Co. Ltd, Japan) who helped to formulate and calculate the models based on the resistor network and percolation. Among collaborating researchers, Dr. H. Hatta, Institute of Aeronautical Science, Japan, deserves a similar recognition as he and the author were the first to extend the framework of the Eshelby model from elasticity problems to uncoupled thermophysical problems. Special acknowledgement goes to Professor Yasuo Kuga, Electrical Engineering department, University of Washington, who provided a summary of his lecture notes on the fundamentals of electromagnetic waves, which are incorporated as a part of Chapter 3. Professor Kuga also helped proofread the sections related to electromagnetic waves. Some of my recent graduate students (Dr. Jong-Jin Park, Mr. Hsiu-Hung Chen, Dr. Marie Le Guilly, Mr. Hiroshi Suzuki and Mr. Chiyuan Wang) contributed to drawing figures and helped access journal papers and webpages. Professor S. K. Lee, visiting scholar, has contributed to the Fresnel formulation of electromagnetic wave propagation applied to surface plasmon resonance, outlined in Section 7.6.2.

The author also acknowledges the following funding agencies, which supported the subject areas of electronic composites over the last 10 years. They are US Government agencies: NSF (National Science Foundation), Air Force Office of Scientific Research (AFOSR), Darpa, Office of Naval Research (ONR); Japanese Government agency: New Energy Development Organization (NEDO); and US and Japanese companies: Boeing, Honeywell, Intel, Toray, Honda, and Hitachi Powder Metals.

Finally, the author is very thankful to his secretaries, Ms. Jennifer Ha and Mr. Samuel J. Yi, who typed all chapters of the book and made extensive corrections. Without their continuous efforts, this book would not have been finished. He is also very thankful to Dr. Brian Watts for his helpful copy-editing of the manuscript.

Publishers and authors

The author would like to acknowledge permissions to reproduce the figures and photographs used in this book, granted by the following publishers:

American Institute of Physics
American Society of Mechanical Engineers
Annual Reviews
American Ceramic Society
Elsevier Ltd
The Institute of Electrical and Electronic Engineers, Inc.
Japan Institute of Metals
John Wiley & Sons
Kluwer/Plenum Publishers
Materials Research Society
The McGraw-Hill Companies
Pearson Education
Sage Publications
Society for Experimental Mechanics, Inc.
Taylor and Francis Ltd

He expresses special thanks to the following people who sent original photographs and figures:

Professor Bontae Han, University of Maryland, College Park
Dr. L. T. Kubota, Instituto de Quimica, Brazil
Professor J. F. Li, Tsingha University, China
Dr. Y. Matsuhisa, Toray, Japan

Dr. M. Matsumoto, Nippon Telegraph and Telecommunication, Japan
Professor Dennis Polla, University of Minnesota
Mr. K. Saijo, Toyokohan, Japan
Professor K. Sasagawa, Hirosaki University, Japan
Professor Ronald A. Sigel, University of Minnesota
Dr. J. J. Sniegowski and Dr. M. P. deBoer, Saudia National Laboratories
Dr. Y. Sutou, Tohoku University, Sendai, Japan
Ms. T. Takei, Mitsubishi Electric Corporation, Sagamihara, Japan

1

Introduction

1.1 What is an electronic composite?

"Composite" is a well-accepted word, generally referring to structural components with enhanced mechanical performance. There are a number of textbooks and review articles on these types of composites (e.g., Kelly, 1973; Tsai and Hahn, 1980; Hull, 1981; Chawla, 1987; Clyne and Withers, 1993; Daniel and Ishai, 1994; Gibson, 1994; Hull and Clyne, 1996).

Historically, composites with enhanced mechanical performance have been in existence from ancient-Egyptian time, *c.* 2000 BC, when bricks were made of mud, soil and straw (*Exodus*, Chapter 5, verse 7). Structural composites are designed primarily to enhance the mechanical properties of a matrix material by introducing reinforcement; the primary mechanical properties to be enhanced are strength, stiffness, and fracture resistance.

Normally, a composite consists of a matrix material and one kind of filler, but sometimes more than one kind of filler can be used, forming a "hybrid composite". Depending on the matrix material, one can group composites into three basic types: polymer matrix composites (PMCs), metal matrix composites (MMCs), and ceramic matrix composites (CMCs). Among these, PMCs are the most popular type for electronic composites due to their low processing temperatures and associated cost-effectiveness.

An "electronic composite" is defined as a composite that is composed of at least two different materials and whose function is primarily to exhibit electromagnetic, thermal, and/or mechanical behavior while maintaining structural integrity. Thus, "electronic" should not be interpreted narrowly as referring only to electronic behavior, but instead be understood in much broader terms, including physical and coupling behavior. In this sense, electronic composites are distinguished from structural composites whose primary function is to enhance mechanical properties.

Among various applications, electronic composites have been extensively used as the major component materials in electronic packaging: printed circuit boards (PCBs), thermal interface materials (TIMs), encapsulants, etc., most of which are polymer-based composites providing ease of fabrication and cost-effectiveness. As an extension of electronic packaging, electronic composites are used now in micro-electromechanical systems (MEMS) and BioMEMS, where their functions are multi-fold: active, sensing and housing materials.

The properties of electronic composites can be tailored to meet specific applications. Thus, prediction of the composite properties at an early stage of designing electronic composites is a key task. Normally the composite property is expected to fall between those of the matrix and filler, following the law-of-mixtures type formula, and depends greatly highly on the microstructural parameters: volume fraction, filler shape and size distributions; the properties of the matrix and filler; and also the properties of the matrix–filler interface. Sometimes, the property of an electronic composite becomes quite different from those of the matrix material and filler, and is far from that based on the law-of-mixture type formula. Such a unique property of the composite can be designed purposely or found accidentally in the course of development of a functional composite; it is termed a "cross product" exhibiting "coupling behavior" (Newnham *et al.*, 1978). Since coupling behavior between various physical properties is included in the definition of electronic composites, composites with such coupling behavior are often referred to as "smart composites" or "multi-functional materials," which are the key materials systems for use in smart structures and devices ranging from bio-micro-electromechanical systems (BioMEMS) through sensors to actuators. Therefore, construction of accurate models for the macro-property–microstructure (or –nanostructure) relation is strongly desired. These models are multi-scale, i.e., covering nano-, micro-, meso-, and macro-levels. If these models at different scale levels are interconnected smoothly, one can establish a hierarchical model which will be useful for many scientists and engineers who want to design new smart (or intelligent) materials, MEMS and BioMEMS devices, and multi-functional structures. The main body of this book is devoted to presenting a number of such models. In the remainder of this chapter, we shall review earlier models of electronic composites.

1.2 Early modeling of electronic composites

Modeling of electronic composites in the nineteenth century and early part of the twentieth century focused on the prediction of the dielectric constant ε and electrical conductivity σ or resistivity ρ of a composite composed of spherical fillers with ε_f, σ_f (ρ_f) and matrix with ε_m and σ_m (ρ_m) where the subscripts f and m denote filler and matrix, respectively. Landauer (1978) made an extensive literature survey of early models for the electrical conductivity of composites that were proposed in the nineteenth century through to the mid twentieth century. We shall review some of the early models used to predict the electromagnetic properties of composites, i.e., (1) Lorentz sphere problem, (2) demagnetization in a ferromagnetic body, and (3) concepts of thermal, electric, and magnetic circuits. The first two provide the background for modeling based on the effective medium theory, the last for the resistor network model. Both models will be discussed in detail in later chapters.

1.2.1 Lorentz sphere problem

Consider a dielectric material with dielectric constant ε which is subjected to a uniform electric field **E**, Fig.1.1(a). At a macroscopic level, the dielectric material is considered to consist of a uniform electric dipole moment with polarization **P** (per unit volume). At the atomic level, one can find a free space with ε_0 dielectric constant between lattice points (atoms) or molecules that constitute the dielectric medium. If we consider a spherical domain of radius r_0 between the atoms or molecules, Fig. 1.1(b), a layer of electric charges (positive and negative) is distributed on the inner surface of the

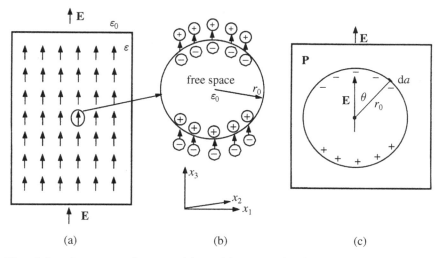

Fig. 1.1 Lorentz sphere problem: (a) macro-level model in a dielectric material subjected to uniform electric field resulting in uniform polarization **P**, (b) nano-level model where free space is polarized by pairs of positive and negative charges. (c) Lorentz sphere problem to idealize (b). (After Ishimaru, 1991, with permission from Pearson Education, Inc.)

sphere, which is called the "Lorentz sphere", Fig. 1.1(c). The net (or total) electric field $\mathbf{E}^t$ in the sphere is expected to be larger than the applied field by an amount $\mathbf{E}_p$, i.e.,

$$\mathbf{E}^t = \mathbf{E} + \mathbf{E}_p. \tag{1.1}$$

We shall compute the magnitude of $\mathbf{E}^t$ using a model developed by four pioneers in the modeling of electric composites, Mossotti (1850), Clausius (1879), Lorenz (1880) and Lorentz (1909), which is summarized by Ishimaru (1991).

The magnitude of the charge P_r in the radial direction on the surface element da is given by

$$P_r = \mathbf{P} \cdot \mathbf{da} = P\cos\theta \ \mathrm{d}a, \tag{1.2a}$$

where

$$\mathrm{d}a = 2\pi r \sin\theta \ r \ \mathrm{d}\theta, \tag{1.2b}$$

and P is the magnitude of $\mathbf{P}$.

This charge P_r induces an electric field dE_p in the radial direction given by the following formula

$$\mathrm{d}E_p = \frac{P_r}{4\pi r_0^2 \varepsilon_0}. \tag{1.3}$$

Thus, the magnitude of the electric field, E_p, in the x_3-direction, contributed by the layer of distributed charge on the entire inner surface of the sphere, is obtained as

$$
\begin{aligned}
E_p &= \int_s \frac{P_r \cos\theta}{4\pi r_0^2 \varepsilon_0} \mathrm{d}a \\
&= \int_0^\pi \frac{P\cos^2\theta}{4\pi r_0^2 \varepsilon_0} 2\pi r_0 \sin\theta \ r_0 \mathrm{d}\theta = \frac{P}{2\varepsilon_0} \int_0^\pi \cos^2\theta \ \sin\theta \ \mathrm{d}\theta \\
&= \frac{P}{3\varepsilon_0}.
\end{aligned}
\tag{1.4a}
$$

Since the components of $\mathbf{E}_p$ along the x_1- and x_2-directions are zero, the result of Eq. (1.4a) can be written in vector form as

$$\mathbf{E}_p = \frac{\mathbf{P}}{3\varepsilon_0}. \tag{1.4b}$$

Macroscopically, the dielectric medium is governed by

$$\mathbf{D} = \varepsilon_0 \mathbf{E} + \mathbf{P}, \tag{1.5}$$

where $\mathbf{D}$ is the electric flux density vector (coulombs [C]/m^2). The first term in Eq. (1.5) is the flux density in free space (i.e., if there were no atoms, or molecules) and the second term is the electric polarization vector resulting from electric dipole moments that exist in the dielectric medium. The electric flux density vector in a medium with dielectric constant ε is also proportional to the applied electric field, i.e.,

$$\mathbf{D} = \varepsilon \mathbf{E}. \tag{1.6}$$

For isotropic materials, $\mathbf{P}$ is proportional to the field $\mathbf{E}$ as

$$\mathbf{P} = \varepsilon_0 \chi_e \mathbf{E}, \tag{1.7}$$

where χ_e is the electric susceptibility. From Eqs. (1.5)–(1.7)

$$\begin{aligned} \varepsilon &= \varepsilon_0(1 + \chi_e) \\ \text{or} \quad \varepsilon_r &= \frac{\varepsilon}{\varepsilon_0} = 1 + \chi_e, \end{aligned} \tag{1.8}$$

where ε_r is the relative dielectric constant (non-dimensional) and ε_0 is the dielectric constant of free space, see Appendix B1.

At the microscopic level, the polarization vector $\mathbf{P}$ is composed of a number N of elemental dipole moments $\mathbf{p}$ which are in turn considered to be proportional to the net local field $\mathbf{E}^t$ in the sphere, i.e.,

$$\mathbf{P} = N\mathbf{p} = N\alpha \mathbf{E}^t, \tag{1.9}$$

where α is the polarizability. From Eqs. (1.1), (1.4), and (1.9), we have

$$\begin{aligned} \mathbf{E}^t &= \mathbf{E} + \frac{\mathbf{P}}{3\varepsilon_0} \\ &= \mathbf{E} + \frac{N\alpha \mathbf{E}^t}{3\varepsilon_0}. \end{aligned} \tag{1.10}$$

Equation (1.10) provides the relation between $\mathbf{E}^t$ and $\mathbf{E}$,

$$\mathbf{E}^t = \frac{\mathbf{E}}{\left(1 - \dfrac{N\alpha}{3\varepsilon_0}\right)}, \tag{1.11}$$

which can be rewritten in terms of ε_r by using Eqs. (1.7), (1.8) and (1.10):

$$\mathbf{E}^t = \frac{(\varepsilon_r + 2)}{3} \mathbf{E}. \tag{1.12}$$

Equating the right-hand side of Eq. (1.11) to that of Eq. (1.12) and using Eq. (1.8), we obtain

$$\chi_e = \frac{\dfrac{N\alpha}{\varepsilon_0}}{1 - \left(\dfrac{N\alpha}{3\varepsilon_0}\right)} \tag{1.13a}$$

and

$$\varepsilon_r = \frac{1 + 2\left(\dfrac{N\alpha}{3\varepsilon_0}\right)}{1 - \left(\dfrac{N\alpha}{3\varepsilon_0}\right)}. \tag{1.13b}$$

Equations (1.13) provide the relation between polarizability α and relative dielectric constant ε_r:

$$\alpha = \frac{3\varepsilon_0 (\varepsilon_r - 1)}{N (\varepsilon_r + 2)}. \tag{1.14}$$

The above formulation, established by the four pioneering physicists named above, is known as the "Clausius–Mossotti" formula or "Lorentz–Lorenz" formula. Among these physicists, Lorentz summarized the formulae of his predecessors, and the model of Fig. 1.1 is called the "Lorentz sphere."

1.2.2 Other models for dielectric constants

Let us extend the case of Fig.1.1 to that of a filler material with dielectric constant ε_2 embedded as spheres in a matrix material with constant ε_1, Fig. 1.2. The effective dielectric constant ε_c of the composite is given by modifying Eq. (1.13b):

$$\frac{\varepsilon_c}{\varepsilon_1} = \frac{1 + \dfrac{2N\alpha}{3\varepsilon_1}}{1 - \dfrac{N\alpha}{3\varepsilon_1}}, \tag{1.15}$$

where the polarizability α is now replaced by

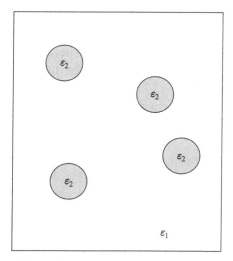

Fig. 1.2   Filler with dielectric constant ε_2 embedded in a matrix with constant ε_1.

$$\alpha = \frac{3\varepsilon_1(\varepsilon_2 - \varepsilon_1)}{(\varepsilon_2 + 2\varepsilon_1)} V \qquad (1.16)$$

and V is the total volume of filler material in a sphere. If the spheres occupy a volume fraction f and N is interpreted as the number of the spheres per unit volume,

$$f = NV. \qquad (1.17)$$

Note that $NV = 1$ in Eq. (1.14) where, effectively, the spheres occupied the entire space.

After substituting Eqs. (1.16) and (1.17) into Eq. (1.15), the composite dielectric constant ε_c is obtained as

$$\varepsilon_c = \varepsilon_1 \frac{1 + 2f \dfrac{(\varepsilon_2 - \varepsilon_1)}{(\varepsilon_2 + 2\varepsilon_1)}}{1 - f \dfrac{(\varepsilon_2 - \varepsilon_1)}{(\varepsilon_2 + 2\varepsilon_1)}} \qquad (1.18)$$

which can be rewritten as

$$\frac{(\varepsilon_c - \varepsilon_1)}{(\varepsilon_c + 2\varepsilon_1)} = f \frac{(\varepsilon_2 - \varepsilon_1)}{(\varepsilon_2 + 2\varepsilon_1)}. \qquad (1.19)$$

The formula Eq. (1.18) was first derived by Maxwell (1904), who considered the case of concentric spheres where an inner sphere with electric conductivity

σ_1 is embedded in an outer sphere with conductivity σ_2. By replacing ε_i by σ_i in Eq. (1.18), one can obtain a formula to predict the electric conductivity σ_c of a composite, which is called the "Maxwell–Garnett mixing formula." In Eqs. (1.18) and (1.19), one can recover special cases, i.e., if $f = 0$, $\varepsilon_c = \varepsilon_1$ (matrix dielectric constant) and, if $f = 1$, $\varepsilon_c = \varepsilon_2$ (filler dielectric constant). Even though the Maxwell–Garnett mixing formula appears to cover the entire range $0 \leq f \leq 1$, the model is based on the assumption of a small volume fraction.

We can further extend our reasoning to the case of two materials (ε_1, ε_2) embedded in a composite matrix material (ε_c), Fig 1.3. Then, in the background composite material (ε_c), each phase (ε_1, ε_2) can be viewed as an inclusion having an electric dipole moment with its polarizability (α_1, α_2). The polarization $\mathbf{P}$ created by these two different dielectric phases is given by

$$\mathbf{P} = (N_1\alpha_1 + N_2\alpha_2)\mathbf{E}^t. \tag{1.20}$$

If $\mathbf{P}$ is integrated over the entire composite domain, it vanishes because the electric field in the background composite material is the same as the local total field $\mathbf{E}^t$. This requires

$$N_1\alpha_1 + N_2\alpha_2 = 0, \tag{1.21}$$

where α_i is the polarizability of the ith dielectric material (ε_i) with respect to the composite medium (ε_c). Thus, from Eq. (1.16), α_i ($i = 1, 2$) are given by

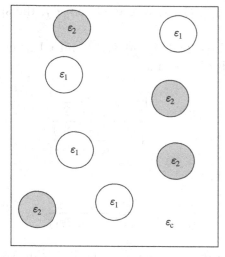

Fig. 1.3   Materials with dielectric constants ε_1 and ε_2 embedded in a background composite with constant ε_c.

$$\alpha_1 = \varepsilon_c \frac{3(\varepsilon_1 - \varepsilon_c)}{(\varepsilon_1 + 2\varepsilon_c)} V_1, \tag{1.22a}$$

$$\alpha_2 = \varepsilon_c \frac{3(\varepsilon_2 - \varepsilon_c)}{(\varepsilon_2 + 2\varepsilon_c)} V_2 \tag{1.22b}$$

and N_i, the number of spheres of the ith dielectric material, is related to the volume fraction f_i of the ith material, by

$$N_1 V_1 = f_1, \tag{1.23a}$$

$$N_2 V_2 = f_2. \tag{1.23b}$$

V_i in the above equations is the volume of each sphere of the ith dielectric material. Upon substituting Eqs. (1.22) and (1.23) into (1.21), and canceling a common factor ε_c, we obtain the formula

$$f_1 \frac{(\varepsilon_1 - \varepsilon_c)}{(\varepsilon_1 + 2\varepsilon_c)} + f_2 \frac{(\varepsilon_2 - \varepsilon_c)}{(\varepsilon_2 + 2\varepsilon_c)} = 0. \tag{1.24}$$

This formula is called "Bruggeman's symmetric formula" (Bruggeman, 1935) since the interchange of dielectric materials 1 and 2 gives the same formula.

We can consider a simpler model than those of Figs. 1.2 and 1.3, i.e., one which we call the "law of mixtures," having two cases: (1) the parallel model, Fig. 1.4(a), and (2) the series model, Fig. 1.4(b).

In the parallel model, Fig. 1.4(a), a composite is composed of material 1 with ε_1 and material 2 with ε_2 which are aligned in a parallel manner, and the

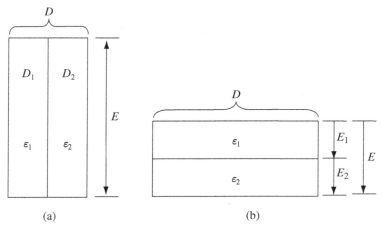

Fig. 1.4   Law-of-mixtures model: (a) parallel, (b) series.

electric field E is applied to the top and bottom surfaces of the composite in such a way that it is common to the two materials, i.e.,

$$D_1 = \varepsilon_1 E, \tag{1.25a}$$

$$D_2 = \varepsilon_2 E. \tag{1.25b}$$

The total electric flux density D_c of the composite is the volume-averaged sum of the electric flux density in each material.

$$D_c = (1 - f)D_1 + fD_2, \tag{1.26}$$

where f is the volume fraction of material 2 which is treated as "filler" here. The composite must obey the following constitutive equation:

$$D_c = \varepsilon_c E. \tag{1.27a}$$

A substitution of Eqs. (1.25) into Eq. (1.26), which is then substituted into Eq. (1.27a), gives the law-of-mixtures formula based on the parallel model,

$$\varepsilon_c = (1 - f)\varepsilon_1 + f\varepsilon_2. \tag{1.27b}$$

If the composite of Fig. 1.4(a) is subjected to an applied electric field E in the transverse direction, then one can create the series model, Fig. 1.4(b). In this case the electric flux density is continuous through materials 1 and 2; thus in the two materials

$$D_c = \varepsilon_1 E_1, \tag{1.28a}$$

$$D_c = \varepsilon_2 E_2. \tag{1.28b}$$

The applied electric field E is the volume-averaged sum of materials 1 and 2,

$$E = (1-f)E_1 + fE_2. \tag{1.29}$$

From Eqs. (1.27) – (1.29), one can derive the law-of-mixtures formula based on the series model,

$$\frac{1}{\varepsilon_c} = \frac{(1 - f)}{\varepsilon_1} + \frac{f}{\varepsilon_2}. \tag{1.30}$$

The law-of-mixtures formulae of Eqs. (1.27) and (1.30) are interpreted as the longitudinal and transverse dielectric constants, respectively, of a continuous fiber composite.

Let us compare the composite dielectric constants predicted by the four models above. To this end we use the following input data:

$$\frac{\varepsilon_1}{\varepsilon_0} = 2, \qquad \frac{\varepsilon_2}{\varepsilon_0} = 50. \qquad (1.31)$$

The numerical results of $\varepsilon_c/\varepsilon_0$ are plotted as a function of volume fraction f of filler (material 2) in Fig. 1.5 and indicate that the predictions of the law of mixtures based on parallel and series models serve as the upper and lower bounds on the composite dielectric constant, while the predictions of Bruggeman's symmetric model exceed those of Maxwell's model.

McLachlan *et al.* (1990) reviewed the above models and proposed a combined model between Bruggeman's symmetric model and a percolation model. The combined model proposed by McLachlan *et al.* will be discussed in Section 6.6.

1.2.3 Demagnetization in a ferromagnetic body

When a ferromagnetic body of finite size is subjected to an applied magnetic field **H** it is magnetized with magnetization vector **M**, and a demagnetization field **H**$_d$ is induced in the ferromagnetic body, with the direction of **H**$_d$ opposite to that of **H**, given by (Chikamizu, 1964)

$$\mathbf{H}_d = -N_d \mathbf{M}. \qquad (1.32)$$

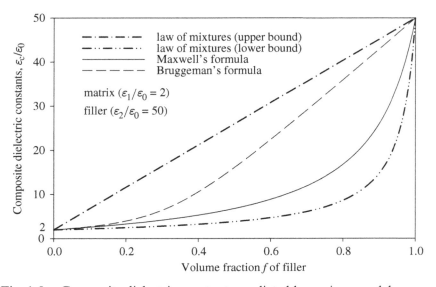

Fig. 1.5   Composite dielectric constants predicted by various models.

N_d is called the "demagnetization factor" and depends on the shape of the ferromagnetic body. The demagnetization vector $\mathbf{H_d}$ remains constant (uniform) if the shape of the body is ellipsoidal.

Let us consider a thin ferromagnetic plate with $\mathbf{M}$ perpendicular to the plate thickness direction, Fig. 1.6, in which magnetic poles $(+, -)$ are also illustrated as well as the demagnetization vector $\mathbf{H_d}$. The surface density of the magnetic poles is $\mu_0 M$ (webers [Wb]/m^2) where M is the scalar value. Applying Gauss' theorem along the dashed closed domain, Fig. 1.6,

$$\int H_d \, \mathrm{d}s = MS, \tag{1.33}$$

where $\mathrm{d}s$ is the surface element, S is the surface within the dashed frame, and H_d is the magnetic field perpendicular to the surface S. The left-hand side of Eq. (1.33) is equal to SH_d, thus Eq.(1.33) provides

$$SH_d = MS \tag{1.34}$$

and, therefore,

$$H_d = M. \tag{1.35}$$

From Eqs. (1.32) and (1.35), the demagnetization factor of a thin ferromagnetic plate is $N_d = 1$. Therefore, the net magnetization of a thin ferromagnetic

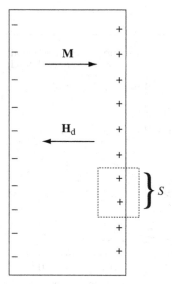

Fig. 1.6   Demagnetization field $\mathbf{H_d}$ induced as a result of magnetization vector $\mathbf{M}$ along the thickness direction of plate.

plate in the direction perpendicular to the plate thickness is $M - H_d = 0$. Similarily, the demagnetization factor of a thin ferromagnetic plate in the plate direction can be found as $N_d = 0$. Osborn (1945) and Stoner (1945) calculated the demagnetization factor of various types of ellipsoidal shapes. For the case of a prolate spheroid where the direction of magnetization is along the longer axis, N_d is given by

$$N_d = \frac{1}{(\beta^2 - 1)} \left\{ \frac{\beta}{(\beta^2 - 1)} \ln\left(\beta + \sqrt{\beta^2 - 1} \right) - 1 \right\}, \tag{1.36}$$

where β is the aspect ratio of the longer to the shorter axis of the prolate spheroid. For a spherical ferromagnetic body, N_d is given by 1/3. The above demagnetization vector $\mathbf{H}_d$ can be examined in a slightly different way. The constitutive equation relating the magnetic flux intensity $\mathbf{B}$ for a ferromagnetic body with magnetization vector $\mathbf{M}$ subjected to applied field $\mathbf{H}$ is

$$\mathbf{B} = \mu_0(\mathbf{H} + \mathbf{M}). \tag{1.37}$$

If the demagnetization vector is taken into account, the second term in the right-hand side of Eq. (1.37) is modified with an added term $\mathbf{H}_d$. Hence,

$$\mathbf{B} = \mu_0(\mathbf{H} + \mathbf{M} + \mathbf{H}_d)$$

$$= \mu_0(\mathbf{H} + \mathbf{M} - N_d\mathbf{M}), \tag{1.38}$$

where Eq. (1.32) was used; μ_0 is the magnetic permeability of free space. The demagnetization field $\mathbf{H}_d$ is considered as a disturbance field due to the ferromagnetic body of finite size located in an infinite (vacuum or air) domain subjected to a far-field applied field of $\mathbf{H}$. Then $\mathbf{H}_d$ plays the same role as the disturbance electric field $\mathbf{E}_p$, see Eq. (1.1) in Subsection 1.2.1, where the shape of the dielectric filler (or inclusion) was assumed to be spherical. For a spherical inclusion problem, the disturbance field in a spherical body with $\mathbf{P}$ (polarization vector) in the dielectric material and $\mathbf{M}$ (magnetization vector) in the ferromagnetic material can be treated in the same manner, see Eqs. (1.4) and (1.32) with $N_d = 1/3$. The difference between the dielectric and ferromagnetic cases is that in the former a factor 1/3 is associated with ε_0 (the dielectric constant in vacuum or air) while in the latter the factor is $-1/3$. Please note that the magnetization vector $\mathbf{M}$ is sometimes written without μ_0 in Eq. (1.37) in older textbooks, i.e.,

$$\mathbf{B} = \mu_0\mathbf{H} + \mathbf{M}. \tag{1.39}$$

If this type of equation is used instead of Eq. (1.37), then the role of μ_0 is the same as that of ε_0. Hence, the demagnetization factor $1/3$ for a spherical shape should be treated exactly the same as in the dielectric case, i.e., Eq. (1.4), except for the sign. In this book, we shall use the Institute of Electrical and Electronic Engineering (IEEE) convention, i.e., Eq. (1.37), to express the relation between magnetic flux intensity (**B**), magnetic field (**H**), and magnetization (**M**).

Next, we shall consider an infinite ferromagnetic body with uniform magnetization **M**, which has an ellipsoidal cavity, Fig. 1.7(a). The problem of finding the demagnetization vector $\mathbf{H_d}$ in an ellipsoidal cavity can be broken down into two parts, Fig. 1.7(b) and (c). If the ellipsoidal cavity is interpreted as a special case of a magnetic inhomogeneity with μ_f embedded in a matrix

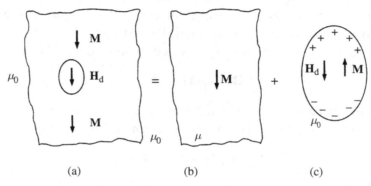

(a)   (b)   (c)

Fig. 1.7   (a) Magnetic body with ellipsoidal cavity is superposition of two problems: (b) uniform magnetization vector **M** in an infinite magnetic body, and (c) ellipsoidal magnetic body with **M** and demagnetization vector $\mathbf{H_d}$.

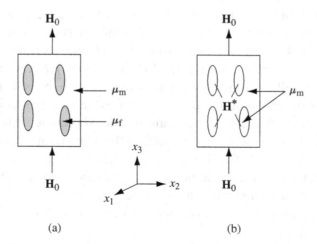

(a)   (b)

Fig. 1.8   Magnetic composite model. (a) Actual magnetic composite, (b) Eshelby model.

of ferromagnetic material with μ_m, Fig. 1.8(a), one can formulate a magnetic composite model by extending the above model. Then Fig. 1.8(a) is converted to the Eshelby model with eigen-magnetic-field $\mathbf{H}^*$ defined in the domain of ferromagnetic fillers, as shown in Fig. 1.8(b). This will be discussed in Chapter 4 as a part of the Eshelby formulation of inclusion problems, where the demagnetization factor N_d is found to be equal to a component of the Eshelby tensor $\mathbf{S}$.

1.3 Concept of electric, magnetic, and thermal circuits

Four basic physical behaviors that are key to electronic composites are electric flux, electric conduction, magnetic flux, and thermal conduction. These behaviors can be cast into a single circuit model. Let us examine these in terms of governing equations.

1.3.1 Electric flux in a dielectric body

In the absence of true electric change in a dielectric medium with constant ε,

$$\nabla \cdot \mathbf{D} = 0, \tag{1.40}$$

where ∇ is the del operator (or divergence [see Chapter 3]) and $\mathbf{D}$ is the electric flux density vector, which is related to electric field $\mathbf{E}$ by

$$\mathbf{D} = \varepsilon\mathbf{E} \tag{1.41}$$

and $\mathbf{E}$ is derivable from the electric potential Φ_e by

$$\mathbf{E} = -\nabla\Phi_e. \tag{1.42}$$

Assuming the dielectric constant ε is uniform in a dielectric body, one can obtain, from Eqs. (1.40)–(1.42),

$$\nabla^2\Phi_e = 0 \tag{1.43}$$

which is Poisson's equation without source charge, i.e., the Laplace equation. (The definition of ∇^2 is given in Section 3.1.)

1.3.2 Electric conduction in a conductive body

The electric current density vector $\mathbf{J}$ in a conductive medium with uniform conductivity σ satisfies

$$\nabla \cdot \mathbf{J} = 0 \tag{1.44}$$

where $\mathbf{J}$ satisfies Ohm's law

$$\mathbf{J} = \sigma \mathbf{E} \tag{1.45}$$

and $\mathbf{E}$ is related to electric potential Φ_e by Eq. (1.42). Therefore, from Eqs. (1.44), (1.45), and (1.42), we obtain Eq. (1.43).

1.3.3 Magnetic flux in a ferromagnetic body

In a ferromagnetic medium with uniform permeability μ, the magnetic field vector $\mathbf{H}$ is derivable from the magnetic potential Φ_m by

$$\mathbf{H} = -\nabla \cdot \Phi_m. \tag{1.46}$$

Inside a ferromagnetic medium with uniform magnetization $\mathbf{M}$, the magnetic field $\mathbf{H}$ is derived from Eq. (1.37) as

$$\mathbf{H} = \frac{\mathbf{B}}{\mu_0} - \mathbf{M} \tag{1.47}$$

where $\mathbf{B}$ is the magnetic flux density vector, satisfying

$$\nabla \cdot \mathbf{B} = 0. \tag{1.48}$$

A substitution of (1.46) and (1.47) into (1.48) results in

$$\nabla^2 \Phi_m = -\nabla \cdot \mathbf{M}. \tag{1.49}$$

For uniform magnetization vector $\mathbf{M}$ in a ferromagnetic medium, the right-hand side of (1.49) vanishes, resulting in

$$\nabla^2 \Phi_m = 0. \tag{1.50}$$

Magnetic potential Φ_m in the above equations plays the same role as electric potential Φ_e, satisfying the Laplace equation.

1.3.4 Heat flow in a conductive body

In the absence of heat sources and sinks, the heat flux density vector $\mathbf{q}$ must satisfy

$$\nabla \cdot \mathbf{q} = 0 \tag{1.51}$$

where $\mathbf{q} \ (= \mathbf{Q}/A)$ follows the conduction law:

Table 1.1 *Comparison of thermophysical behavior where A is the cross-section of the circuit line.*

Physical behavior	Field–Gradient relation	Potential (V)	Current (I) or Flow	Resistance (R)	Ohm's law
electric	$\mathbf{E} = -\nabla\Phi_e$	$V = \Phi_e$	$Aj = I$	$R = \int(\mathrm{d}s/\sigma A)$	$V = RI$
dielectric	$\mathbf{E} = -\nabla\Phi_e$	$V = \Phi_e$	$Q_d = AD$	$R_d = \int(\mathrm{d}s/\varepsilon A)$	$V = R_d(AD)$
magnetic	$\mathbf{H} = -\nabla\Phi_m$	$V_m = NI^*$ or $\dfrac{B_r\ell_p}{\mu_r}{}^{**}$	$\Phi = BA$	$R_m = \int(\mathrm{d}s/\mu A)$	$V_m = R_m\Phi$
thermal	$\mathbf{q} = -\mathbf{K}\nabla T$	T	$Q = qA$	$R_t = \int(\mathrm{d}s/kA)$	$T = R_t Q$

*N is number of turns of an electromagnetic coil
**B_r is the magnetic remanence, and μ_r the magnetic permeability of B–H curve of a permanent magnet in the second quadrant (see Fig. 3.21), and ℓ_p is the length of the permanent magnet along the line of magnetic flux.

$$\mathbf{q} = -\mathbf{K}\cdot\nabla T, \tag{1.52}$$

where $\mathbf{Q}$ is the heat vector (watts (W)), A is the cross-sectional area of the conductive body and T is temperature. A substitution of (1.52) into (1.51) leads to

$$\nabla^2 T = 0, \tag{1.53}$$

which is the Laplace equation.

The fact that the above four cases converge to the same Laplace equation suggests that for all four thermophysical properties, problems associated with inhomogeneities (fillers, voids, etc.) embedded in a matrix can be treated by the same model, "Eshelby's equivalent inclusion method," the details of which we shall discuss in Chapter 4.

In the above four cases of physical behavior the potential function satisfying a Laplace equation is well defined as a common basis. Another common feature arising from these four physical behaviors is a conduction law, e.g., Ohm's law in an electric circuit (or Kirchhoff's second law), i.e.,

$$V = RI. \tag{1.54}$$

Equation (1.54) is also applicable to the other three thermophysical behaviors. Table 1.1 summarizes the roles of potential V, current I or flow, and resistence R in the four cases.

Table 1.1 demonstrates that two relations are common to all four types of thermophysical behavior: (i) field (or flux density)–(gradient of potential) relation, in column 2 and (ii) circuit relation in the columns 3, 4 and 5. The former is the basis for the Eshelby effective medium model (Chapter 4), while the latter provides the key basis for the resistor network model (Chapter 5).

2

Applications of electronic composites

We shall review applications of electronic composites in this chapter, which is subdivided into four areas: electronic packaging, MEMS, composites for sensors and actuators, and control of electromagnetic waves.

2.1 Electronic packaging

Electronic composites have been used extensively for electronic packaging materials, which often have conflicting requirements that are difficult to meet – high thermal conductivity, low electrical conductivity, and low coefficient of thermal expansion (CTE). In addition, low density and low cost are normally required, particularly for commercial packaging applications. Thus polymer matrix composites are the most popular choice, while some military and space applications require the use of high-performance electronic composites, thus sometimes justifying the use of metal matrix composites and ceramic matrix composites. A detailed review of electronic packaging materials is summarized in a book by Chung (1995). Zweben (1995) made an extensive review of electronic composites for use in electronic packaging, which include all types of composites, polymer matrix composites (PMCs), metal matrix composites (MMCs), ceramic matrix composites (CMCs) and carbon/carbon (C/C) composites.

Applications of electronic composites are quite extensive, covering computers, appliances, generators, and transformers, to name a few. Depending on their physical functions, one can categorize them into several types – housing, insulator, resistor, and conductor – which are discussed below. It is to be noted here that the housing material can be an insulator, a conductor, or an electromagnetic wave absorber, depending on the specific application.

Housing. Most electronic composites in electronic packaging applications are used in housing, which protects the electromagnetic functions of the key

Table 2.1 *Requirements of electronic composites for electronic packaging housing.*

Electronic packaging application	Main requirements
Computers: cover plate and housing	Shielding from heat, dust, moisture, UV, vibration and shock, and EMIs Mechanical stiffness and strength, high specific stiffness, low CTE, high thermal conductivity; and electrical insulator
VTR: mechatronic components in video deck	High specific stiffness with high-precision 3D geometry; electrical insulator
Floppy disk drive	High specific stiffness with high-precision clearance; electrical insulator
CD pick-up device	High specific stiffness with extra-high-precision clearance; electrical insulator

components from various environments: heat, dust, UV, moisture, mechanical vibration, and incoming electromagnetic waves which would induce Electromagnetic Interactions (EMIs), as summarized in Table 2.1.

Insulator. Electronic composites as insulating materials have been used extensively for a long time, because the polymer matrix is a lightweight and cost-effective insulator. Examples are printed circuit boards (PCBs), connectors, encapsulants, and rotor arms in automobile engine distributors. These electronic composites are processed by affordable processing routes, such as injection molding, sheet molding, and bulk molding (see the details of processing in Section 8.1). Even though the polymer is an electrical insulator, it may exhibit electric breakdown under high voltage. To avoid this, fillers with high dielectric constant, such as mica, are often used to form a composite which provides extra-high shielding from high-voltage breakdown. This is particularly so for the distributor/rotor-arm design in automobile engines.

Conductor. A conductor is normally conductive both electrically and thermally. Conductive composites, both MMC- and PMC-based, are used as interconnection materials for both thermal and electrical applications, as well as the housing material for EMI shielding. For some applications, a material that is an electrical insulator, but also a thermal conductor, is required, for example, diamond and AlN. Table 2.2 summarizes the applications of conductive composites.

Resistor. A resistor controls the flow of electric current. Two designs of resistor are commercially available: fixed, and variable. A typical resistor

Table 2.2 *Conductive composites and examples of their applications.*

Conductive composites	Applications
Polymer-based composites:	
Carbon/PVC[a]	EMI housing
Silver/PC[b] *or*	Thermal interface material including die attach
Graphite/PC silver/epoxy	material for thermal conduction
Ceramic-based composite	Electrode for electronic packaging and fuel cell
Carbon/carbon, C/C	Bonding material
Metal-based composite, Cu/C	Interconnecting materials, heat spreader and
	heatsink

[a] Polyvinyl chloride; [b] polycarbonate

material is composed of a conductor (such as carbon) and an insulator (polymer), thus its microstructure is composite. Some resistors are made of monolithic or alloyed metals. There exist three different designs of fixed resistor: rod, film, and wirewound. For microelectronic applications, the most popular type is a film resistor with two versions: chip resistor, composed of a thin resistance coating and a ceramic substrate; and a resistor network composed of the resistor material and a number of metal terminals.

2.1.1 Examples of electronic packages

We shall show a few examples of electronic packages whose microstructures are often complicated, and may need simplification to a set of standard electronic-composite microstructures for their analysis.

Figure 2.1 shows one classical design of computer chip, a small outline gull-wing package (SOP) where (a) is an overview, (b) a vertical cross-sectional view across line A–A', and (c), (d), and (e) show, respectively, the horizontal cross-sectional view along lines B–B', C–C' and D–D' shown in Fig. 2.1 (b).

Figure 2.2 shows the microstructure of an Intel® Pentium® II chip along vertical and selected horizontal cross-sections.

Next, the packaging of a recent Intel® Pentium® 4 Processor chip is shown in Fig. 2.3, wherein most of the components are electronic composites, chip (die), encapsulant, printed circuit board (PCB), and thermal interface material (TIM).

In order to see those components that are used for better thermal management, including TIM, the package of the Intel® Pentium® 4 Processor that is vertically integrated is shown in Fig. 2.4, where the function and microstructure of the TIM are emphasized. The above TIM will be discussed in detail in Subsection 8.2.2, where measurements of thermal resistance are considered.

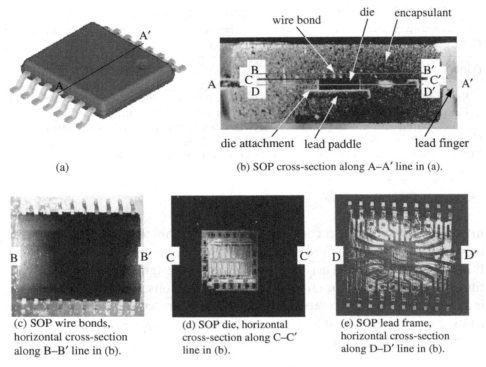

(a)

(b) SOP cross-section along A–A' line in (a).

(c) SOP wire bonds, horizontal cross-section along B–B' line in (b).

(d) SOP die, horizontal cross-section along C–C' line in (b).

(e) SOP lead frame, horizontal cross-section along D–D' line in (b).

Fig. 2.1   Small outline gull-wing package (SOP).

Even though a TIM exhibits anisotropic thermal conductivity if the arrangement of conductive fillers is made along a specific direction, most TIMs are isotropic. Modeling of the electrical and thermal properties of isotropic conductive materials was attempted by Li and Morris (1997) and Kim (1998). In parallel with the development of TIMs, anisotropically conductive polymers have emerged as electrically interconnected materials (Chang *et al.*, 1993). A typical anisotropic conductive polymer is composed of electrically conductive fillers and a polymer matrix which is to be sandwiched between a chip with bumps and a substrate with electrode pads. Conductive fillers in an anisotropic conductive polymer are chosen which can deform easily under the compressive force exerted by a pair of bump and pad, thus forming a z-directional (along the thickness direction) electronic connection. Shiozawa *et al.* (1995) designed such a conductive adhesive and reported its reliability data. This anisotropic conductive polymer has been used as a key interconnect material in a liquid crystal display (LCD) driver used in large-scale integration (LSI) (Watanabe *et al.*, 1996).

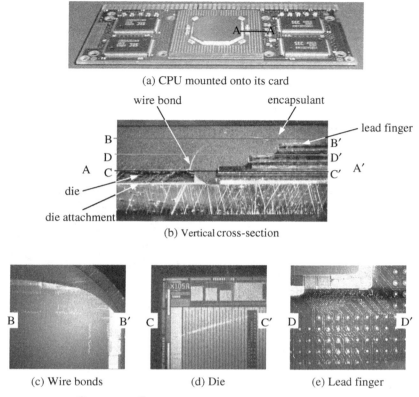

(a) CPU mounted onto its card

(b) Vertical cross-section

(c) Wire bonds   (d) Die   (e) Lead finger

Fig. 2.2   Intel® Pentium® II chip: (a) overall view, (b) vertical cross-section along the A–A′ line, (c)–(e) horizontal cross-sectional view along lines B–B′, C–C′, and D–D′ in (b).

2.2 Micro-electromechanical systems (MEMS) and BioMEMS

Micro-electromechanical systems (MEMS) is a collective name for packages that are micromachined using silicon-based integrated circuit (IC) processing. The functions of MEMS are as sensors and actuators. This MEMS technology started in the late 1980s and has expanded during the 1990s. During the expansion period, MEMS technology evolved into several branches: (i) nano-electromechanical systems (NEMS), by down-sizing the MEMS packages and their components; (ii) BioMEMS, with application to biosensors; and (iii) biomimetics (or bio-inspired design). Design and fabrication of MEMS is based on "top down" manufacturing, i.e., miniaturization from larger- to smaller-sized components, while Biomimetics aims to borrow or be inspired by design and fabrication concepts observed in biological cells and their functions, often associated with "nano-sized components." Whatever the design and fabrication

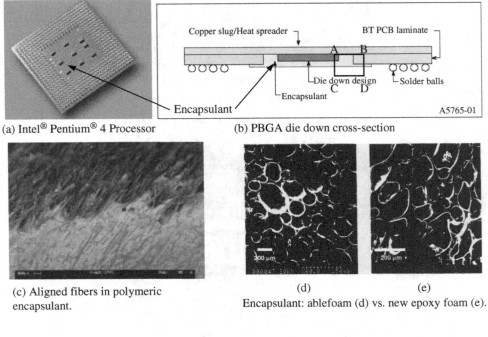

(a) Intel® Pentium® 4 Processor		(b) PBGA die down cross-section

(c) Aligned fibers in polymeric encapsulant.

(d)						(e)
Encapsulant: ablefoam (d) vs. new epoxy foam (e).

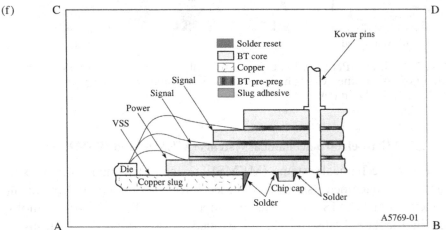

Fig. 2.3	Intel® Pentium® 4 Processor in 423-pin package (*source*: ftp://download.intel.com/design/packtech/ch_14.pdf): (a) overview,(b) profile in a 2-level package with encapsulant, PCB, heat spreader and solder balls, (c)–(e) microstructures of three different encapsulant materials. (f) Enlarged vertical cross-section within ABCD, see (b), but shown upside down.

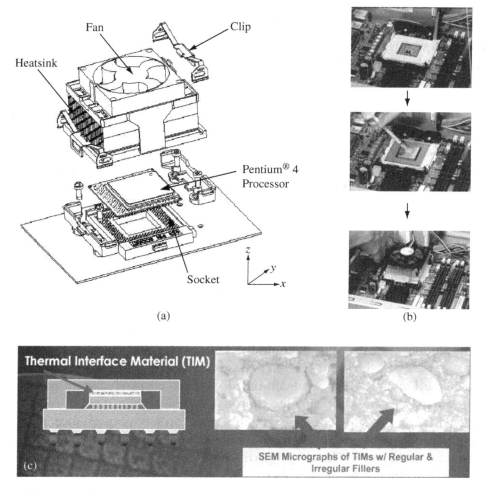

Fig. 2.4 Thermal management components: (a) expanded view, (b) sequence of packaging – installation of socket, applying TIM, and installation of fan, heatsink, and clips, (c) microstructure of TIMs.

concept may be called, the structures of MEMS, NEMS and biomimetics have a common feature, namely an inhomogeneous structure which can be considered as a "micro- or nano-composite." The key properties of these composites are thermomechanical, electromagnetic and electrochemical. Therefore, they may be classified as "electronic composites." Knowing with accuracy the properties of the electronic composites is a prerequisite in designing the optimum components of MEMS (NEMS), BioMEMS and biomimetics-based devices.

In the following Subsections, we shall review representative MEMS and BioMEMS, as they are well developed and some of them are already marketed as commercial products.

2.2.1 MEMS

Polysilicon surface machining was started by researchers at Westinghouse Research Laboratory (Nathanson *et al.*, 1967). Howe and Muller (1983) processed a cantilever beam made of polysilicon by using a metal–oxide–semiconductor (MOS) planar process, a technique later adopted by many researchers in the 1990s (Howe *et al.*, 1996), leading to rapid expansion of polysilicon-based MEMS technology.

Sniegowski (2001) defined four different groups of MEMS technology based on polysilicon surface micromachining – 2-level, 3-level, 4-level and 5-level MEMS, Fig. 2.5, where 1-level denotes the silicon substrate base. The 2-level MEMS structures are able to perform restricted motion, such as cantilever beams attached to substrate vibrate in resonance mode. The 3-level MEMS have a moving component such as a free-spinning rotating gear attached to the substrate with a free-spinning pin. The 4-level MEMS have interconnected elements between moving gears and other functioning components. Finally, 5-level MEMS have higher hierarchy structures, i.e., up to five levels of components are placed on mobile platforms which are translated over the substrate. Examples of 5-level MEMS components are shown in Fig. 2.6.

In parallel with the above definition, Sniegowski (2001) redefined MEMS into four classes shown in Fig. 2.7; classes I and II are already commercial products. Examples of polysilicon-based commercial products are pressure sensors. Esashi *et al.* (1998) made a thorough review of three types of vacuum-sealed MEMS pressure sensors: capacitive, based on servo capacitance; micro-diaphragm, based on piezoresistive polysilicon; and resonant, based on several different mechanisms. The key properties required for MEMS are also listed in Fig. 2.7, from mechanical properties for lower-level MEMS to frictional properties for higher-level MEMS, as some components in the latter have moving parts.

Sniegowski (2001) identified engineering problems associated with the above four classes of MEMS devices, listed in Table 2.3.

Sniegowski and de Boer (2000) reviewed polysilicon-based MEMS which are better suited to the processing of higher-level MEMS. A flowchart of the processing of the moving gear component of a 4-level MEMS is shown in Fig. 2.8 where Pi and Si are polysilicon and sacrificial layer, respectively, used in the ith process step.

MEMS components are exposed to air without any passivation (whereas most electronic packages are encapsulated by plastic passivation materials), and higher-level (or higher-class) MEMS components move during service use, so that durability becomes a serious issue. In MEMS devices, free-standing

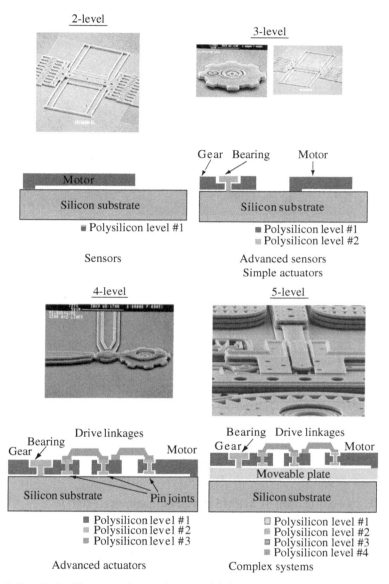

Fig. 2.5   Polysilicon surface micromachining nomenclature (Sniegowski, 2001).

components have a small gap between themselves and the substrate. This "gap" geometry poses a serious problem during processing and service use, called "stiction."

In processing MEMS, liquid solutions are used, which need to be removed at the end of processing, but which can be easily trapped in, say, a small gap

Fig. 2.6   Moving components of 5-level MEMS (Sniegowski, 2001).

Class I *No Moving Parts*	Class II *Moving Parts,* *No Rubbing or* *Impacting Surfaces*	Class III *Moving Parts,* *Impacting Surfaces*	Class IV *Moving Parts, Impacting* *and Rubbing Surfaces*

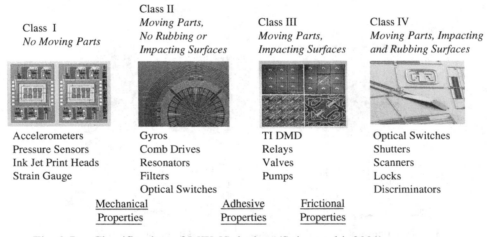

Accelerometers Pressure Sensors Ink Jet Print Heads Strain Gauge	Gyros Comb Drives Resonators Filters Optical Switches	TI DMD Relays Valves Pumps	Optical Switches Shutters Scanners Locks Discriminators

<u>Mechanical</u>   <u>Adhesive</u>   <u>Frictional</u>
Properties   Properties   Properties

Fig. 2.7   Classification of MEMS devices (Sniegowski, 2001).

between a free-standing cantilever and the substrate. The stiction force that arises between two adjacent components with a small gap stems from various mechanisms: (1) capillary force, (2) hydrogen bonding, (3) electrostatic force, and (4) van der Waals force, with (1)–(4) arranged in order from strongest to weakest. All of these forces influence the stiction of MEMS. Sniegowski and de Boer (2000) studied in detail the mechanical and surface properties of materials used in MEMS based on polysilicon. They suggested several approaches to overcome the stiction problem – for example, use of a tungsten (W) coating on the polysilicon surface of micromachined MEMS moving components such as gears. Figure 2.9 demonstrates the effectiveness of a W-coated gear in increasing durability; after 1 billion cycles it is less worn than the uncoated gear

Table 2.3 *Engineering problems associated with four classes of MEMS defined in Fig. 2.6* (Sniegowski, 2001).

MEMS type	Class I (No moving parts)	Class II (Moving parts, no rubbing or impacting surfaces)	Class III (Moving parts, impacting surfaces)	Class IV (Moving parts, impacting and rubbing surfaces)
Engineering problems	Particle contamination Shock-induced stiction	Particle contamination Shock-induced stiction Mechanical fatigue	Particle contamination Shock-induced stiction Stiction Mechanical fatigue Impact damage	Particle contamination Shock-induced stiction Stiction Mechanical fatigue Friction wear

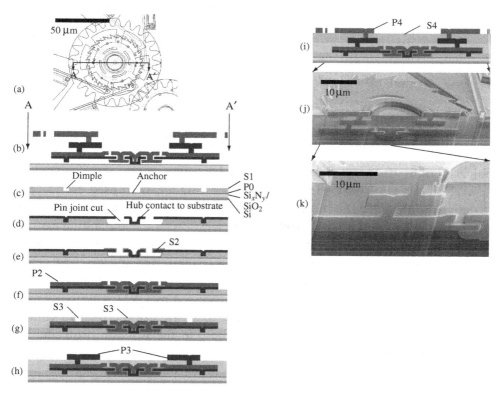

Fig. 2.8 (a) Plan view of gear-ratio component of 5-level MEMS. (b) Vertical cross-section along A–A′ line in (a). (c)–(k) Sequence of processing in terms of deposition of polysilicon layers (P*i*, *i* = 0–4) and sacrificial oxide layers (S*i*, *i* = 1–4). (Sniegowski and de Boer, 2000.)

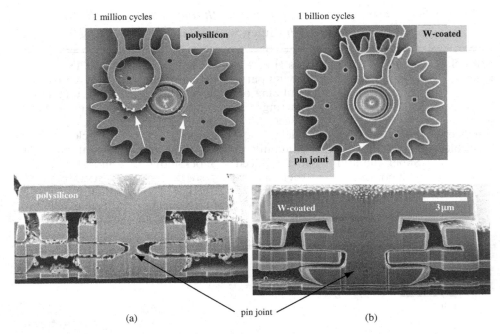

Fig. 2.9   Durability of gears after many cycles: (a) polysilicon gear, (b) W-coated gear. (Mani *et al.*, 2000, with permission from Mater. Res. Soc.)

viewed at 1 million cycles (Mani *et al.*, 2000). The W-coated layer, whose elastic modulus is higher than that of polysilicon, improves the wear resistance of the gear. An improvement in wear resistance can also be achieved by using a polycrystalline diamond coating that can be deposited by chemical vapor deposition (CVD). This CVD diamond processing has been used extensively to design a variety of MEMS: active cooling (Björkman *et al.*, 1999a,b), optical components (Björkman *et al.*, 1999a,b), acceleration sensor and ink-jet ejector (Kohn *et al.*, 1999), micro-switch (Ertl *et al.*, 2000), micro-tweezer and atomic force microscope (AFM) tip (Shibata *et al.*, 2000), and motors and gears (Hunn *et al.*, 1994; Aslam and Schulz, 1995; Mao *et al.*, 1995). The surface roughness of the polycrystalline diamond processed by CVD is large, $\sim$1 mm rms, which is a disadvantage when seeking to down-size the MEMS components, and also the rough surfaces contribute to the striction problem. To overcome this, Sullivan *et al.* (2001) proposed to use amorphous diamond (aD) as a key MEMS material as this provides a much smoother surface and also residual stress in the as-deposited aD films can be reduced by adjusting the deposition condition. Figure 2.10 shows three examples of aD-based MEMS components: (a) electrostatically activated comb drive unit, (b) micro-tensile specimens, and (c) cantilever beams.

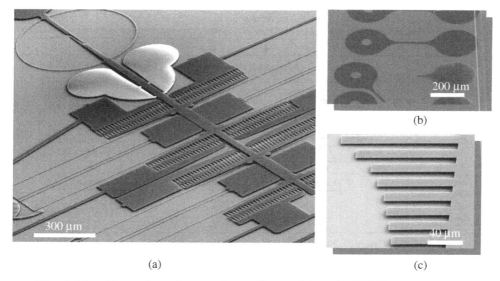

Fig. 2.10   Examples of amorphous-diamond-based MEMS components:
(a) comb drive, (b) tensile specimens, and (c) cantilever beams. (Sullivan *et al.*,
2001, with permission from Mater. Res. Soc.)

Akiyama *et al.* (1997) designed a "scratch drive" actuator based on polysilicon which provides a 100-μm displacement and a vertical force of 10 μN
with net work of 1 nJ. Bourouina *et al.* (2002) designed a MEMS optical
scanner based on a magnetostrictive actuator and a piezoresistive detector.

2.2.2 BioMEMS

Most MEMS devices are made from hard materials (silicon and glass, etc.),
while biomedical devices are often composed of soft materials, i.e., polymers,
which are more compatible with biological tissues and reduce irritation and
scarring. Miniaturization of biomedical devices has marked advantages, such
as being implantable in a human body, so BioMEMS-based soft materials
have emerged as a new technology. These soft materials are used as passivation
material, and as active and sensing materials.

BioMEMS can be categorized into three groups: (1) diagnostic, (2) surgical,
and (3) therapeutic. Polla *et al.* (2000) reviewed BioMEMS for these three
areas. Among these, diagnostic BioMEMS have been studied most extensively
and are reviewed first below, followed by other types of BioMEMS.

Diagnostic BioMEMS

There are many diagnostic BioMEMS components, with designs based on a
variety of mechanisms. Mello and Kubota (2002) made a thorough review on

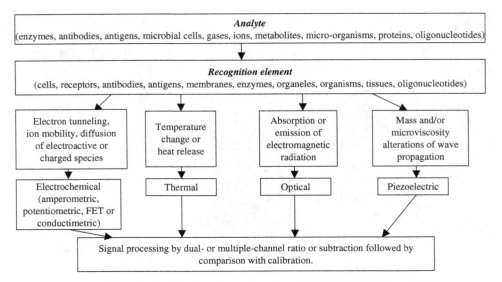

Fig. 2.11   Flowchart of sensing procedure for BioMEMS (Mello and Kubota, 2002, with permission from Elsevier Ltd). FET is field-effect transistor.

biosensors. Figure 2.11 illustrates a flowchart of an analyte/recognition-element method of analysis.

A diagnostic BioMEMS device normally is composed of micro-fluidics and a sensor. The former is to pump fluid (blood and body fluids) into a sensor chamber, while the latter is to analyze the sample fluid by a known sensing mechanism. A key component of the micro-fluidics is a valve which moves or vibrates so as to push and displace the fluid sample. This valve can be actuated by several mechanisms, e.g., thermal (bimetallic laminated composite, shape memory alloy based laminate), and piezoelectric and electro-active (laminated composite – monomorph, bimorph). A typical micro-pump based on a piezo-electric monomorph actuator is illustrated in Fig. 2.12. A piezoelectric constant d_{13} plays a key role in providing a bending motion to the top plate of the micro-pump unit. The definition of d is the strain induced in the direction (x_1-axis) of the piezoelectric plate under the action of an electric field applied across the plate thickness (x_3-axis). The constitutive equations of piezoelectric materials and examples of the bending mode of piezoelectric actuators are detailed in Chapters 3 and 7. Polla *et al.* (2000) reviewed types of micro-pumps composed of multiple pumps used in drug-delivery systems.

There are two types of biosensing for the pair analyte/recognition element: (i) bio-affinity recognition, and (ii) biometabolic recognition (Dewa and Ko, 1994). The former relies on a strong bonding between the pairs, such as

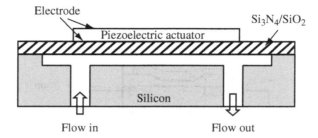

Fig. 2.12   A micro-pump based on a monomorph piezoelectric actuator.

receptor–ligand and antibody–antigen. In the latter, the analyte and any reactants are changed to other product molecules; examples are enzyme–substrate reaction, and the metabolism of molecules with micro-organisms.

Biosensors based on an electrochemical principle have three subsets, depending on what signal is measured – impedance (impedometric sensor), potential (potentiometry), and current (amperometry). The impedance method measures the frequency response of a small-signal impedance between the electrochemical electrode and the biological sensing element; it can measure very small signals accurately. However, this requires large equipment, so its miniaturization to BioMEMS remains to be explored. The impedometric sensor can measure the concentration of specific molecules, micro-organisms, and other foreign species in the sample solution. Potentiometry can detect ion concentration and pH. Potentiometry that is miniaturized is based on an immuno-field-effect transistor (IFET), an example of which is shown in Fig. 2.13.

Amperometric biosensors measure the electric current resulting from the chemical reaction of an electroactive species under an applied electric potential; the measured current is proportional to the concentration of the electroactive species. Amperometric biosensors are extensively used for detection of glucose, galactose, lactose, sucrose, acetic acid, and others. Miniaturization of amperometric biosensors has been accelerated recently, for example, Petrou *et al.* (2002) designed a microdialysis probe amperometer.

Optical biosensors are able to detect the change in optical absorption or emission of molecules. Classical equipment used for this measurement method includes infrared (IR) and ultraviolet–visible (UV–Vis) spectroscopes, but these are relatively larger sized and suited to a laboratory, not portable for field measurements. Optical fibers are better suited for use in portable biosensors. There are two methods based on optical-fiber biosensors: (i) direct measurement via fiber-end modified with sensing dye or fluorophore surrounded by a membrane; and (ii) the middle of an optical fiber modified with an inserted sensor unit composed of bound receptors surrounded by membrane.

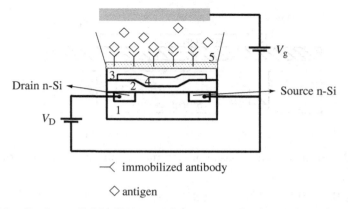

$-\!\!\prec$ immobilized antibody

$\diamond$ antigen

Fig. 2.13   Immuno-field-effect-transistor sensor for formation of antibody–antigen complex in solution (Mello and Kubota, 2002, with permission from Elsevier Ltd). V_D and V_g are drain and gate voltages, respectively.

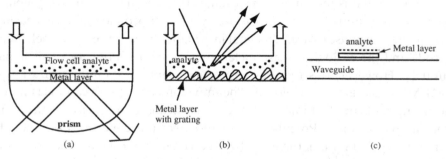

Fig. 2.14   Surface plasmon resonance sensors: (a) ATR-prism, (b) grating, and (c) optical fiber.

Another optical biosensor is based on surface plasmon resonance (SPR). Homola *et al.* (1999) made an extensive review of SPR biosensors. Three different concepts have been proposed and put into practical SPR sensors: (i) a prism complex with attenuated total reflection (ATR); (ii) a grating complex; and (iii) an optical waveguide. These three concepts are illustrated in Fig. 2.14. The ATR prism-based SPR system has been extensively used since its design by Kretschmann and Raether (1968) and Otto (1968). An optical wave incident on the prism with a specific angle θ_i is reflected and a surface plasmon wave propagates at the interface between the metal layer and the analyte layer if the propagation constant k_{SPR} of such a surface plasmon wave is given by

$$k_{SPR} = k_0 \sqrt{\frac{\varepsilon_m n_{a^2}}{\varepsilon_m + n_a^2}}, \qquad (2.1)$$

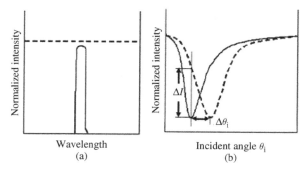

Fig. 2.15   SPR based on incident angle (angle modulation mode).

where k_0 is the free-space wavenumber, equal to $2\pi/\lambda_0$ (λ_0 is the free-space wavelength), n_a is the refractive index of the analyte, and ε_m is a complex-valued dielectric constant of the analyte, $\varepsilon_m = \varepsilon'_m - j\varepsilon''_m$, where $j = \sqrt{(-1)}$. The wave-vector corresponding to the incident wave along the interface (x-axis) is given by

$$k_x = k_0 n_{glass} \sin(\theta_{in}).\tag{2.2}$$

where n_{glass} is the refractive index of the prism glass.

When k_x is equal to k_{SPR}, a surface plasmon wave is generated, resulting in a reduction in the intensity of the reflected wave. There are two methods of analysis based on an ATR prism-based SPR system: angle modulation, and wavelength modulation modes. In the former design, a single wave such as a collimated laser wave, Fig 2.15(a), is used as an incident wave and the reflected intensity is measured as a function of the incident angle θ_i, Fig. 2.15(b). If a sensor chip is mounted on an unknown analyte layer, the reflected-intensity vs. θ_i relation is shifted as shown by the dashed line in Fig. 2.15(b). This shift can be measured in terms of the change, $\Delta\theta_i$, in θ_i, or the change in the intensity, ΔI, which can be measured more accurately than $\Delta\theta_i$. In the case of wavelength modulation using white light with an infinite number of wavelengths, Fig. 2.16(a), the intensities of the reflected multiple waves exhibit the lowest intensity at a certain wavelength of incident waves, the solid curve in Fig. 2.16(b). The details of SPR analysis based on Maxwell's equations and transmission-line theory is given in Subsection 7.6.2.

If an unknown analyte is mounted on the surface of the SPR sensor chip, the intensity–wavelength curve is shifted to the dashed line in Fig. 2.16(b), providing the change $\Delta\lambda$ in the wavelength. Cahill *et al.* (1997) designed an SPR sensor based on the wavelength modulation mode and the use of a prism with retro-reflection, Fig. 2.17. An advantage of using this retro-reflector SPR is the portability of the probe portion, which can be attached to any analyte medium. Akimoto *et al.* (2000) examined the sensitivity of this retro-reflection type SPR.

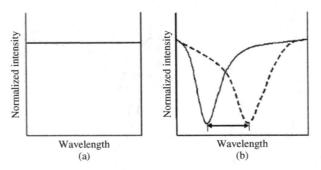

Fig. 2.16   SPR based on multiple incident waves (wavelength modulation mode) (Cahill *et al.*, 1997).

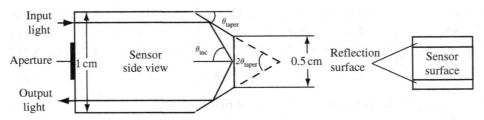

Fig. 2.17   Retro-reflector SPR system (Cahill *et al.*, 1997).

Akimoto *et al.* (2003) proposed an SPR design based on a step-like sensor surface which provides an integrated reference surface. Rich and Myszka (2002) made an extensive literature survey (700 papers published in 2001) of biosensors based on optical sensing methods, most of which are based on commercial SPR systems: Biocore AB (Uppsala, Sweden), Affinity Sensors (Franklin, MA, USA), Nippon Laser and Electronics (Nagoya, Japan), Texas Instruments (Dallas, TX, USA), IBIS Technologies (Enschede, The Netherlands), and Analytical μ-Systems (Woodinville, WA, USA).

Surgical BioMEMS

BioMEMS actuators are considered useful tools in surgical applications. A miniaturized piezoelectric inchworm actuator is a key component of surgical actuators. The concept of a piezoelectric inchworm actuator is illustrated in Fig. 2.18, where the key active component is the piezoelectric actuator with top and bottom electrodes. Under an applied voltage, the piezoelectric actuator expands by constant d_{13}. By repeating this process at a high-frequency voltage, the piezoelectric plate can move a sizeable distance.

The displacement u of the piezoelectric inchworm actuator per unit pulse is given by

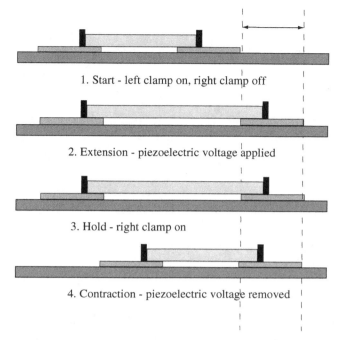

1. Start - left clamp on, right clamp off

2. Extension - piezoelectric voltage applied

3. Hold - right clamp on

4. Contraction - piezoelectric voltage removed

Fig. 2.18   Four-step operation of piezoelectric inchworm actuator. (1) Back foot clamped while front foot released, (2) piezoelectric plate extension, (3) front foot clamped, then back foot released, (4) piezoelectric plate contraction. It is to be noted that both front and back feet are permanently bonded to the piezoelectric plate while clamping (by electrostatic force) and release take place between the lower surfaces of feet and substrate. (After Polla *et al.*, 2000.)

$$u = \frac{L d_{13} V}{t} \tag{2.3}$$

where L and t are the length and thickness of the piezoelectric plate, V is the applied voltage and d_{13} is the piezoelectric constant. An example of the application of a piezoelectric inchworm actuator to ophthalmology is shown in Fig. 2.19 where the silicon lens attached to the top of the actuator moves in a step-wise fashion.

Therapeutic BioMEMS

Patients with a chronic disease can be benefited by a drug-delivery system that can be either implantable or transdermal (through skin).

An implantable drug-delivery system requires miniaturized BioMEMS. Polla *et al.* (2000) designed such a MEMS micro-pump based on silicon micromachining and a phosphorus zinc titanate (PZT) piezoelectric actuator.

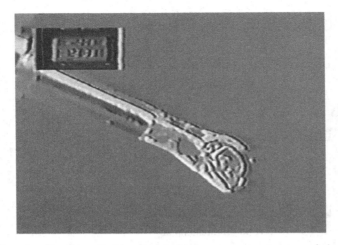

Fig. 2.19   Application of piezoelectric inchworm actuator to ophthalmology
(Polla *et al.*, 2000).

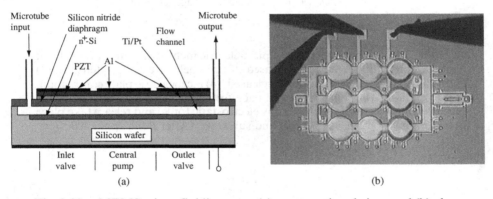

(a)   (b)

Fig. 2.20   MEMS micro-fluidic pump: (a) cross-sectional view, and (b) plan-
view photograph (Polla *et al.*, 2000).

Figure 2.20 (a), (b) illustrates the profile and plan view of the MEMS micro-
pump. The key actuator in the micro-pump is based on a PZT piezoelectric
dish with key piezoelectric constant d_{13}. Then the volume (ΔVol) to be
displaced and the maximum pressure ($P_{\max}$) needed can be estimated as

$$\Delta\mathrm{Vol} = \frac{3a^4(5 + 2\nu)(1 - \nu)d_{13}V}{3h^2(3 + 2\nu)}, \tag{2.4}$$

$$P_{\max} = \frac{6Ehd_{13}V}{(3 + \nu)a^2}, \tag{2.5}$$

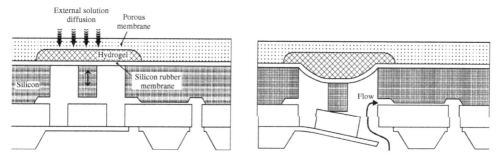

Fig. 2.21   Glucose-sensitive micro-valve based on a hydrogel (Ziaie *et al.*, 2004, with permission from Elsevier Ltd).

where a, h are the radius and the thickness of a diaphragm made of PZT and Si_3N_4, E and ν are the averaged Young's modulus and Poisson's ratio of PZT and Si_3N_4, and V is the applied voltage. If the pump is driven by 1 KHz AC, the flow rate to be pumped is estimated as $1\,\mu\ell/s$.

An alternative to a piezoelectric material for use in a micro-pump is hydrogel as the actuator material whose volume change upon stimulus (applied voltage, pH change, or temperature change) is much larger than any other active material. Hydrogels are also advantageous due to their flexibility and can be implantable in a human body if the hydrogel is sealed within a biocompatible membrane. Ziaie *et al.* (2004) reviewed several designs of hydrogel-based micro-fluidic systems with application to a glucose-sensitive insulin delivery system. Figure 2.21 illustrates a micro-valve design based on a hydrogel that can swell or shrink depending on pH, which is in turn sensitive to glucose.

2.3 Composites for sensors and actuators

The key components in actuators and sensors are active and sensing materials respectively, which include materials that exhibit coupled behavior between mechanical, thermal, electromagnetic, optical, and chemical properties, for example, ferroelectric, piezoelectric, pyroelectric, and magnetostrictive materials, conducting polymers, and shape memory alloys. These materials are often composed of a material exhibiting coupled behavior and a host matrix material. This is because, to achieve optimum performance, the composite structure of the material with coupled behavior is usually brittle and a host matrix material is needed to enclose the brittle phase. In the following Subsections we shall review examples of piezoelectric, magnetostrictive and magnetic composites.

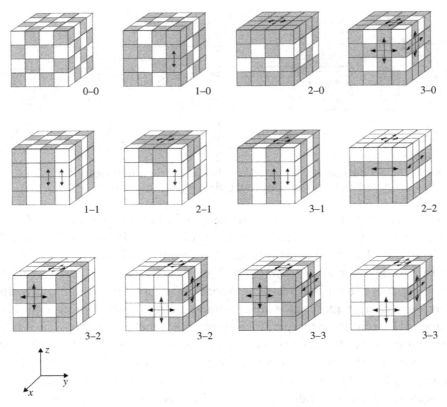

Fig. 2.22   Connectivity pattern of a piezoelectric composite. (Newnham *et al.*, 1978, with permission of Mater. Res. Soc.)

2.3.1 *Piezoelectric composites*

The development of piezoelectric composites has its roots in the early work of Newnham *et al.* (1978). Newnham laid out a terminology used to describe the microstructure of piezoelectric composites in terms of their connectivity, where a two-phase composite is denoted by an $X–Y$ system in which X and Y refer to the connectivity of the piezoelectric and non-piezoelectric phases, respectively, Fig. 2.22. By this definition, a 0–3 composite denotes the case of piezoelectric particles (no connectivity, i.e., isolated islands) embedded in a non-piezoelectric matrix, such as a polymer matrix (continuous in three dimensions), while 3–0 denotes a composite composed of a piezoelectric matrix (continuous in three dimensions) with spherical voids (isolated islands, thus no connectivity). Composites of 1–3 connectivity are those composed of piezoelectric rods or fibers (connectivity in one dimension) and a polymer matrix (connectivity in three dimensions). These are often fabricated by embedding continuous piezoelectric rods in a polymer matrix, while 3–1 composites are made by boring holes in a solid piezoelectric ceramic and impregnating them with a soft polymer. Newnham's group (Newnham *et al.*, 1978; Klicker *et al.*, 1981) advanced the understanding of the physical behavior of piezoelectric composites through the development of *mechanics-of-materials*-type parallel and series models. This understanding, particularly of 1–3 composites, was significantly enhanced by Smith and co-workers (Smith *et al.*, 1985; Smith, 1989; Smith, 1990; Smith and Auld, 1991; Smith, 1993) in the analysis of continuous fiber-reinforced piezoelectric composites. In these approaches, a simplifying assumption of either a constant stress or a constant strain component in the composite has been made which then leads to the application of Voigt- and Reuss-type estimates of electroelastic constants. More recent developments in the modeling of piezoelectric composites based on rigorous micromechanics modeling are those of Dunn and his co-workers (Dunn and Taya, 1993a,b; Dunn and Wienecke, 1997). These include the dilute, self-consistent, Mori–Tanaka, and differential scheme models. The key to each of these methods is the use of stress and strain concentration factor tensors obtained through the solution of a single particle embedded in an infinite medium, while the goal in the analysis of composite materials is the determination of the effective properties of the homogenized piezoelectric composite, which are functions of the electroelastic properties of the constituent phases and the microstructure.

Uchino *et al.* (1987) fabricated a monomorph made from semiconductive piezoelectric ceramics. A rainbow-type actuator was developed by Hartling (1994) by reducing one surface of a PZT wafer that creates a non-uniform electric field when voltage is applied. Wu *et al.* (1996) fabricated a piezoelectric actuator where the electrical resistivity was graded through the plate thickness, leading to a graded poling of the material. A piezoelectric bimorph produces a high bending displacement at the cost of a high stress concentration at the interface between the top and bottom layers, reducing its lifetime (Gamano, 1970). The use of epoxy resin to bond the top and bottom layers in the bimorph piezoelectric actuator made the interface fragile and fatigue prone (Uchino, 1998). In order to reduce the high stress concentration at the mid-interface, the concept of functionally graded microstructure (FGM) has been introduced, where a piezo-actuator of the bimorph type is composed of a number of laminae with each being a piezoelectric composite, and the electroelastic properties of the laminae are graded through the thickness direction such that mirror symmetry with respect to the mid-plane is maintained. In order to identify the optimum microstructure of FGM piezoelectric laminated composites, Taya and his co-workers have developed two-stage modeling: (1) homogenization of each FGM layer (lamina) based on the Eshelby

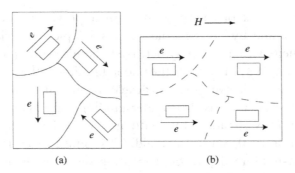

(a)   (b)

Fig. 2.23   Strain axis (shown by arrows) associated with the crystal lattice in each magnetic domain rotates from (a) demagnetized state to (b) magnetized state at saturation during increased magnetization. (Chikamizu, 1964.)

model (Dunn and Taya, 1993a,b); and (2) the classical laminate theory (Almajid *et al.*, 2001; Almajid and Taya, 2001). A detailed account of this modeling is given in Chapter 7.

2.3.2 *Magnetostrictive composites*

Magnetostriction is the change of shape or strain induced in a ferromagnetic material during its magnetization. Under an applied magnetic field, the crystal lattice in each magnetic domain deforms spontaneously in the direction of the domain magnetization. For a small magnitude of the applied magnetic field, the directions of magnetization and strain in each domain, which are more or less coincident, rotate toward the applied field, see Fig. 2.23.

If we assume that the spontaneous strain e in each magnetic domain is independent of the crystallographic direction of magnetization, the magnetostrictive strain λ_{dem} (or macroscopic strain of a ferromagnetic material) in its demagnetized state, Fig. 2.23(a), is related to e as

$$\lambda_{\text{dem}} = \int_0^{\pi/2} e \cos^2\phi \sin\phi \, d\phi = \frac{e}{3}, \qquad (2.6)$$

where the direction of spontaneous strain makes an angle ϕ with respect to the macroscopic specimen (observation) axis. In the saturated state $\lambda_{\text{s}} = e$. Therefore, the net strain λ at saturation magnetization to be observed between the initially demagnetized state and the saturated state is obtained as

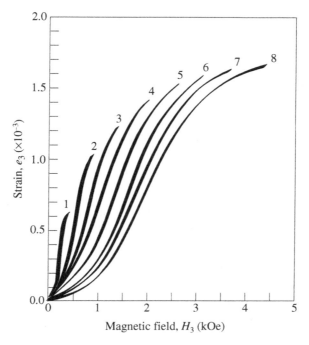

Fig. 2.24   Strain vs. applied magnetic field for Tefenol-D ($Tb_{0.3}Dy_{0.7}Fe_2$) under various compressive stress levels where numbers 1–8 denote constant stresses of 6.9, 15.3, 23.6, 32, 40.4, 48.7, 57.1, and 65.4 MPa, respectively (after Moffett *et al.*, 1991).

$$\lambda = \lambda_s - \lambda_{\text{dem}} = \frac{2}{3}e. \qquad (2.7)$$

Magnetostriction is the coupling behavior between mechanical and magnetic behavior where the strain λ induced is proportional to the square of the applied magnetic field (H). The largest magnetostrictive strain that has been reported is for the alloys of iron (Fe) with the rare-earth elements, terbium (Tb) and dysprosium (Dy). One of these alloys is Tefenol-D (Clark, 1993); a typical curve of the strain e and magnetic field H of Tefenol-D ($Tb_{0.3}Dy_{0.7}Fe_2$) is illustrated in Fig. 2.24 (Moffett *et al.*, 1991) where the e–H relation exhibits non-linear behavior (approximated as $e \propto \lambda H^2$) and also low hysteresis. The order of strain is as high as 1.5×10^{-3}. This high strain and low hysteresis at fast actuation speed makes this Tefenol-D very attractive for applications in transducers. However, Tefenol-D is known to be brittle and electrically conductive (thus resulting in higher eddy-current loss), limiting its use at higher frequencies. To overcome these difficulties, composites made of

Tefenol-D fillers and polymer matrix have been designed, which are mechanically tougher and whose eddy current loss at high frequencies can be minimized by using a polymer insulator matrix.

Sandlund *et al.* (1994) designed two kinds of magnetostrictive composites, isotropic and anisotropic composites. In the isotropic composites, Tefenol-D particles are randomly distributed, while in the anisotropic composites the Tefenol-D particles are magnetically oriented during processing. They measured the elastic modulus, magnetostriction, and piezo–magnetic coupling coefficient of these composities to conclude that these properties of the anisotropic composite are superior to those of the isotropic composite. Kim *et al.* (1998a) used amorphous Tb–Fe–B ribbons as a magnetostrictive filler to design magnetostrictive composites whose magnetostriction reached the value of 500×10^{-6} at $H = 1.1\,\mathrm{kOe}$ while maintaining good compressive strength of the composite. They optimized the processing parameters to obtain higher magnetostriction (536×10^{-6} at $H = 1.1\,\mathrm{kOe}$), piezo–magnetic coupling coefficient and compressive strength (Lim *et al.*, 1999).

The University of California at Los Angeles (UCLA) group (McKnight and Carman, 2001; Nersessian and Carman, 2001) has designed magnetostrictive composites whose magnetostriction at saturation is of order 1×10^{-3}. If the magnetostrictive fillers are made of Tefenol-D particles oriented along [112], the magnetostriction λ_c of the composite is increased to 1.6×10^{-3}, and it is expressed as

$$\lambda_c = \frac{3}{2}\lambda_s \left(\frac{M}{fM_s}\right)^2, \tag{2.8}$$

where λ_s is the saturation magnetostriction along the prescribed crystal axis, M and M_s are the applied and saturation magnetization, respectively, of Tefenol-D, and f is the volume fraction of the Tefenol-D fillers.

Since magnetostriction is the coupling behavior between mechanical and magnetic behavior, the elastic (Young's) modulus of the polymer matrix is expected to have an impact on the overall magnetostrictive strain λ. Chen *et al.* (1999) studied the effect of the matrix elastic modulus E_m on λ both experimentally and theoretically. Figure 2.25 (a) and (b) show the comparison between the experimental data of λ and E_m, and volume fraction f of Tefenol-D filler, respectively.

The model proposed by Chen *et al.* (1999) is based on two types of "laws of mixture": (1) uniform strain model; and (2) uniform stress model, see Fig. 4.1. For the uniform strain model, the composite stress σ_c is the sum of the stress σ_f in the filler with its volume fraction f and the stress σ_m in the matrix with its volume fraction $(1-f)$,

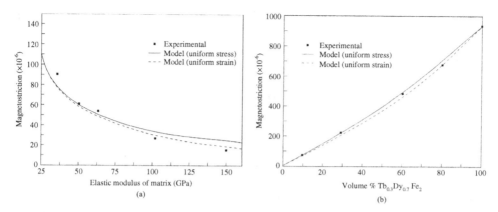

Fig. 2.25   (a) Magnetostriction vs. Young's modulus E_m of polymer matrix and (b) magnetostriction vs. volume % Tefenol-D (Tb$_{0.3}$Dy$_{0.7}$Fe$_2$). (Chen *et al.*, 1999, with permission from Amer. Inst. Phys.)

$$\sigma_c = (1-f)\sigma_m + f\sigma_f. \tag{2.9}$$

The elastic Hooke law holds for each phase, composite (c), filler (f), and matrix (m), hence

$$\sigma_c = E_c e_c, \tag{2.10a}$$

$$\sigma_m = E_m e_m, \tag{2.10b}$$

$$\sigma_f = E_f e_f. \tag{2.10c}$$

From Eqs. (2.9) and (2.10) and using the uniform strain condition, i.e.,

$$e_c = e_m = e_f, \tag{2.11}$$

the composite Young modulus is obtained as

$$E_c = (1-f)E_m + fE_f. \tag{2.12}$$

For the uniform stress model, the composite strain is the sum of the strain in the filler with volume fraction f and the strain in the matrix with volume fraction $(1-f)$, i.e.,

$$e_c = (1-f)e_m + fe_f. \tag{2.13}$$

Substituting Hooke's law of Eqs. (2.10) into Eq. (2.11) and using uniform stress conditions, i.e.,

$$\sigma_c = \sigma_m = \sigma_f, \tag{2.14}$$

we obtain the composite Young modulus as

$$E_c = \frac{1}{\dfrac{(1-f)}{E_m} + \dfrac{f}{E_f}}. \tag{2.15}$$

It is assumed in the model of Chen *et al.* (1999) that the stress induced in a magnetostrictive composite is solely due to the stress in the magnetostrictive filler; and the composite stress, when the composite is magnetically saturated, is the product of the volume fraction f of the magnetostrictive filler and its stress, i.e.,

$$E_c\lambda_s = fE_f\lambda_{sf} \tag{2.16}$$

where λ_s is the magnetostriction of the composite at saturation, and λ_{sf} is that of the magnetostrictive filler. When E_c is substituted from Eq. (2.12) or Eq. (2.15) into Eq. (2.16), one can obtain the magnetostriction at saturation of the composite based on the uniform strain model (λ_{se}) and the uniform stress model ($\lambda_{s\sigma}$) respectively,

$$\lambda_{se} = \frac{fE_f\lambda_{sf}}{(1-f)E_m + fE_f}, \tag{2.17a}$$

$$\lambda_{s\sigma} = f\lambda_{sf}E_f\left(\frac{1-f}{E_m} + \frac{f}{E_f}\right). \tag{2.17b}$$

The predictions of Eqs. (2.17) are shown by the solid line (uniform stress model) and the dashed line (uniform strain model) in Fig. 2.25. Both models predict well the experimental results.

2.3.3 *Magnetic composites*

More-compact actuators are highly desirable in a number of applications. This demands even more-compact electric motor and electromagnet systems as sources of driving high-density power where both stator magnets and moving rotors are made of soft and hard magnets with the requirement of high magnetic properties. The requirements of a soft magnet are high permeability μ_r, high magnetization B_s at saturation, and small hysteresis or equivalently low coercivity H_c at use frequencies f. Those of a permanent magnet are high remanent flux B_r or equivalently high B_s, high coercivity H_c, and high magnetic energy $(BH)_{max}$. Some of the soft and hard magnets are used as moving parts, such as the rotor, which then requires additional properties – lighter weight and mechanical toughness – to render them usable for longer time. Morita *et al.* (1985) designed Fe powders reinforced with epoxy resin and Kevlar® fiber composites for use in the magnetic wedge in an induction motor.

The key constituent magnetic materials for the above magnetic composites are nano-crystalline Nd–Fe–B alloys. Thus, we shall discuss the magnetic characteristics of these first, followed by the detailed design and process issues of magnetic composites.

Nano-crystalline Nd–Fe–B alloys

Since Matumoto and Maddin processed the first amorphous metal in 1970, a number of amorphous metals, particularly magnetic alloys, have been studied. Early studies on amorphous magnetic materials centered on Fe–Co alloys (Fujimori *et al.*, 1974) which are extensively used as commercial magnetic materials but whose price remains still high. The high price of the Fe–Co system prompted researchers over the last 12 years to turn their attentions to less-expensive alloys. Yoshizawa *et al.* (1988) reported first a successful soft magnetic material based on Fe–Si–B–Nb–Cu nano-crystals of size 10 nm. Coehoorn *et al.* (1988) reported an $Nd_2Fe_{14}B/Fe_3B$ nano-composite that exhibited large B_s (=1.3 tesla [T]) and large H_c (= 0.3 mega-ampere/meter [MA/m]). The Tohoku University group then found other nano-crystalline soft magnetic materials based on Fe–M–B where M can be Zr, Hf, or Nb (Suzuki *et al.*, 1990 Makino *et al.*, 1994, Makino *et al.*, 1995) and its variations (Makino *et al.*, 1997). By careful arrangement of a soft phase (t-Fe_3B, Fe_3B, α-Fe), a hard phase ($Nd_2Fe_{14}B$), and an amorphous phase, one can design both hard and soft nano-composites, which are illustrated schematically in Fig. 2.26(a) and (b), respectively.

It has been confirmed by the above researchers and others that soft (α-Fe, t-Fe_3B) and hard ($Nd_2Fe_{14}B$) nano-crystals have very high B_s ($\approx$1.6 T), and if they are formed as nano-composites, they exhibit strong exchange coupling where the soft phase plays the role of a 3d transition metal while the hard phase's function is that of a 4f rare-earth element. By virtue of the exchange coupling, the nano-composite can exhibit a wide range of magnetic characteristics, from soft to hard, depending on the composition ratio. Another feature of the nano-composite is its reversible behavior in the second quadrant of the B–H curve, i.e., "spring-back" (Kneller and Hawig, 1991) where the soft phase is forced to change its magnetization direction into that of the applied

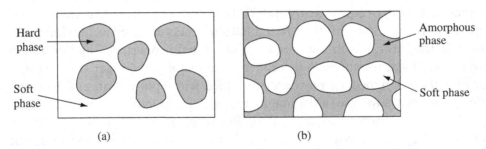

Fig. 2.26   Nano-composites: (a) hard magnet, and (b) soft magnet.

demagnetization field, but its movement is constrained by the hard phase which acts as a bias spring, so that the local B–H curve becomes reversible.

Hinomura *et al.* (1997) reported that Nd–Fe–B ribbons with a specific composition which were processed by the spin-melting technique can be tailored from soft to hard nano-composites by careful annealing treatments. The higher the annealing temperature, the harder the nano-crystal Nd–Fe–B becomes, see Fig. 2.27.

Hard magnetic composites

Hard magnetic composites are composed of hard magnet fillers and bonding (matrix) materials (usually polymer) where the hard magnet fillers are required to have high B_s and H_c which are now available from nano-magnetic composites such as the Nd–Fe–B system.

Composites made of hard ferromagnetic particles and polymer matrix are called "bond magnets" whose magnetic performance is less than for "sintered" magnets. For the magnetization process to convert the "green" composite body to a permanent magnet with strong and anisotropic magnetization requires simultaneous sintering and magnetization at high temperature, which cannot be used for bond magnets as the polymer phase cannot survive at a high temperature. Despite their lower magnetic performance, bond magnets are used extensively due to their low cost and ease of forming into any 3D shape.

Ferromagnetic shape memory alloy (FSMA) composites

An unusual type of ferromagnetic material is emerging as an actuator material. This is the "ferromagnetic shape memory alloy" (FSMA), such as Fe–Pd and Ni–Mn–Ga alloys, which possesses both ferromagnetic and martensitic transformation properties (Sohmura *et al.*, 1980; Webster *et al.*, 1984; Ullakko *et al.*, 1996; James and Wuttig, 1998; O'Handley, 1998; Liang *et al.*, 2000; Liang *et al.*, 2001; Kato *et al.*, 2002a,b; Sakamoto *et al.*, 2003; Wada *et al.*, 2003a). There are

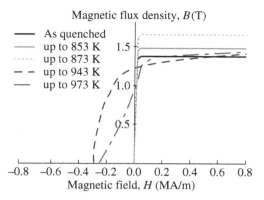

Fig. 2.27   Demagnetization curves measured at room temperature for $Nd_{4.5}Fe_{77}B_{18.5}$ ribbons as quenched, and annealed at various temperatures (after Hinomura *et al.*, 1997, with permission from Japan Inst. of Metals).

three mechanisms of actuation associated with FSMAs under an applied magnetic field, which can be used as the driving force for fast responsive actuator materials: (i) magnetic-field-induced phase transformation, (ii) martensite variant rearrangement, and (iii) hybrid mechanism.

The first mechanism is based on the phase change from austenite to martensite under an increasing magnetic field, or the reverse phase transformation under a decreasing magnetic field. If we construct a three-dimensional phase-transformation diagram with stress (σ)–temperature (T)–magnetic field (H) axes, see Fig. 2.28, the phase transformation of austenite to martensite under a modest magnetic field requires that the T–H phase boundary surface be inclined toward the T-axis, otherwise the increasing H loading would not intersect the T–H phase boundary surface. Kato *et al.* (2002a) made a preliminary estimate of the magnetic energy needed to induce a phase transformation based on a thermodynamic model, and concluded that a large H-field is required for the phase change in both Ni–Mn–Ga and Fe–Pd. Therefore, this mechanism is not suited for use in designing compact actuators, which may need a small and portable electromagnetic system as a driving unit.

The second mechanism is to induce the strain in an FSMA with 100% martensite phase subjected to a constant H-field which acts on the magnetic moments in the magnetic domains that exist in the martensite phase so as to rotate them along the easy axis, i.e., the c-axis in the case of Ni–Mn–Ga and Fe–Pd. The strain induced by the mechanism is a function of the c/a ratio of the FSMA, i.e., the order of the shear strain, given by $a/c - c/a$. Thus, the smaller the c/a ratio, the larger the shear strain that can be induced by this mechanism. The c/a ratio of Ni–Mn–Ga is reported to be 0.94, which could

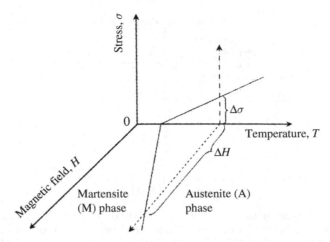

Fig. 2.28 Three-dimensional phase transformation diagram for a ferromagnetic shape memory alloy (FSMA) where loading is composed of three axes, stress σ, temperature T, and magnetic field H.

provide 6% or more strain. Even though the strain induced by the second mechanism is very large, the corresponding stress remains modest, just several MPa, under a modest applied magnetic flux density (1T). The variant rearrangement mechanism (the second mechanism) is still useful for micro-electromechanical systems (MEMS)-based actuators as the order of force and stroke in a typical MEMS-based actuator is modest but the fast speed of actuation is attractive to MEMS designers. Variant rearrangement in martensite has been also reported in Ni–Mn–Ga (Ullakko *et al.*, 1996; Sozinov *et al.*, 2001; O'Handley *et al.*, 2003), Fe–Pt (Sakamoto *et al.*, 2003; Kakeshita *et al.*, 2000), Fe–Pd (Kato *et al.*, 2002a,b; Yamamoto *et al.*, 2004), and Ni–Mn–Al systems (Fujita *et al.*, 2000). The order of reversible strain induced by a magnetic field was only 0.2% in Ni_2MnGa in an earlier study (Ullakko *et al.*, 1996), but it increased to 6% in an Ni_2MnGa single crystal with off-stoichiometric composition (Murray *et al.*, 2000; Murray *et al.*, 2001). This remarkable improvement in the magnetic field-induced strain was acheived with an external bias stress which helps converted variants to return to initial ones upon removal of the magnetic field.

The third mechanism, which we call a "hybrid mechanism," is based on a set of chain reactions, first applied magnetic flux (or field) gradient, magnetic force, stress-induced martensite phase transformation (see Fig 2.28), resulting in the phase change from a stiff austenite phase to a soft martensite phase, leading to a large displacement. The advantages of this are a large stress (hundreds of MPa in the case of Fe–Pd), a modest-to-intermediate strain, and a fast actuation time. This phase change can be applied by bringing close to the FSMA specimen a compact and portable magnet which provides a large magnetic field gradient, and thus is suited for use in designing compact actuators with large force capability.

The above FSMA systems are rather difficult to process and the cost of the materials is also high. To make more affordable FSMA systems, one can use the concept of a composite, i.e., use two different starting materials, a ferromagnetic material and a shape memory alloy of superelastic grade, to form an FSMA composite. If this FSMA composite is activated using the hybrid mechanism, one can induce large displacement while mantaing a large stress and a high actuation speed. There are two routes for designing such FSMA composites, one is a laminated FSMA composite, the other is a particulate FSMA composite. Kusaka and Taya (2004) made an analytical study of a laminated FSMA composite, with the aim of determining the optimum microstructure for use as a bending actuator and also as an axial spring actuator.

One can design an FSMA composite composed of a ferromagnetic material and a shape memory alloy whose cost may be less than those of Fe–Pd and Ni–Mn–Ga (Kusaka and Taya, 2004). Taya and his co-workers have designed several actuators based on FSMA composites (Liang *et al.*, 2003; Wada *et al.*, 2003b; Taya *et al.*, 2003).

2.4 Control of electromagnetic waves

The effective complex refractive index, which is given by the square root of the product of relative permittivity ε_r and permeability μ_r, determines the EM wave propagation characteristics, such as phase velocity, in composite materials. Materials with a very high refractive index are extremely useful for microwave applications because the effective wavelength inside the material will be much smaller than that of free space. Most materials are non-magnetic; therefore, the high refractive index of materials such as ceramic composites is due solely to their high ε_r values. However, it has been shown that ε_r can be controlled by adding magnetic dipoles of certain orientation into the host materials. By using a composite concept one can increase both ε_r and μ_r. One way to increase ε_r is by the conductor loading technique, though techniques to increase ε_r of non-magnetic materials are not yet well established.

2.4.1 Very high-ε_r meta-materials

High-ε_r materials such as barium tetratitanate ($\varepsilon_r = 37$ at 6 GHz) are widely used for wireless devices as a substrate of miniature-sized antennae and dielectric resonators. A typical antenna has a dimension such as patch length or line length of approximately $\lambda/2$ to $\lambda/4$ where λ is the wavelength inside the

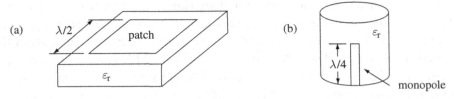

Fig. 2.29   Examples of patch antenna (GPS application, 1.5 GHz) and resonator-backed antenna (Global Star application, 2.5 GHz) (Carver and Mink, 1981).

substrate. Because most wireless systems operate at a lower microwave region with a fairly long wavelength, it is important to fabricate antennae on a high-ε_r substrate to reduce the physical dimensions. For example, the Global Star transmitting antenna for the Qualcomm unit uses two monopole antennae attached to a dielectric resonator which has ε_r of 36. The length of each monopole is only 10 mm, which is much less than the free-space $\lambda_0/4$ (Global Star: 1.6 GHz transmitter and 2.5 GHz receiver). To design small antennae for wireless systems such as the global positioning system (GPS) and Global Star, therefore, the substrate materials must have very high ε_r, and the loss tangent must be small to maximize the efficiency. Figure 2.29 shows typical patch and dielectric resonator-backed antennae.

There have been significant US Department of Defense (DoD) and industry efforts to reduce the size of microwave antennae in recent years. Small, rugged, low-profile antennae can be mounted on military uniforms, helmets, and vehicle surfaces. The antenna size reduction can be done by two approaches. The first one is a clever design of the antenna geometry. The second approach is to develop new, very high-ε_r materials. Typical high-ε_r substrate materials are composites based on titania ($\varepsilon_r = 96$ at 6 GHz) whose ε_r ranges from 30 to 100. These dielectric materials also have low loss at microwave frequency. Although materials such as barium titanate ($BaTiO_3$) and strontium titanate ($SrTiO_3$) have very high ε_r (2000 to 4500 within a limited temperature range), they have a limited use as antenna substrates.

It is known that when dielectric materials are loaded with conducting particles, both real and imaginary parts of ε_r will increase as the fractional volume of filler increases (Lee *et al.*, 2000b). This technique has been used for designing lossy dielectric materials such as beryllia–silicon carbide (BeO–SiC) for microwave tube applications. Table 2.4 shows the measured complex dielectric constant of BeO(60%)–SiC(40%) at different frequencies. These values are much higher than that of the undoped beryllia (BeO), which is close to 6.4 in the microwave region. BeO(60%)–SiC(40%) is designed to

Table 2.4 *Measured relative complex dielectric constant of* BeO(60%)–SiC(40%) *composite.* BeO *without* SiC *has a dielectric constant of* 6.4–j0.002 *at* 10 GHz.

Frequency (GHz)	$\varepsilon_r = \varepsilon_r' - j\varepsilon_r''$	
	Real, ε_r'	Imaginary, ε_r''
0.4	90	11
0.9	86.6	15.5
1.95	80.5	21.9
2.99	78.6	28.9
8.5	33	13.5
10.0	31	13
12.0	30	13.2

absorb microwave energy and one of the requirements is a high loss tangent, which can be controlled by the amount of SiC.

Through careful experimentation using AlN, Lee *et al.* (2000a) found that the rate of increase of both real and imaginary parts of ε_r is related not only to particle concentrations but also to the particle size, shape and orientation. Figure 2.30 shows microstructures of $BaTiO_3$–Pt composites with varying levels of Pt conducting particles. Lee and co-workers have also conducted numerical simulations and confirmed that the change of ε_r is related to the particle size and shape. This suggests that it is possible to increase the real part while keeping the increase of the imaginary part to a minimum by selecting the type of inclusion and manufacturing process. However, the simple approach of mixing conducting particles with a host will not produce the desired characteristics. One needs to custom-design material compositions and processing methods. The material processing must be guided by theoretical bases and extensive numerical analysis.

2.4.2 Non-magnetic meta-material with ε_r greater than 1

The effective relative permeability μ_r can be modified to be either less or greater than 1 even for non-magnetic materials. The split-ring resonator is one approach. Another approach is periodically located conducting rings, as shown in Fig. 2.31(a). We are interested in the frequency region in which the size of the ring is much less than the wavelength. Kuga *et al.* (2005) have conducted preliminary numerical simulations using the finite element method (FEM) (HFSS from Ansoft). The material consists of layers of conducting rings placed inside the X-band waveguide. It is assumed that the substrate

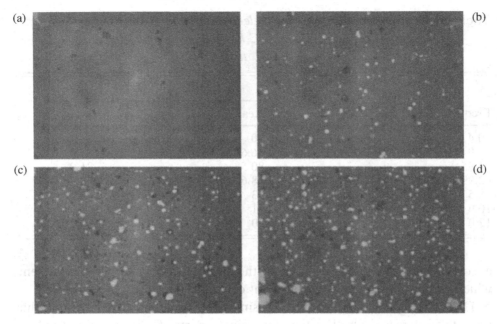

Fig. 2.30   Microstructure of BaTiO$_3$–Pt composites with (a) 0, (b) 3, (c) 5, and (d) 10 vol.% Pt (Kuga *et al.*, 2005).

which holds the rings is a thin styrofoam ($\varepsilon_r = 1.05$). The reflection S_{11} and transmission S_{21} coefficients are computed using HFSS at microwave frequency. This provides four real numbers which can be related to two complex values ε_r and μ_r. The inversion method is used for estimating the effective ε_r and μ_r. Figure 2.31(b)–(d) shows numerical results for an electronic composite with a periodic array of conducting rings (Kuga *et al.*, 2005).

2.4.3 *Active meta-materials with negative to large positive ε_r and μ_r*

Pendry and his group (1996, 1998) proposed a composite with negative ε_r and μ_r which is made of a non-magnetic, metal-based split-ring resonator (SRR) and an insulator housing plate, Fig. 2.32(a). This work was immediately followed by the UCSD group (Smith *et al.*, 2000), who measured transmission power as a function of frequency, and claimed that a composite with negative ε_r and μ_r is indeed realized. The above researchers studied theoretically the effects of several parameters on the effective values of ε_r and μ_r, such as radius, gap between inner and outer split ring, and the thickness of these split rings. Additional key parameters influencing strongly the effective values of ε_r and μ_r

Non-magnetic meta-material with $\varepsilon_r > 1$

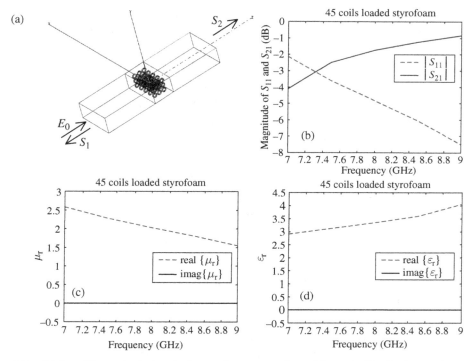

Fig. 2.31   Numerical simulations of a composite with periodically located conducting rings. (a) Structure, (b) magnitude of S_{11} and S_{21} responses, (c) effective μ_r, and (d) effective ε_r. (After Kuga *et al.*, 2005.)

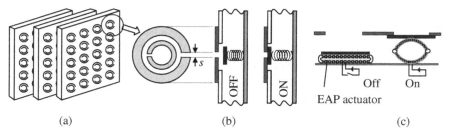

Fig. 2.32   Tunable split-ring resonator board: (a) entire view, (b) gap s can be opened or closed by a mechanical switch based on a shape memory alloy actuator, (c) electro-active polymer (EAP) mechanical switch.

are the split gap(s) and also the electric properties (ε and μ) of the housing plate. Figure 2.32(b) and (c) illustrate the design of such an active antenna system in the control of opening/closing of the split-ring gap, where (b) illustrates active control by use of a shape memory alloy (SMA) spring (via temperature change) and (c) that by an electro-active polymer switch (via an applied voltage of 1V).

Active composites will be useful for many applications. One example is a reflective surface which excites surface waves called plasmons. It is well known that for transverse magnetic (TM) incident polarization the surface plasmons can be excited for a surface with negative ε_r. By solving the wave equations at the boundary of free space and the dielectric layer, one can obtain the condition $\varepsilon_r P_0 = -P$ where, Fig. 2.33, P_0 and P are complex propagation constants normal to the boundary in free space and in the dielectric material, respectively (Ishimaru, 1991). The imaginary part of P_0 and P which gives an attenuating wave must be negative to satisfy the radiation condition. From this equation, it can be shown that a condition to excite plasmons is when ε_r becomes negative for non-magnetic material ($\mu_r = 1$). There will be very little reflection from the surface when the plasmons are excited. For the transverse electric (TE) case, one can obtain a similar expression. From this, one can show that there is no plasmon excitation for $\mu_r > 1$. However, as in the TM case, the surface plasmons can be excited if ε_r becomes negative for the TE incident wave.

Optical sensors which use surface plasmons have been applied in many areas (Mansuripur and Li, 1997). However, there has been little effort to use surface plasmons for microwave applications because of a lack of materials with negative ε_r and/or μ_r. The active SRR will exhibit negative μ_r when the gap is open, within a limited frequency range. One can demonstrate the existence of plasmons for the TE case using a split ring. By combining structures which create negative ε_r, such as a dipole, one can develop surfaces which can support plasmons for both TE and TM incident waves. By tuning the gaps, one can change material properties which, in turn, excite or suppress plasmons on the surfaces. Radar cross section (RCS) and radar polarimetric signature of military targets are closely related to the surface reflectivity of materials. Using active composites, it will be possible to control and disguise the target characteristics to enemy radars.

Another application of composites with negative ε_r and/or μ_r is in near-field microwave imaging systems. The traditional near-field microscopes use the evanescent wave in an optical fiber whose diameter is much less than the wavelength. Although there is no propagating mode in a small optical fiber, the evanescent mode still exits and it can be used for optical imaging. The spatial resolution is determined by the tip size of the optical fiber and it can be much smaller than the wavelength. A similar approach has also been applied to microwave imaging with tapered waveguides. Pendry *et al.* (1996, 1998) proposed an optical imaging technique which also uses the evanescent wave in a thin gold layer ($\varepsilon_r < 0$) (*IEEE Spectrum*, Jan. 2001). With the proposed meta-materials which have negative ε_r and/or μ_r, one can construct a micro-wave/millimeter-wave

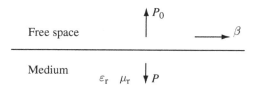

Fig. 2.33   Definitions of complex propagation constants (P_0 and P) near a boundary. Here $\beta\,(=2\pi/\lambda)$ is the phase constant.

imaging system which has a spatial resolution much less than the wavelength. This system will be useful for non-destructive testing of materials.

2.4.4 Wave-absorbing materials

Housing materials that can absorb electromagnetic waves at quasi-microwave frequencies (1–3 GHz) are increasingly important for wireless telecommunication applications such as cellular phones, personal computers (PCs), and local access networks (LANs). Operational interference can be minimized if one can use a housing material with high absorbing capacity. To design thinner wave absorbers, one needs to use materials with high absorption, i.e., low reflection loss where the reflection loss Γ of a wave-absorbing material backed by a metal plate is defined by

$$\Gamma\ \text{(dB)} = -20\ \log\left|\frac{Z\tanh(\gamma d) - Z_0}{Z\tanh(\gamma d) + Z_0}\right|. \tag{2.18}$$

Here Z, γ, and d are characteristic impedance, propagation constant, and thickness, respectively, of the wave absorber and they are defined by

$$Z = \sqrt{\frac{\mu}{\varepsilon}};\ Z_0 = \sqrt{\frac{\mu_0}{\varepsilon_0}}, \tag{2.19a}$$

$$\gamma = 2\pi f\sqrt{\mu\varepsilon} \tag{2.19b}$$

where μ and ε are complex permeability and permittivity (dielectric constant), given by

$$\mu = \mu' - j\mu'', \tag{2.20a}$$

$$\varepsilon = \varepsilon' - j\varepsilon''. \tag{2.20b}$$

The first and second terms on the right-hand side of Eq. (2.20) are real and imaginary parts, respectively, and $j = \sqrt{-1}$. The requirements of a wave absorber are to minimize reflection from the absorber material. To this end,

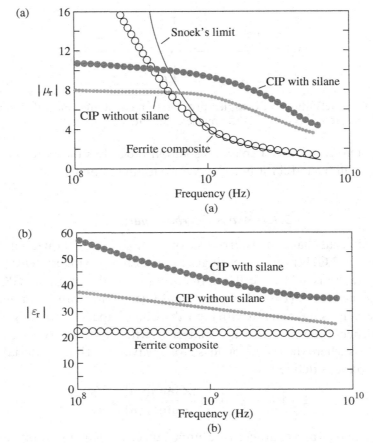

(a)

(b)

Fig 2.34　Frequency dependence of wave-absorbing composites: (a) relative permeability μ_r, and (b) relative permittivity ε_r. (Matsumoto and Miyata, 1999b.)

characteristic impedance Z is made close to Z_0. In addition, to make the wave absorber thinner, the attenuation of waves must be maximized, which can be realized by the maximization of the values of the imaginary relative permeability μ_r'' and the imaginary relative permittivity ε_r'' for a given use frequency. Matsumoto and Miyata (1999b) designed a high-performance wave-absorbing material which is composed of carbonyl-iron particles (CIP) and a polymer matrix. Figure 2.34(a) and (b) illustrate the absolute values of the complex relative permeability μ_r and permittivity ε_r of a CIP/polymer composite as a function of frequency. The figure demonstrates the effectiveness of using a CIP/polymer composite as a wave absorber as compared with the conventional ferrite particle/polymer composites. If a silane coupling agent is used to coat the

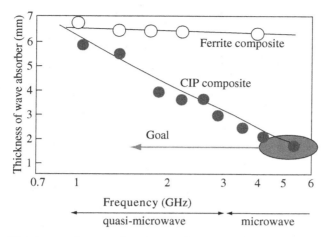

Fig. 2.35 Thickness of wave absorber vs. frequency (Matsumoto and Miyata, 1999b).

CIP, both $|\mu_r|$ and $|\varepsilon_r|$ increase, thus contributing to the design of a thinner wave absorber. Figure 2.35 illustrates the effectiveness of using CIP fillers to make thinner wave absorbers, although there still exists the challenge of designing an even thinner CIP composite plate for the quasi-microwave region.

Model for laminated composite

Matsumoto and Miyata (1997) designed two types of composite: (1) a laminated composite composed of a magnetic layer with permeability μ and polymer matrix with permittivity ε, Fig. 2.36(a); and (2) a composite composed of the same constituents but with their spatial arrangement random, Fig. 2.36(b).

Matsumoto and Miyata proposed a simple model to predict the composite ε and μ for the above two types of composite geometry. For the laminated composite, the magnetic field propagates along the plane of lamination (y-axis in Fig. 2.36(a)), thus the magnetic flux density B_c in the composite is the sum of that in each consititiuent multiplied by its volume fraction, v_1, and v_2 with $v_1 + v_2 = 1$, i.e.,

$$B_c = v_1 B_1 + v_2 B_2 \tag{2.21}$$

$$B_1 = \mu_1 H_1 \tag{2.22a}$$

$$B_2 = \mu_2 H_2 \tag{2.22b}$$

$$B_c = \mu_c H_c, \tag{2.22c}$$

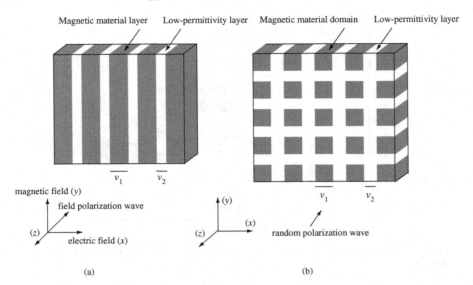

Fig. 2.36 Two designs of wave-absorbing materials: (a) laminated composite for fixed polarization wave, and (b) random composite (similar to 1–3 type defined by Newnham *et al.* (1978), see Fig. 2.22). (Matsumoto and Miyata, 1997, © 1997 IEEE. Reproduced with permission from IEEE.)

where B_i, H_i, and μ_i are, respectively, magnetic flux density, magnetic field, and magnetic permeability of the ith phase material with $i = 1$ (magnetic material), 2 (low-permittivity material) and c (composite). Since the magnetic field in the composite is equal to that in each material, we have

$$H_c = H_1 = H_2. \tag{2.23}$$

A substitution of Eqs. (2.22) into Eq. (2.21) and use of Eq. (2.23) yields

$$\mu_c = v_1\mu_1 + v_2\mu_2. \tag{2.24}$$

As to the propagation of the E-wave, it is along the direction perpendicular to the plane of lamination (x-axis in Fig. 2.36(a)). The total electric field E_c of the composite is the sum of that in each material with its volume fraction,

$$E_c = v_1E_1 + v_2E_2, \tag{2.25}$$

while the electric flux density D is uniform throughout the composite,

$$D_c = D_1 = D_2. \tag{2.26}$$

For each material,

$$D_c = \varepsilon_cE_c, \tag{2.27a}$$

$$D_1 = \varepsilon_1 E_1, \tag{2.27b}$$

$$D_2 = \varepsilon_2 E_2. \tag{2.27c}$$

A substitution of Eqs. (2.27) into Eq. (2.25) and use of Eq. (2.26) provides another law-of-mixtures formula for the composite permittvity ε_c as

$$\frac{1}{\varepsilon_c} = \frac{v_1}{\varepsilon_1} + \frac{v_2}{\varepsilon_2}. \tag{2.28}$$

Model for random composite

For non-fixed polarization waves, Matsumoto and Miyata (1997) proposed a model that accounts for 3D distribution of the magnetic material and the low-permittivity material, Fig. 2.36(b). They considered first that both materials are continuous along the z-direction, thus the average permeability along the z-direction is given by Eq. (2.24), i.e., $\mu_c = v_1 \mu_1 + v_2 \mu_2$. Then they considered the magnetic field along the x- or y-axis, for which the magnetic field is additive and magnetic flux density is uniform. Therefore, using the same concept as applied to the derivation of the composite permittivity for Fig. 2.36(a), they obtained the composite permeability μ_c as

$$\mu_c = \frac{1}{\dfrac{v_1}{(v_1 \mu_1 + v_2 \mu_2)} + \dfrac{v_2}{\mu_2}}. \tag{2.29}$$

Similarly, the permittivity of the random composite is obtained as

$$\varepsilon_c = \frac{1}{\dfrac{v_1}{(v_1 \varepsilon_1 + v_2 \varepsilon_2)} + \dfrac{v_2}{\varepsilon_2}} \tag{2.30}$$

By using the above simple model for the laminated composite of Fig. 2.36(a) and Eq. (2.18), Matsumoto and Miyata explained the experimental results of reflection vs. frequency for laminated composites with several thicknesses (t) of low-permittivity layer, see Fig. 2.37 where both predictions and experimental data for the case of a thin wave absorber with $t = 3$ mm are compared, resulting in good agreement. For random composites, they used the predictions based on Eqs. (2.29) and (2.30) and plotted these as functions of the volume fraction v_2 of free space (porosity), see in Fig. 2.38 where the input data of μ, ε are used at 2 GHz. The predictions in Fig. 2.38 imply that the values of ε decrease with increase of the free-space volume fraction, while those of μ remain almost constant. If this random composite is used as a wave absorber, 20 dB or

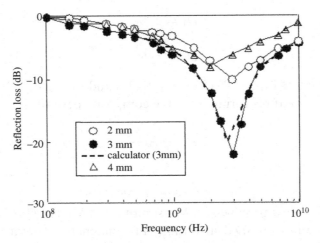

Fig. 2.37   Reflection loss of wave absorber of laminated composite type of Fig. 2.36(a) mounted in a coaxial sample holder (Matsumoto and Miyata, 1997, © 1997 IEEE. Reproduced with permission from IEEE).

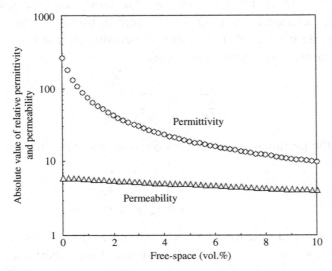

Fig. 2.38   Absolute values of ε_r and μ_r vs. volume fraction (vol.%) of free space v_2 predicted by the model for wave absorber of random composite type of Fig. 2.36(b) (Matsumoto and Miyata, 1997, © 1997 IEEE. Reproduced with permission from IEEE).

over of reflection loss for a frequency range of 2–3 GHz is expected for a wave absorber with thickness 2.5–3.5 mm.

The above model requires the input data of constituents at a given frequency, i.e., ε and μ at 1 GHz, and it can treat the frequency dependency. Matsumoto

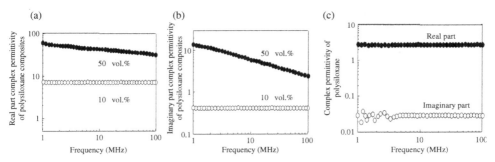

Fig. 2.39   Relative complex permittivity of iron particle/polysiloxane composites: (a) real part (ε'_{cr}), (b) imaginary part (ε''_{cr}), and (c) ε'_{pr} and ε''_{pr} of polysiloxane, as functions of frequency. (Matsumoto and Miyata, 1999a, © 1999 IEEE. Reproduced with permission from IEEE.)

and Miyata (1999a) proposed another simple model for the predictions of the complex permittivity of electronic composites made of metal particles and polymer matrix, where two different cases of particle loading are considered, low volume fractions (10–20%) and high volume fractions (30–50%). Their model is based on an equivalent circuit model. Before discussing their model, let us examine the trend of the experimental data for the complex permittivity of carbonyl-iron spherical particles (of diameter 2.3 mm) in a polysiloxane matrix, with volume fractions of particles of 10 and 50%. Figure 2.39(a) and (b) illustrate the real part (ε'_{cr}) and the imaginary part (ε''_{cr}) of the relative permittivity of the iron particle/polysiloxane composite as functions of frequency. The complex relative permittivities (ε'_{pr}, ε''_{pr}) of the polymer matrix (polysiloxane) are also shown as functions of frequency, Fig. 2.39(c).

The experimental results of Fig. 2.39 reveal that the relative complex permittivity of composites with 10% volume fraction of particles does not depend on frequency while that with 50% volume fraction of particles decreases with frequency despite the fact that the relative complex permittivity of the polysiloxane matrix material is itself independent of frequency. Matsumoto and Miyata found experimentally that the complex permittivity of metal particle/polymer matrix composites (ε_c) is proportional to that of the polymer matrix (ε_p) for low volume fractions of metal particle, i.e.,

$$\varepsilon_c = k\varepsilon_p, \qquad (2.31)$$

where k (see Eq. (2.37) below) is related to the volume fraction v_m of metal particles and k increases with v_m. For high volume fractions, they approximate the experimental data of the composite permittivity by

$$\varepsilon_c = k_1\varepsilon_p + F(f), \qquad (2.32)$$

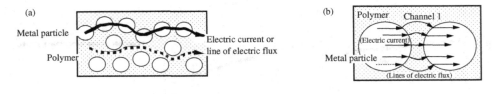

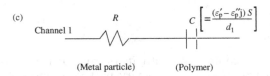

(Metal particle)   (Polymer)

Fig. 2.40   Equivalent circuit model for metal particle/polymer matrix composite, with low volume fraction of particles (after Matsumoto and Miyata, 1999a, © 1999 IEEE. Reproduced with permission from IEEE).

where k_1 increases with v_m and $F(f)$ is the term that depends only on frequency f. They proposed a simple model to explain the experimental data . The model for low volume fractions of metal particles is illustrated in Fig. 2.40. The electric current is presumed to flow mainly through channel 1 (solid line in(a)), a series connection of metal particles and polymer matrix sandwiched between the particles.

Assuming that the impedance of the composite in the equivalent circuit model of Fig. 2.40(c) consists only of the polymer capacitance, i.e.,

$$Z_c = -\frac{j}{2\pi f C},\qquad(2.33a)$$

where

$$C = \frac{\varepsilon_p S}{d_1} = \frac{(\varepsilon_p' - j\varepsilon_p'')S}{d_1},\qquad(2.33b)$$

and S and d_1 are the surface area of the composite specimen and total length of the polymer region in channel 1, respectively. Therefore, the composite admittance Y_c is obtained from Eq. (2.33a) as

$$Y_c = 2\pi f \left[\frac{(\varepsilon_p'' + j\varepsilon_p')S}{d_1}\right]\qquad(2.34)$$

and Y_c is connected to the composite capacitance, C_c, as

$$Y_c = 2\pi j f C_c\qquad(2.35a)$$

where

$$C_c = \frac{(\varepsilon'_c - j\varepsilon''_c)S}{d} \qquad (2.35b)$$

with d the specimen thickness.

From the above equations, the complex permittivity of the composite is obtained as

$$\varepsilon_c = \varepsilon'_c - j\varepsilon''_c = \frac{d}{2\pi f S j} Y_c, \qquad (2.36a)$$

and

$$\varepsilon'_c - j\varepsilon''_c = \left(\frac{d}{d_1}\right)\left(\varepsilon'_p - j\varepsilon''_p\right). \qquad (2.36b)$$

The ratio of specimen thickness d to the total length of polymer region in channel 1 (d_1) is defined by

$$k = \frac{d}{d_1}. \qquad (2.37)$$

Equations (2.36b) and (2.37) lead to Eq. (2.31), the key equation for low volume fractions. The predictions of Eq. (2.36b) or Eq. (2.31) are verified by the experimental data, as shown in Fig. 2.41 where ε_c is observed to be proportional to ε_p with its interception at the origin. It is seen from Fig. 2.40 that the slope (gradient, k value in Eq. (2.31)) of the $\varepsilon_c - \varepsilon_p$ lines increases with volume fraction of particles, as shown in Fig. 2.42.

It is to be noted in Fig. 2.42 that the gradient (k value) for polymer composites measured at different frequencies leads to the conclusion that the complex permittivity of polymer composites with low volume fractions of particles is independent of frequency in accord with the results of Fig. 2.39.

Matsumoto and Miyata (1999a) proposed another equivalent circuit model to explain the experimental data of the complex permittivity of metal particle/polymer matrix composites with high volume fractions of particles. The model is illustrated in Fig. 2.43 where the electrical conduction mechanism is based on two channels, channel 1 (similar to the model of Fig. 2.39), and channel 2 for the path through interconnected metal particles.

The key formula based on the equivalent circuit model for high volume fractions of particles is expressed by Eq. (2.32), which is derived from the model of Fig. 2.43(c) and verified by the experimental results of the polymer matrix, Fig. 2.44.

It follows from Fig. 2.44 that the lines of the $\varepsilon_c - \varepsilon_p$ relation intercept at positive values of ε_c (ε'_c and ε''_c) which correspond with the second right-hand

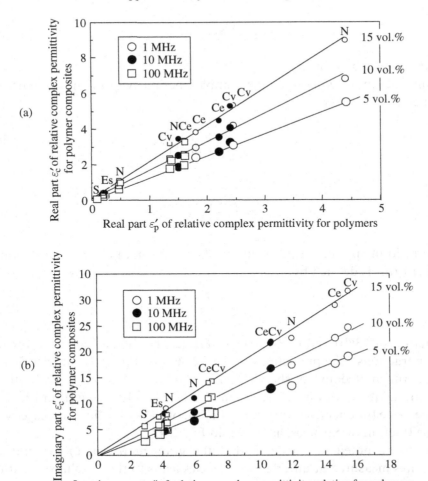

Fig. 2.41   Relative permittivity of polymer composites as a function of that of the polymer matrix: (a) real part, (b) imaginary part. (After Matsumoto and Miyata, 1999a, © 1999 IEEE. Reproduced with permission from IEEE.) S: polysiloxane; Es: unsaturated polyester; N: poly(acrylonitrile butadiene); Ce: cyanoethyl-poly(hydroxyethyl-cellulose); Cv: cyanoethyl-poly (vinylalcohol).

term in Eq. (2.32). This second term in Eq. (2.32) is dependent on frequency, as is also supported by their experimental data.

A further study on the effects of the polymer matrix on the complex permittivity of metal particle/polymer matrix composites with higher volume fractions of particles (55–60%) was reported by Matsumoto and Miyata (2002). They concluded that the optimum composite for wave absorbers should have a high real part and a low imaginary part of the composite permittivity.

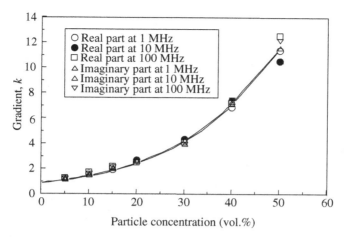

Fig. 2.42 The gradient of the ε'_c vs. ε'_p lines of Fig. 2.41 as a function of vol.% particles (Matsumoto and Miyata, 1999a, © 1999 IEEE. Reproduced with permission from IEEE).

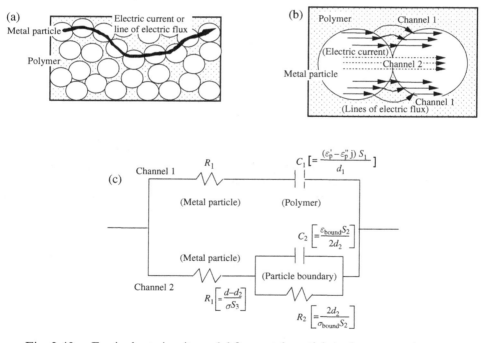

Fig. 2.43 Equivalent circuit model for metal particle/polymer matrix composites with high volume fraction of particles (after Matsumoto and Miyata, 1999a, © 1999 IEEE. Reproduced with permission from IEEE).

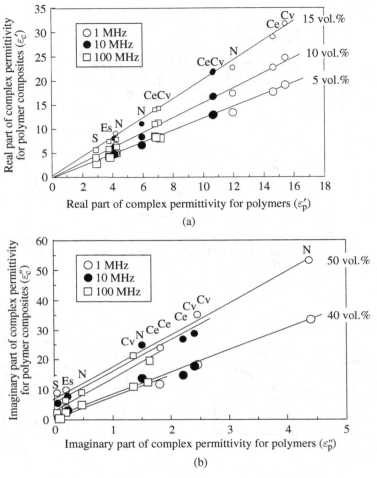

Fig. 2.44   Relative permittivity of polymer composites with high particle concentrations as a function of polymer permittivity: (a) real part, and (b) imaginary part. (After Matsumoto and Miyata, 1999a, © 1999 IEEE. Reproduced with permission from IEEE.) See Fig. 2.41 caption for meaning of letters.

3

Foundations of modeling

3.1 Introduction

The microstructure–macro-property relation of an electronic composite is a prerequisite in designing the electronic composite with unique properties. In order to establish the microstructure–macro-property relation, one must know a priori the governing equations which must be satisfied in the domain of the composite as well as in the domain of each constituent. By using these governing equations and appropriate boundary conditions, one can develop a composite model. Before discussing composite models we shall review the governing equations of mechanical, thermal, electromagnetic, and coupled behavior of materials.

3.1.1 Vector and tensor notation

In this chapter we discuss mostly mathematical equations, by using two types of expression: index formulation, and symbolic formulation. We introduce briefly the concepts of vector and tensor. A vector $\mathbf{A}$ (symbolic form) can be expressed in index form as

$$\mathbf{A} = A_i \mathbf{e}_i = A_1 \mathbf{e}_1 + A_2 \mathbf{e}_2 + A_3 \mathbf{e}_3 \tag{3.1}$$

where A_i is the ith component of vector $\mathbf{A}$ and $\mathbf{e}_i$ is the ith base vector based on Cartesian coordinates and a repeated index is to be summed over i. The dot (inner) product between two base vectors, $\mathbf{e}_i$ and $\mathbf{e}_j$, is

$$\mathbf{e}_i \cdot \mathbf{e}_j = \delta_{ij} = \begin{cases} 1 & \text{if } i = j, \\ 0 & \text{if } i \neq j, \end{cases} \tag{3.2}$$

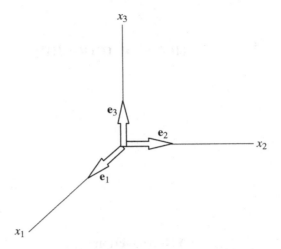

Fig. 3.1   Orthogonal coordinates x_1, x_2, and x_3 with corresponding base vector $\mathbf{e}_i$ with $|\mathbf{e}_i| = 1$

where δ_{ij} is called Kronecker's delta. The vector (cross) product is

$$\mathbf{e}_i \times \mathbf{e}_j = e_{ijk}\mathbf{e}_k, \tag{3.3}$$

$$e_{ijk}\, e_{mnk} = \delta_{im}\delta_{jn} - \delta_{in}\delta_{jm}, \tag{3.4}$$

where e_{ijk} is a permutation symbol (third-order tensor) and takes a value $+1$ for an even permutation $(i \to j \to k)$, -1 for an odd permutation $(k \to j \to i)$, and 0 if any two indices coincide. In this book, we consider only orthogonal coordinates, Fig. 3.1.

Another important vector operator is the partial derivative operator, called the "del operator," defined by

$$\nabla = \frac{\partial}{\partial x_i}\mathbf{e}_i = \frac{\partial}{\partial x_1}\mathbf{e}_1 + \frac{\partial}{\partial x_2}\mathbf{e}_2 + \frac{\partial}{\partial x_3}\mathbf{e}_3. \tag{3.5}$$

Next we define tensors. Most important tensors that appear often in the book are second- through fourth-order tensors. For example, a second-order tensor $\boldsymbol{B}$ and fourth-order tensor $\boldsymbol{C}$ can be expressed in index form as

$$\boldsymbol{B} = B_{ij}\,\mathbf{e}_i\,\mathbf{e}_j, \tag{3.6a}$$

$$\boldsymbol{C} = C_{ijkl}\,\mathbf{e}_i\,\mathbf{e}_j\,\mathbf{e}_k\,\mathbf{e}_l. \tag{3.6b}$$

In this book, we sometimes use only components for simplicity. When the del operator ∇ is applied to scalar, vector, and itself with inner and vector products, the following are the expressions in both symbolic and index forms:

$$\nabla T = \frac{\partial T}{\partial x_i}\mathbf{e}_i = \frac{\partial T}{\partial x_1}\mathbf{e}_1 + \frac{\partial T}{\partial x_2}\mathbf{e}_2 + \frac{\partial T}{\partial x_3}\mathbf{e}_3$$
$$= T_{,i}\mathbf{e}_i = T_{,1}\mathbf{e}_1 + T_{,2}\mathbf{e}_2 + T_{,3}\mathbf{e}_3, \tag{3.7}$$

where subscript ",i" denotes $\partial/\partial x_i$;

$$\nabla \cdot \mathbf{A} = \left(\frac{\partial}{\partial x_i}\mathbf{e}_i\right) \cdot (A_j\,\mathbf{e}_j) = \frac{\partial(A_j\delta_{ij})}{\partial x_i}$$
$$= \frac{\partial A_i}{\partial x_i} = A_{i,i} = A_{1,1} + A_{2,2} + A_{3,3}; \tag{3.8}$$

$$\nabla \times \mathbf{A} = \left(\frac{\partial}{\partial x_i}\mathbf{e}_i\right) \times (A_j\,\mathbf{e}_j)$$
$$= \frac{\partial A_j}{\partial x_i}\,e_{ijk}\mathbf{e}_k$$
$$= \left\{\frac{\partial A_3}{\partial x_2} - \frac{\partial A_2}{\partial x_3}\right\}\mathbf{e}_1 + \left\{\frac{\partial A_1}{\partial x_3} - \frac{\partial A_3}{\partial x_1}\right\}\mathbf{e}_2 + \left\{\frac{\partial A_2}{\partial x_1} - \frac{\partial A_1}{\partial x_2}\right\}\mathbf{e}_3; \tag{3.9}$$

$$\nabla \cdot \nabla = \left(\frac{\partial}{\partial x_i}\mathbf{e}_i\right) \cdot \left(\frac{\partial}{\partial x_j}\mathbf{e}_j\right) = \frac{\partial}{\partial x_i}\frac{\partial}{\partial x_i} = \frac{\partial^2}{\partial x_1^2} + \frac{\partial^2}{\partial x_2^2} + \frac{\partial^2}{\partial x_3^2}. \tag{3.10}$$

The inner product of second-order tensors $\mathbf{B}$ and $\mathbf{D}$ can be expressed as

$$\mathbf{B} \cdot \mathbf{D} = \left(B_{ij}\mathbf{e}_i\mathbf{e}_j\right) \cdot (D_{mn}\mathbf{e}_m\mathbf{e}_n)$$
$$= B_{ij}D_{ij}$$
$$= B_{11}D_{11} + B_{22}D_{22} + B_{33}D_{33} + B_{12}D_{12} + B_{21}D_{21} + B_{13}D_{13}$$
$$+ B_{31}D_{31} + B_{23}D_{23} + B_{32}D_{32}, \tag{3.11}$$

where the repeated indices i and j are summed over 1, 2, and 3 (Einstein convention) and Eqs. (3.2) and (3.3) are used to arrive at the above equations.

3.1.2 Integral theorems

Here we shall discuss two integral theorems, Gauss' divergence theorem and Stokes' theorem.

Gauss' divergence theorem

Let $\mathbf{A}$ be a vector-valued or tensor-valued continuous function in a finite Euclidean space $(\mathbf{x})$, the volume V of which is closed. Gauss' divergence

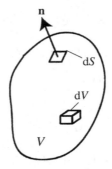

Fig. 3.2   Closed volume V that contains both surface element dS with outward unit normal vector **n** and volume element dV.

theorem is a relation between the volume and surface integrals which are related by

$$\int_V \nabla * \mathbf{A}\, dV = \int_S \mathbf{n} * \mathbf{A}\, dS,\qquad\qquad (3.12)$$

where ∇ is defined by Eq. (3.5), **A** is either a vector or a tensor, **n** is a unit vector normal to the surface S of a finite volume V pointing outward, dV and dS are volume and surface elements, respectively, see Fig. 3.2. The symbol "$*$" can be either a dot ($\cdot$), a cross product ($\times$), or have empty meaning (i.e., can be ignored). Therefore, the following three equations are explicitly given as Gauss' divergence integral equations:

$$\int_V \nabla\varphi\, dV = \int_S \varphi\,\mathbf{n}\, dS,\qquad\qquad (3.13a)$$

$$\int_V \nabla\cdot\mathbf{A}\, dV = \int_S \mathbf{n}\cdot\mathbf{A}\, dS,\qquad\qquad (3.13b)$$

$$\int_V \nabla\times\mathbf{A}\, dV = \int_S \mathbf{n}\times\mathbf{A}\, dS,\qquad\qquad (3.13c)$$

where φ is a scalar field, $\varphi(\mathbf{x})$, and **A** is a vector or tensor field, $\mathbf{A}(\mathbf{x})$. For mechanical problems, **A** can be a stress $\boldsymbol{\sigma}$ or strain e tensor while, in electromagnetic problems, **A** can be an electric field vector **E** or a magnetic field vector **H**.

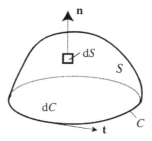

Fig. 3.3   Open surface *S*, with outward unit normal vector **n**, bounded by closed curve *C* with unit tangent vector **t** such that traveling along *C* in the positive direction of **t** following the right-handed screw is coincident with positive **n**.

Stokes' theorem

Consider now an open surface *S*, with unit outward normal vector **n**, that is bounded by a closed curve *C* where the positive unit vector **t** tangent to the curve is along *C* and such that the circular path along positive **t** makes right-handed screw direction coincident with positive **n**, Fig. 3.3. Stokes' theorem is given by

$$\int_S (\mathbf{\nabla} \times \mathbf{A}) \cdot \mathbf{n} \, dS = \oint_C \mathbf{A} \cdot \mathbf{t} \, dC, \tag{3.14}$$

where dS and dC are surface and line elements, respectively, and **A** can be either a vector field or a tensor field, $\mathbf{A}(\mathbf{x})$. In Eqs. (3.13) and (3.14), $\mathbf{n} \, dS$ can be replaced by d**S**.

In summary, we shall use both symbolic and index formulations throughout this book, for example in Eq. (3.11) $\boldsymbol{B} \cdot \boldsymbol{D}$ represents the symbolic form, while $B_{ij} D_{ij}$ is the index form.

We shall state the governing equations related to the behavior of electronic composites – mechanical, thermal, electromagnetic, and coupling behavior. Due to the use of many variables to cover a variety of behaviors, we shall employ those symbols summarized in Table 3.1, where some symbols are used to denote different parameters, depending upon the type of behavior. Hopefully, this will not confuse the reader.

3.2 Mechanical behavior

The governing equations for the mechanical behavior of solids are equilibrium equations, strain–displacement relations (or equivalent compatibility equations), and constitutive equations. The first two equations are given by

Table 3.1 *Symbols representing selected mechanical, thermal, and electromagnetic field and flux variables.*

Behavior	Flux and Field Vector and Tensor	Component	SI unit
Mechanical	displacement	u_i	m
	stress	σ_{ij}	N/m^2 (Pa)
	strain	e_{ij}	–
Thermal	temperature	θ	K
	entropy	σ	J/K
	heat flux	q_i	J/m^2
	thermal conductivity	K_{ij}	W/(m·K)
	thermal resistance	R_t	K/W
Electromagnetic	electric charge	Q	C
	capacitance	C	F
	electric potential	Φ_e	V
	magnetic potential	Φ_m	A
	electric field	E_i	V/m
	magnetic field	H_i	A/m
	electric flux density	D_i	C/m^2
	electric polarization	P_i	C/m^2
	magnetic flux density	B_i	Wb/m^2 (T)
	magnetization	M_i	A/m
	electric current density	J_i	A/m^2
	electric current	I	A
	electric conductivity	σ_{ij}	S/m
	dielectric constant	ε_{ij}	F/m
	magnetic permeability	μ_{ij}	H/m
	electric resistance	R	Ω
	electric inductance	L	H

$$\sigma_{ij,j} + f_i = \rho \ddot{u}_i, \qquad (3.15)$$

strain–displacement:

$$e_{ij} = \frac{1}{2}\left(u_{i,j} + u_{j,i}\right), \qquad (3.16a)$$

compatibility of strain:

$$R_{ij} = e_{ipr}e_{jqs}e_{rs,pq} = 0, \qquad (3.16b)$$

where f is the force per unit volume on the body, ρ is the body's mass density and R_{ij} is the strain compatibilty tensor. As to the constitutive equations, we first consider elastic materials, then elastic–plastic materials, and finally materials exhibiting creep.

3.2.1 Elastic materials

Here we consider only linear elastic solids. The constitutive equations of linear elastic solids can be expressed in general form:

$$\sigma_{ij} = C_{ijkl} e_{kl} \tag{3.17a}$$

or

$$e_{ij} = S_{ijkl} \sigma_{kl} \tag{3.17b}$$

where C_{ijkl} and S_{ijkl} are elastic stiffness and compliance tensors, respectively, (of rank 4). Due to symmetry of the stress and strain tensors, and the positiveness of the elastic strain energy density $\left(E = \frac{1}{2}\sigma_{ij}e_{ij} > 0\right)$, the numbers of independent components of C_{ijkl} and S_{ijkl} are reduced from 81 to 21. This number is reduced further to 9 if the characteristic directions of the elastic solid are made coincident with the three orthogonal Cartesian coordinates x_1, x_2 and x_3, which is then called an orthotropic elastic solid. The strain–stress relation of an orthotropic elastic material is

$$\mathbf{e} = \mathbf{S}\boldsymbol{\sigma}, \tag{3.18}$$

where

$$\mathbf{e} = \left\{ \begin{array}{c} e_{11} \\ e_{22} \\ e_{33} \\ 2e_{23} \\ 2e_{31} \\ 2e_{12} \end{array} \right\}, \quad \boldsymbol{\sigma} = \left\{ \begin{array}{c} \sigma_{11} \\ \sigma_{22} \\ \sigma_{33} \\ \sigma_{23} \\ \sigma_{31} \\ \sigma_{12} \end{array} \right\}, \tag{3.19}$$

and

$$\mathbf{S} = \begin{bmatrix} \frac{1}{E_1} & -\frac{\nu_{12}}{E_2} & -\frac{\nu_{13}}{E_3} & & & \\ -\frac{\nu_{21}}{E_1} & \frac{1}{E_2} & -\frac{\nu_{23}}{E_3} & & \mathbf{0} & \\ -\frac{\nu_{31}}{E_1} & -\frac{\nu_{32}}{E_2} & \frac{1}{E_3} & & & \\ & & & \frac{1}{G_{23}} & & \\ & \mathbf{0} & & & \frac{1}{G_{13}} & \\ & & & & & \frac{1}{G_{12}} \end{bmatrix}, \tag{3.20}$$

where E_i and G_{ij} are Young's and the shear modulus, respectively, of the orthotropic material.

Poisson's ratio ν_{ij} used in Eq. (3.20) is defined as

$$\nu_{ij} = \left| \frac{e_{ii}}{e_{jj}} \right|, \tag{3.21}$$

where index i is the direction of observed strain under loading along the x_j-axis, and repeated indices in Eq. (3.21) are not to be summed. Twelve constants in Eq. (3.20) are used and they are inter-related by the requirement of symmetry of the compliance matrix $[S]$:

$$\frac{\nu_{12}}{E_2} = \frac{\nu_{21}}{E_1},$$
$$\frac{\nu_{13}}{E_3} = \frac{\nu_{31}}{E_1}, \tag{3.22}$$
$$\frac{\nu_{23}}{E_3} = \frac{\nu_{32}}{E_2}.$$

Hence, we have only nine independent constants for orthotropic elastic materials. If the property is symmetric with respect to the x_1–x_2 plane, we have

$$E_T \equiv E_1 = E_2, \quad \nu_{12} = \nu_{21},$$

$$G_{23} = G_{13} \equiv G_L;$$

$$\text{set } G_{12} = G_T, \quad \text{set } E_3 = E_L.$$

Here, subscript T (transverse) means perpendicular to and L (longitudinal) means in the direction of the fiber axis.

Then Eq. (3.20) is reduced to

$$S = \begin{bmatrix} \frac{1}{E_T} & -\frac{\nu_{12}}{E_T} & -\frac{\nu_{13}}{E_L} & & & \\ -\frac{\nu_{12}}{E_T} & \frac{1}{E_T} & -\frac{\nu_{13}}{E_L} & & 0 & \\ -\frac{\nu_{31}}{E_T} & -\frac{\nu_{31}}{E_T} & \frac{1}{E_L} & & & \\ & & & \frac{1}{G_L} & & \\ & 0 & & & \frac{1}{G_L} & \\ & & & & & \frac{1}{G_T} \end{bmatrix}. \tag{3.23}$$

This is for a transversely isotropic elastic material with axis of symmetry along the x_3-axis. There are seven elastic constants in Eq. (3.23), which will be reduced to only five constants by using the following relations:

$$G_T = \frac{E_T}{2(1 + \nu_{12})},$$

$$\frac{\nu_{31}}{E_T} = \frac{\nu_{13}}{E_L}. \tag{3.24}$$

If an elastic material is isotropic, i.e., the property does not depend on the material texture (direction), then we have

$$E_1 = E_2 = E_3 = E,$$
$$\nu_{12} = \nu_{13} = \nu_{31} = \nu, \tag{3.25}$$
$$G_L = G_T = G = \mu,$$

$$S = \begin{bmatrix} \frac{1}{E} & -\frac{\nu}{E} & -\frac{\nu}{E} & & & \\ -\frac{\nu}{E} & \frac{1}{E} & -\frac{\nu}{E} & & 0 & \\ -\frac{\nu}{E} & -\frac{\nu}{E} & \frac{1}{E} & & & \\ & & & \frac{1}{G} & & \\ & 0 & & & \frac{1}{G} & \\ & & & & & \frac{1}{G} \end{bmatrix} \tag{3.26}$$

where the following relation holds for E, G and ν –

$$2G = 2\mu = \frac{E}{1 + \nu} \tag{3.27}$$

– and where G (or μ) and E are, respectively, the shear and Young's modulus with dimensions of pascals (Pa) or newtons (N)/m^2, and ν is Poisson's ratio, which is non-dimensional.

The above three cases of orthotropic, transversely isotropic, and isotropic materials are illustrated in Fig. 3.4(a), (b), and (c) respectively, where the orientations of isotropic fillers are used to illustrate the anisotropy of a composite. Please note that the anisotropy of a composite is realized even if isotropic fillers are in an isotropic matrix. Electronic composites are often processed by using short fibers in a polymer matrix, which is cost-effective. Then the macroscopic properties are strongly dependent, not only on the properties of the short fibers, but also on the distributions of the fiber aspect ratio, and on the orientation angles (θ, ϕ in spherical coordinates), Fig. 3.4(d).

3.2.2 Elastic–plastic materials

Here we confine ourselves to describing the constitutive equations of isotropic elastic–plastic materials. The case for anisotropic elastic–plastic materials can

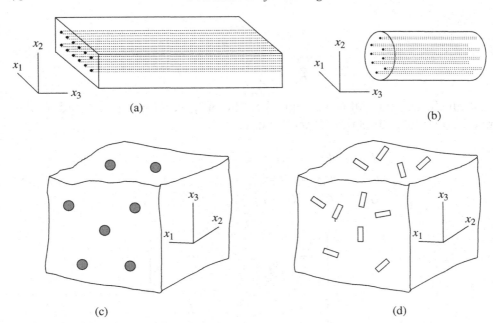

Fig. 3.4   Composite property depends on the orientation of fillers or fibers: (a) orthotropic (continuous filler aligned along the x_3-axis, periodically), (b) transversely isotropic (continuous filler aligned along the x_3-direction but randomly orientated in the x_1–x_2 plane), (c) isotropic (spherical fillers, orientated randomly), and (d) anisotropic (short fibers, randomly orientated or with given distribution functions of orientation angles (θ, ϕ).

be found in Hill (1950). Following Prandtl (1924) and Reuss (1929), the constitutive equations of an isotropic elastic–plastic material is given, for a work-hardening material, by

$$\mathrm{d}e'_{ij} = \frac{3\sigma'_{ij}\mathrm{d}\bar{\sigma}}{2\bar{\sigma}H'} + \frac{\mathrm{d}\sigma'_{ij}}{2G}, \tag{3.28a}$$

$$\mathrm{d}e_{ii} = \frac{(1-2\nu)}{E}\,\mathrm{d}\sigma_{ii}; \tag{3.28b}$$

and for non-work-hardening material, by

$$\mathrm{d}e'_{ij} = \frac{3\sigma'_{ij}}{2\sigma_y}\,\mathrm{d}\bar{e}_{\mathrm{p}} + \frac{\mathrm{d}\sigma'_{ij}}{2G}, \tag{3.29a}$$

$$\mathrm{d}e_{ii} = \frac{(1-2\nu)}{E}\,\mathrm{d}\sigma_{ii}, \tag{3.29b}$$

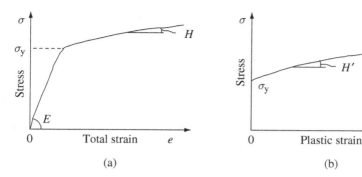

Fig. 3.5   (a) Stress–strain curve where Young's modulus E, yield stress σ_y, and work-hardening rate H are defined. (b) Stress–plastic-strain curve where yield stress σ_y and work-hardening rate H' are defined.

where the work-hardening rate H' of the stress–plastic-strain region and yield stress σ_y are defined in Fig. 3.5. In Eqs. (3.28) and (3.29) de'_{ij} and $d\sigma'_{ij}$ are the incremental deviatoric strain and stress components, respectively, $d\bar{e}_p$ and $d\bar{\sigma}$ are the incremental equivalent plastic strain and incremental equivalent stress, respectively, and $\bar{\sigma}$ is defined by Eq. (3.34) below. Deviatoric strain and stress are defined in Eq. (3.36) below.

The yield criterion of an isotropic elastic–plastic material follows the J'_2 flow rule (Hill, 1950),

$$\frac{1}{2}\sigma'_{ij}\sigma'_{ij} = k^2. \tag{3.30}$$

The instantaneous stiffness tensor of an isotropic elastic–plastic material, C^{ep}_{ijkl}, has been used for finite-element programs in conjuction with updated Lagrangian formulation (Nemat-Nasser and Taya, 1980). Equation (3.30) is called the yield surface criterion. If the yield criterion is satisfied, then a material undergoes plastic deformation following Eqs. (3.28) or (3.29). The parameter k in Eq. (3.30) is the yield stress in shear and is related to the normal (uniaxial) yield stress σ_y depending on the yield criterion used:

$$k = \begin{cases} \frac{1}{2}\sigma_y & \text{for Tresca,} \\ \frac{1}{\sqrt{3}}\sigma_y & \text{for von Mises.} \end{cases} \tag{3.31}$$

The yield surface for a state of biaxial stress, such as uniaxial tension and torsional shear stress τ, is expressed, following von Mises' criterion, as

$$\sigma^2 + 3\tau^2 = \sigma_y^2. \tag{3.32}$$

The work-hardening rate H' defined in terms of the equivalent stress $(\bar{\sigma})$–equivalent plastic strain $(\bar{e}_p)$ relation is related to the slope (H) of the uniaxial stress–strain curve by

$$H' = \frac{H}{1-\left(\frac{H}{E}\right)}, \tag{3.33}$$

where H and H' are defined in Fig. 3.5(a) and (b), respectively. In the above equations $\bar{\sigma}$ is the equivalent stress defined by

$$\bar{\sigma} = \left(\frac{2}{3}\sigma'_{ij}\sigma'_{ij}\right)^{1/2}. \tag{3.34}$$

The equivalent plastic strain increment $d\bar{e}_p$ is defined by $\left(\frac{3}{2} de^p_{ij} de^p_{ij}\right)^{1/2}$. When a material experiences unloading, the above constitutive equations are all reduced to an elastic equation

$$de'_{ij} = \frac{d\sigma'_{ij}}{2G}, \tag{3.35a}$$

$$de_{ii} = \frac{(1-2\nu)}{E} d\sigma_{ii}. \tag{3.35b}$$

Of course, in this case one can omit the incremental symbol "d" in front of stress and strain. In the above equations, σ'_{ij} and e'_{ij} are deviatoric stress and strain components, respectively, and they are related to σ_{ij} and e_{ij} by

$$\begin{aligned}
\sigma'_{ij} &= \sigma_{ij} - \frac{1}{3}\delta_{ij}\sigma_{kk}, \\
e'_{ij} &= e_{ij} - \frac{1}{3}\delta_{ij}e_{kk}
\end{aligned} \tag{3.36}$$

where δ_{ij} is Kronecker's delta defined by Eq. (3.2) and the repeated index k follows the Einstein convention, see Section 3.1. Equations (3.28) and (3.29) can be rewritten in terms of an incremental stress deviator–strain deviator relation:

$$d\sigma'_{ij} = C^{ep}_{ijkl} de'_{kl}, \tag{3.37a}$$

$$de_{ii} = \frac{(1-2\nu)}{E} d\sigma_{ii}, \tag{3.37b}$$

where C^{ep}_{ijkl} is the instantaneous stiffness tensor of an isotropic elastic–plastic material, given by

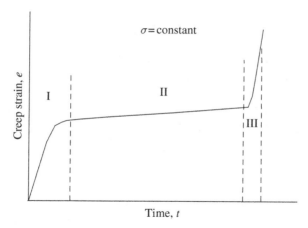

$\sigma = $ constant

I   II   III

Creep strain, e

Time, t

Fig. 3.6   Typical creep strain vs. time curve which exhibits three stages: I (primary), II (secondary), and III (tertiary).

$$C_{ijkl}^{ep} = 2G\left[\delta_{ik}\delta_{jl} - \frac{9\sigma_{ij}'\sigma_{kl}'}{\sigma^2(2H'/G + 6)}\right]. \qquad (3.38)$$

A more rigorous treatment of inelastic behavior of materials is summarized by Nemat-Nasser (2004).

3.2.3 Materials exhibiting creep

Creep is defined here as time-dependent inelastic deformation of a material when it is subjected to a constant stress σ at a given temperature θ. A typical creep strain e vs. time t curve is shown in Fig. 3.6, which exhibits three stages: I (primary), II (secondary or steady state), and III (third or tertiary stage). Usually, the duration of the secondary creep is the longest, thus the assessment of its creep rate $\dot{e}_c$ $(= de_c/dt)$ is the most important task. Among various parameters, stress σ and temperature θ are the major ones influencing $\dot{e}_c$ and here we are concerned with creep due to climb and glide of dislocations. Frost and Ashby (1982) summarized the mechanisms of deformation of a metal under various regimes of temperature and stress, to produce what is known as a "deformation map."

Since creep is a thermally activated process, its temperature dependence is given by an Arrhenius-type equation

$$\dot{e}_c = f(\sigma)\, e^{-Q_a/R\theta} \qquad (3.39)$$

where $f(\sigma)$ is a known function of stress σ, R (joules [J]/(mol·K)) is the gas constant, θ is the temperature in Kelvin, and Q_a is the apparent activation energy. The most popular form of $f(\sigma)$ is a power-law relation given by

$$f(\sigma) = A\left(\frac{\sigma}{G}\right)^n,$$

(3.40)

where A and n are material constants, and G is the shear modulus. Although Eq. (3.39) provides a basic equation for secondary creep, a more rigorous equation is needed if the effects of parameters other than σ and θ are to be examined. For example the Dorn-type equation may be suited for this purpose (Mukherjee *et al.*, 1969):

$$\dot{e}_c = A\left(\frac{\sigma}{G}\right)^n \frac{Gb}{k_B\theta} D_0 e^{-Q/R\theta}$$

(3.41)

where b is the Burgers vector, D_0 is the pre-exponential constant, Q is an activation energy which is constant for self-diffusion, k_B (J/K) is the Boltzmann constant. The values of these constants for most metals are tabulated elsewhere (Mukherjee *et al.*, 1969). It is known that the stress dependence of the creep rate at higher stresses cannot be described by the power-law-type equation. In this case an exponential-type law is better suited; that is, Eq. (3.40) is replaced by

$$f(\sigma) = A\exp(\sigma/\sigma_0)$$

(3.42)

where A and σ_0 are material constants.

3.3 Thermal behavior

3.3.1 Governing equations

The governing equation for heat conduction in an anisotropic solid is

$$\rho c_p \dot{\theta} + q_{i,i} + q_s = 0$$

(3.43)

where ρ, c_p, θ, q_i, and q_s are mass density, specific heat capacity at constant pressure, temperature, heat flux component and heat source per unit volume, respectively, $\dot{\theta}$ is $\partial\theta/\partial t$, where t is time. Heat flux q_i is related to the temperature gradient $\theta_{,j}$ through Fourier's law by

$$-q_i = K_{ij}\theta_{,j}.$$

(3.44)

When Eq. (3.44) is substituted into Eq. (3.43), we obtain

$$\rho c_p \dot{\theta} + q_s = \left(K_{ij}\theta_{,j}\right)_{,i}.$$

(3.45)

In the absence of a heat source ($q_s = 0$) and for a homogeneous material, Eq. (3.45) is reduced to

$$K_{ij}\theta_{,ij} = \rho c_p \dot{\theta}. \tag{3.46}$$

where K_{ij} (W/(m $\cdot$ K)) is the component of the thermal conductivity vector. For steady-state heat conduction, Eq. (3.46) is further reduced to

$$K_{ij}\theta_{,ij} = 0. \tag{3.47}$$

Special cases of K_{ij} given by Eq. (3.47) are examined here. When one plane of reflection symmetry exists, for example the x_3-axis, then K_{ij} is reduced to

$$K_{ij} = \begin{bmatrix} K_{11} & K_{12} & 0 \\ K_{21} & K_{22} & 0 \\ 0 & 0 & K_{33} \end{bmatrix}. \tag{3.48}$$

Though $K_{12} \neq K_{21}$, their values are usually much smaller than the diagonal components (K_{11}, K_{22}, and K_{33}); hence one can omit off-diagonal components. When two planes of reflection symmetry exist and the coordinates are taken along such axes (perpendicular to the planes), the material is called an orthotropic solid and its K_{ij} is given by

$$K_{ij} = \begin{bmatrix} K_{11} & 0 & 0 \\ 0 & K_{22} & 0 \\ 0 & 0 & K_{33} \end{bmatrix}. \tag{3.49}$$

If the axes in an orthotropic solid are interchangeable, or the heat conduction is independent of direction, the material is called isotropic and K_{ij} is reduced to

$$K_{ij} = \begin{bmatrix} K_{11} & 0 & 0 \\ 0 & K_{11} & 0 \\ 0 & 0 & K_{11} \end{bmatrix} = K\delta_{ij} = K \begin{bmatrix} 1 & 0 & 0 \\ 0 & 1 & 0 \\ 0 & 0 & 1 \end{bmatrix}, \tag{3.50}$$

where K is the thermal conductivity of an isotropic material (W/(m $\cdot$ K)).

3.3.2 Thermal resistance

The concept of thermal resistance R_t is defined in an analogous fashion to electrical resistance, as discussed in Section 1.3. Consider two observation points for which temperatures T_1 and T_2 (K) are measured for a given heat flow Q (W), Fig. 3.7. Then the thermal resistance R (K/W) between the two observation points is defined by

$$R_t = \frac{(T_1 - T_2)}{Q} = \frac{\Delta T}{Q}. \tag{3.51}$$

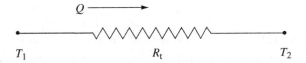

Fig. 3.7   Thermal resistance R_t

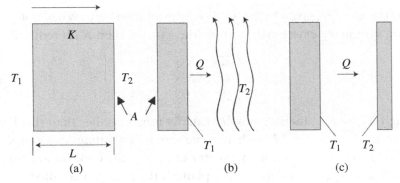

Fig. 3.8   Three different modes of heat flow: (a) heat conduction through solid medium, (b) heat flow by convected coolant, and (c) radiation heat transfer.

Three modes of heat transfer exist with equivalent thermal resistance (Lau *et al.*, 1998): (1) heat conduction through a solid medium, (2) heat flow from a hot surface to a cold coolant (fluid), and (3) radiation heat flow from a hot surface to a cold surface, Fig. 3.8. We shall discuss briefly the formulae for calculating thermal resistance for these three modes below.

Heat conduction through medium

The thermal resistance R_t for heat conduction through a solid medium with thermal conductivity K (watts [W]/(m · K)) is given by

$$R_t = \frac{L}{KA},\qquad(3.52)$$

where L is the thickness and A the cross-sectional area of the medium, see Fig. 3.8(a). Values of thermal conductivity K and coefficient of thermal expansion (CTE) of materials that are often used in electronic packaging are given in Appendix B2. It is to be noted that Eq. (3.52) is derived from Eq. (3.44) by using

$$q_i = \frac{Q_i}{A},\qquad(3.53a)$$

$$\theta_{,j} = \frac{\Delta T}{\Delta x},$$ (3.53b)

where all indices are ignored to indicate the steady-state heat conduction along a one-dimensional (x-axis) direction and $\Delta x = L$.

Heat transfer through coolant

When a hot component (T_1) is subjected to forced flow of coolant (T_2), the equivalent thermal resistance is given by

$$R_t = \frac{1}{hA},$$ (3.54)

where h is the heat transfer coefficient (W/(m$^2 \cdot$ K)) and A is the area of the hot component from which heat flow Q is emitted, Fig. 3.8(b). When a mass of coolant is flowing over the surface of the hot component which emits heat flow Q, the coolant increases its temperature by ΔT given by

$$\Delta T = \frac{Q}{cM}.$$ (3.55)

In view of Eq. (3.51), the equivalent thermal resistance R_t is given by

$$R_t = \frac{1}{cM},$$ (3.56)

where c is the specific heat capacity (J/(kg $\cdot$ K)) of the coolant and M is the mass flow rate of the coolant (kg/s). The specific heat capacities c of common coolant materials are given by $c = 1.16$ (air), 1590 (coolant FC77), and 1000 (water) (Lau *et al.*, 1998). Use of a fluid coolant is very effective in removing heat from a hot component.

When a coolant fluid is forced to flow over the flat surface of electronic components, it develops a "boundary layer" in both laminar and turbulent flows. Then the heat transfer coefficient h defined by Eq. (3.54) needs to be modified to account for this boundary-layer effect. The velocity profile $u(x, y)$ of a fluid flowing over a flat plate is approximated by the following parabolic shape:

$$u(x, y) = U\left(\frac{2y}{\delta} - \frac{y^2}{\delta^2}\right) \quad \text{for } 0 \le y \le \delta(x),$$ (3.57)

where U is the constant velocity in the laminar flow region and δ, the thickness of the boundary layer, is a function of x, Fig. 3.9.

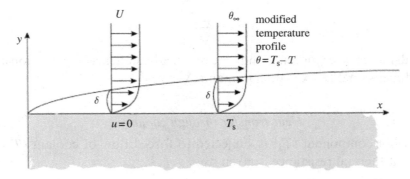

Fig. 3.9   Laminar flow with boundary-layer thickness δ where the velocity u
and modified temperature θ follow parabolic profiles.

If the modified temperature θ is defined by

$$\theta = T_s - T, \tag{3.58}$$

where T and T_s are the temperature at an arbitrary point (x, y) and that of the
surface, respectively, Fig. 3.9, then θ is assumed to follow a parabolic law
similar to Eq. (3.57), i.e.,

$$\theta = \theta_\infty \left(\frac{2y}{\delta} - \frac{y^2}{\delta^2} \right), \tag{3.59}$$

where θ_∞ is the constant temperature in the laminar flow region. The heat flux
q along the y-axis at the surface of the flat plate is given by

$$q = -K_f \left(\frac{\partial T}{\partial y} \right)_{y=0} = K_f \left(\frac{\partial \theta}{\partial y} \right)_{y=0}. \tag{3.60}$$

where K_f is the thermal conductivity of the fluid.
From Eqs. (3.59) and (3.60), we obtain

$$q = \frac{Q}{A} = \frac{2K_f}{\delta} \theta_\infty = \frac{2K_f(T_s - T_\infty)}{\delta}. \tag{3.61}$$

In view of Eq. (3.51),

$$R_t = \frac{(T_s - T_\infty)}{Q} = \frac{1}{(2K_f/\delta)A}. \tag{3.62}$$

A comparison between Eqs. (3.54) and (3.62) provides the heat transfer
coefficient h for the convected coolant flow with a boundary layer,

$$h = \frac{2K_f}{\delta}. \tag{3.63}$$

Equation (3.63) implies that, the higher the thermal conductivity of the coolant and the thinner the boundary layer, the higher the heat transfer coefficient becomes. It is noted here that the boundary-layer thickness δ decreases with increase of the coolant velocity U, but it increases with larger values of x, see Fig. 3.9.

Heat transfer by radiation

When a hot body (T_1) emits heat into a surrounding medium (T_2), even into a vacuum, the radiation heat flux q_r (W/m^2) is given by

$$q_r = \varepsilon\sigma\left(T_1^4 - T_2^4\right), \tag{3.65}$$

where ε is the emissivity of the hot surface and σ is the Stefan–Boltzmann constant, equal to 5.670×10^{-8} W/(m$^2 \cdot$ K^4). The value of ε of a hot medium with a rough surface is 0.91, while that of a medium with a highly polished mirror surface is as small as 0.04 (Lau *et al.*, 1998). For a small temperature difference T_1-T_2,

$$T_1^4 - T_2^4 \approx 4T_2^3(T_1 - T_2). \tag{3.66}$$

From Eqs. (3.51), (3.65), and (3.66),

$$R_t = \frac{1}{(4\varepsilon\sigma T_2^3)A} = \frac{(T_1 - T_2)}{Q}. \tag{3.67}$$

Therefore, we can define the radiation heat transfer h_r using the formula for heat transfer through a coolant, i.e., Eq. (3.54). Then

$$R_t = \frac{1}{h_r A}, \tag{3.68a}$$

where

$$h_r = 4\varepsilon\sigma T_2^3. \tag{3.68b}$$

It is to be noted in Eqs. (3.68) that h_r is not a constant, but dependent on the temperature of the surrounding medium.

3.4 Electromagnetic behavior

Before we examine Maxwell's equations, we shall discuss briefly Ampère's law, Gauss' electric and magnetic laws, and Faraday's law, as they inspired Maxwell to establish unified equations.

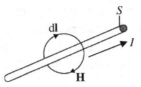

Fig. 3.10   Ampère's circuital law.

Ampère's law due to direct current I

Consider an electric current I running through a straight conductive wire. Then a magnetic field H is generated around the wire, Fig. 3.10. The integral of H along the circular path $\mathbf{l}$ (whose element is d$\mathbf{l}$) is equal to the current I, which in turn is equal to the integral of current density J across the cross-section S of the wire:

$$\oint_C \mathbf{H} \cdot d\mathbf{l} = I = \int_S \mathbf{J} \cdot d\mathbf{S}. \tag{3.69}$$

Using Stokes' theorem (Eq. 3.14), we can write Eq. (3.69) as

$$\int_S (\nabla \times \mathbf{H}) \cdot d\mathbf{S} = \int_S \mathbf{J} \cdot d\mathbf{S}. \tag{3.70}$$

The integrands in the two sides of Eq. (3.70) should be the same. Therefore we get

$$\nabla \times \mathbf{H} = \mathbf{J}. \tag{3.71}$$

The right-hand side of Eq. (3.71) contains the current density J due to the direct current. For a time-varying field, we need to add a time-dependent electric flux density $\partial \mathbf{D}/\partial t$:

$$\nabla \times \mathbf{H} = \mathbf{J} + \frac{\partial \mathbf{D}}{\partial t}. \tag{3.72}$$

Gauss' law for electric field

Consider a finite volume V bounded by a closed surface S, where the total charge inside is Q, Fig. 3.11. The surface integral of electric flux density D over the closed surface S is equal to the electric charge Q inside the surface S:

$$\oint_S \mathbf{D} \cdot d\mathbf{S} = Q. \tag{3.73}$$

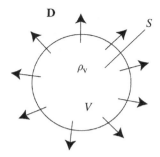

Fig. 3.11   Finite volume V with surface S contains electric charge Q with its volume charge density ρ_v.

Using the divergence theorem (Eq. (3.13b)) we obtain

$$\oint_S \mathbf{D} \cdot d\mathbf{S} = \int_V \boldsymbol{\nabla} \cdot \mathbf{D} \, dV. \tag{3.74}$$

With Eq. (3.73), Eq. (3.74) is rewritten as

$$\int_V \boldsymbol{\nabla} \cdot \mathbf{D} \, dV = Q = \int_V \rho_v \, dV. \tag{3.75}$$

The integrands on the two sides of Eq. (3.75) must be the same, resulting in

$$\boldsymbol{\nabla} \cdot \mathbf{D} = \rho_v \tag{3.76}$$

where ρ_v is the electric charge per unit volume (C/m^3).

Gauss' law for magnetic field

No magnetic charge exists, Fig. 3.12, i.e.,

$$\oint_S \mathbf{B} \cdot d\mathbf{S} = 0. \tag{3.77}$$

Using the divergence theorem, we can write Eq. (3.77) as

$$\oint_S \mathbf{B} \cdot d\mathbf{S} = \int_V \boldsymbol{\nabla} \cdot \mathbf{B} \, dV = 0. \tag{3.78}$$

Therefore,

$$\boldsymbol{\nabla} \cdot \mathbf{B} = 0. \tag{3.79}$$

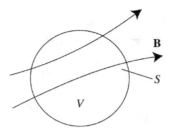

Fig. 3.12   Finite volume V with surface S does not contain any magnetic
charge.

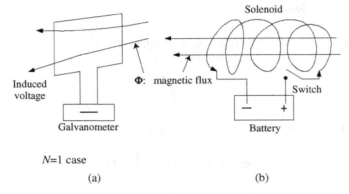

Fig. 3.13   Experimental setup to verify Faraday's law.

Faraday's law

Faraday discovered that electric voltage is induced in a circuit when the
magnetic flux Φ (webers (Wb)) changes with time. Figure 3.13 illustrates an
experimental setup to verify Faraday's law, which is composed of (a) a solenoid
with one turn ($N = 1$) attached to a galvanometer to measure the induced
voltage, and (b) an electromagnet with a switch and battery that generates a
time-varying magnetic flux Φ where the time-dependent Φ is generated by
switching on and off.

The induced voltage V_{emf} due to Φ is given by

$$V_{\text{emf}} = N\frac{\mathrm{d}\Phi}{\mathrm{d}t} = N\frac{\mathrm{d}}{\mathrm{d}t}\int \mathbf{B} \cdot \mathrm{d}\mathbf{S}, \tag{3.80}$$

where N is the number of turns of the induced-voltage side, and V_{emf}, called the
electromotive force (emf), is the voltage around a closed path and is also
related to the line integral of the E-field as

$$V_{\text{emf}} = -\oint_{C} \mathbf{E} \cdot \mathrm{d}\mathbf{l}. \tag{3.81}$$

From these, we obtain

$$\oint_C \mathbf{E} \cdot d\mathbf{l} = -\int_S \frac{\partial \mathbf{B}}{\partial t} \cdot d\mathbf{S} \quad (N = 1). \tag{3.82}$$

Using Stokes' theorem (Eq. (3.14)), we convert the left-hand side of Eq. (3.82) to a surface integral and, putting the derivative inside the integral on the right-hand side of Eq. (3.80), we can rewrite this equation as

$$\int_S (\mathbf{\nabla} \times \mathbf{E}) \cdot d\mathbf{S} = -\int_S \frac{\partial \mathbf{B}}{\partial t} \cdot d\mathbf{S}. \tag{3.83}$$

Since Eq. (3.83) is valid for any arbitrary surface, we obtain

$$\mathbf{\nabla} \times \mathbf{E} = -\frac{\partial \mathbf{B}}{\partial t}. \tag{3.84}$$

If the magnetic flux does not change with time, we obtain the formula for the static case, which will be discussed later.

3.4.1 Maxwell's equations

Maxwell (1904) summarized the above laws into a set of equations that govern electromagnetic phenomena in all ranges of frequencies and size scales above atomic size. The following set is known as the Maxwell equations.

Maxwell's equations in differential form

$$\mathbf{\nabla} \cdot \mathbf{D} = \rho_v, \tag{3.85}$$

$$\mathbf{\nabla} \times \mathbf{E} = -\frac{\partial \mathbf{B}}{\partial t}, \tag{3.86}$$

$$\mathbf{\nabla} \cdot \mathbf{B} = 0, \tag{3.87}$$

$$\mathbf{\nabla} \times \mathbf{H} = \mathbf{J} + \frac{\partial \mathbf{D}}{\partial t}. \tag{3.88}$$

Maxwell's equations in integral form

$$\oint_S \mathbf{D} \cdot d\mathbf{S} = Q, \tag{3.89}$$

$$\oint_C \mathbf{E} \cdot \mathbf{dl} = -\int_S \frac{\partial \mathbf{B}}{\partial t} \cdot \mathbf{dS}, \tag{3.90}$$

$$\oint_S \mathbf{B} \cdot \mathbf{dS} = 0, \tag{3.91}$$

$$\oint_C \mathbf{H} \cdot \mathbf{dl} = \int_S \left(\mathbf{J} + \frac{\partial \mathbf{D}}{\partial \mathbf{T}} \right) \cdot \mathbf{dS}. \tag{3.92}$$

In addition to the above Maxwell equations, the following equations are needed to fully describe the electromagnetic behavior of a given material system.

Lorentz force law

When a charge q is moving with velocity $\mathbf{v}$ in electric field $\mathbf{E}$ and magnetic flux density $\mathbf{B}$, force $\mathbf{F}$ (newtons [N]) is generated:

$$\mathbf{F} = q\{\mathbf{E} + \mathbf{v} \times \mathbf{B}\}. \tag{3.93}$$

Ohm's law

For a conductive medium, electric current density $\mathbf{J}$ is proportional to electric field:

$$\mathbf{J} = \boldsymbol{\sigma} \cdot \mathbf{E} \tag{3.94a}$$

where $\boldsymbol{\sigma}$ is electric conductivity which is generally a second-order tensor but, for isotropic materials, it is reduced to scalar σ. Then (3.94a) is reduced to

$$\mathbf{J} = \sigma \, \mathbf{E}. \tag{3.94b}$$

Convection current law

When a charge density ρ_v (C/m^3) is moving with velocity $\mathbf{v}$, it induces an electric current density $\mathbf{J}$ as

$$\mathbf{J} = \rho_v \mathbf{v}. \tag{3.95}$$

Constitutive equations in dielectric materials and magnetic materials

For a dielectric medium:

$$\mathbf{D} = \boldsymbol{\varepsilon} \cdot \mathbf{E} \tag{3.96}$$

where $\boldsymbol{\varepsilon}$ is a dielectric tensor (farads [F]/m), and $\mathbf{D}$ is electric flux density $(\mathrm{C/m^2})$,

$$\mathbf{D} = \varepsilon_0\mathbf{E} + \mathbf{P}. \tag{3.97}$$

The physical meaning of the polarization vector $\mathbf{P}$ is the electric dipole moment, which consists of pairs of positive and negative charges, per unit volume. For isotropic and linear materials, the electric polarization is proportional to the electric field:

$$\mathbf{P} = \varepsilon_0\chi_e\mathbf{E} \tag{3.98}$$

where χ_e is electric susceptibility. From Eqs. (3.97) and (3.98), Eq. (3.96) can be rewritten as

$$\mathbf{D} = \varepsilon_0\varepsilon_r\mathbf{E}, \tag{3.99}$$

where the relative dielectric constant ε_r is defined by

$$\varepsilon_r = \frac{\varepsilon}{\varepsilon_0} = 1 + \chi_e. \tag{3.100}$$

For a magnetic medium:

$$\mathbf{B} = \boldsymbol{\mu} \cdot \mathbf{H} \tag{3.101a}$$

where $\boldsymbol{\mu}$ is a magnetic permeability tensor, and should be distinguished from the shear modulus (which we denote by G). It is reduced to scalar μ for isotropic materials:

$$\mathbf{B} = \mu\mathbf{H}. \tag{3.101b}$$

For certain magnetic materials (ferrimagnetic and ferromagnetic), a magnetic dipole moment (magnetization) exists either permanently or under an applied electric field. The magnetic dipole moment per unit volume (magnetization vector) $\mathbf{M}$ can be defined by modifying Eqs. (3.101) as

$$\mathbf{B} = \mu_0(\mathbf{H} + \mathbf{M}). \tag{3.102}$$

For isotropic magnetic materials under a moderate magnetic field and not too low a temperature, $\mathbf{M}$ is proportional to the magnetic field:

$$\mathbf{M} = \chi_m\mathbf{H}, \tag{3.103}$$

where χ_m is the magnetic susceptibility. Using Eq. (3.103), Eq. (3.102) can be rewritten as

$$\mathbf{B} = \mu_0\mu_r\mathbf{H}, \tag{3.104}$$

where the relative magnetic permeability μ_r is defined by

$$\mu_r = \mu/\mu_0 = 1 + \chi_m. \tag{3.105}$$

In the above equations, ε_0 and μ_0 are the dielectric constant and magnetic permeability, respectively, for free space and they are given by

$$\varepsilon_0 = 8.854 \times 10^{-12} \text{ F/m}, \tag{3.106a}$$

$$\mu_0 = 4\pi \times 10^{-7} \text{ H/m}. \tag{3.106b}$$

In Eqs. (3.100) and (3.105), ε_r and μ_r are the non-dimensional dielectric constant and magnetic permeability, respectively, of a linear isotropic material.

3.4.2 *Electrostatics and magnetostatics*

If there is no time variation, we can set $d/dt = 0$ in Maxwell's equations. Then Maxwell's equations can be separated into two groups: electrostatic and magnetostatic. The former are related to capacitor equations and the latter to inductor equations.

Electrostatics

$$\nabla \cdot \mathbf{D} = \rho_v, \tag{3.107a}$$

$$\oint_S \mathbf{D} \cdot d\mathbf{S} = Q, \tag{3.107b}$$

$$\nabla \times \mathbf{E} = 0, \tag{3.108a}$$

$$\oint_C \mathbf{E} \cdot d\mathbf{l} = 0, \tag{3.108b}$$

$$C = \frac{Q}{V}, \tag{3.109}$$

where C (F) is electric capacitance.

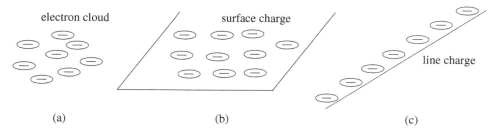

electron cloud   surface charge   line charge

(a)   (b)   (c)

Fig. 3.14   Three different cases of electric charge distribution: (a) volume charge ρ_v, (b) surface charge ρ_s, and (c) line charge ρ_l.

Magnetostatics

$$\nabla \cdot \mathbf{B} = 0, \tag{3.110a}$$

$$\oint_S \mathbf{B} \cdot d\mathbf{S} = 0, \tag{3.110b}$$

$$\nabla \times \mathbf{H} = \mathbf{J}, \tag{3.111a}$$

$$\oint_C \mathbf{H} \cdot d\mathbf{l} = \mathbf{I}, \tag{3.111b}$$

$$L = \frac{\Phi}{I} \tag{3.112}$$

where L (H) is electric inductance.

These two groups show that there is no coupling between magnetic and electric behavior.

Charge and current distributions

There are three cases of electric charge distribution: (a) volume charge ρ_v (C/m^3) in a volume, (b) surface charge on a surface ρ_s (C/m^2), and (c) line charge ρ_l (C/m) along a line. These three cases (a), (b), and (c) are illustrated in Fig. 3.14 (a), (b), and (c), respectively. Integrals of ρ_v, ρ_s, ρ_l over the volume, surface, and line in consideration lead to the total charge Q (C):

$$Q = \int_V \rho_v dV, \tag{3.113a}$$

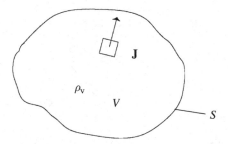

Fig. 3.15   Electric current density **J** due to change with time of volume charge density ρ_v.

$$Q = \int_S \rho_s \mathrm{d}S, \qquad\qquad (3.113b)$$

$$Q = \int_l \rho_l \mathrm{d}l. \qquad\qquad (3.113c)$$

Continuity of charge

The total current moving out from an enclosed area must be equal to the rate of change of charge within the volume, Fig. 3.15:

$$\oint_S \mathbf{J} \cdot \mathrm{d}\mathbf{S} = -\frac{\mathrm{d}}{\mathrm{d}t} \int_V \rho_v \mathrm{d}V. \qquad\qquad (3.114)$$

Using the divergence theorem Eq. (3.13b), we can change the left-hand side to a volume integral and obtain

$$\nabla \cdot \mathbf{J} = -\frac{\partial \rho_v}{\partial t}. \qquad\qquad (3.115)$$

Coulomb's law

Electrical force $\mathbf{F}_{21}$ (N) between two charged bodies (in free space), q_1 and q_2, Fig. 3.16, is given by

$$\mathbf{F}_{21} = \widehat{\mathbf{R}}_{12} \frac{q_1 q_2}{4\pi\varepsilon_0 R_{12}^2}. \qquad\qquad (3.116)$$

Coulomb's law of two charged bodies can be generalized to the electric field **E** induced by an electric charge q, Fig. 3.17.

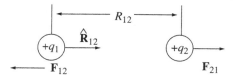

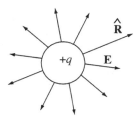

Fig. 3.16   Coulomb's force between two charged bodies, q_1 and q_2

Fig. 3.17   Electric force due to electric charge q.

$$\mathbf{F} = q\mathbf{E}, \tag{3.117a}$$

where

$$\mathbf{E} = \widehat{\mathbf{R}}\frac{q}{4\pi\varepsilon R^2}, \quad \varepsilon = \varepsilon_0\varepsilon_{\mathrm{r}}. \tag{3.117b}$$

From Eqs. (3.116) and (3.117), force (N) can be interconnected in SI units,

$$N = \frac{C^2m}{Fm^2} = \frac{VC}{m}. \tag{3.118}$$

Therefore, we obtain

$$C = FV. \tag{3.119}$$

The electric field due to multiple point charges, Fig. 3.18, is the vector sum of E-fields due to many charges:

$$\mathbf{E}_{\text{total}} = \mathbf{E}_1 + \mathbf{E}_2 + \mathbf{E}_3 + \cdots = \frac{1}{4\pi\varepsilon}\sum_{i=1}^{N}\frac{q_i(\mathbf{R} - \mathbf{R}_i)}{|\mathbf{R} - \mathbf{R}_i|^3}. \tag{3.120}$$

Electric field due to a charge distribution

Due to a volume charge:

using the E-field due to the differential charge

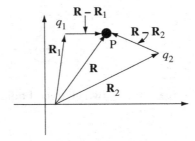

Fig. 3.18   Electric field at point P due to multiple charges.

$$dE = \frac{\widehat{R}\rho_v dV}{4\pi\varepsilon R^2},$$

where $\widehat{R}$ is a unit vector along the E-field vector, we obtain

$$E = \int_v dE = \frac{1}{4\pi\varepsilon} \int_V \widehat{R}\frac{\rho_v \, dv}{R^2}. \qquad (3.121)$$

Due to a surface charge:

$$E = \frac{1}{4\pi\varepsilon} \int_S \widehat{R}\frac{\rho_s dS}{R^2}. \qquad (3.122)$$

Due to a line charge:

$$E = \frac{1}{4\pi\varepsilon} \int_l \widehat{R}\frac{\rho_l dl}{R^2}. \qquad (3.123)$$

Electric scalar potential (voltage)

Potential between P_1 and P_2, Fig 3.19(a), is given by

$$V_{21} = V_2 - V_1 = - \int_{P_1}^{P_2} E \cdot dl. \qquad (3.124)$$

Potential difference does not depend on the path, Fig. 3.19(b), whence we can derive that potential in a closed contour is zero, Fig. 3.19(c):

$$\oint_C E \cdot dl = 0. \qquad (3.125)$$

Potential at point P is given by

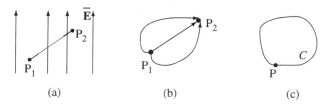

Fig. 3.19   Electric potential.

$$V = -\int_{-\infty}^{P} \mathbf{E} \cdot d\mathbf{l}$$

$$= -\int_{-\infty}^{R} \left(\widehat{\mathbf{R}} \frac{q}{4\pi\varepsilon R^2} \right) \cdot \widehat{\mathbf{R}} dR = \frac{q}{4\pi\varepsilon R} \cdot \qquad (3.126)$$

Using Eq. (3.124) and the gradient, we can write the electric field $\mathbf{E}$ derivable from the potential V for the electrostatic case:

$$\mathbf{E} = -\nabla V. \qquad (3.127)$$

If we combine this with Eq. (3.107a), we obtain Poisson's equation.

$$\nabla \cdot (\nabla V) = -\frac{\rho_v}{\varepsilon},$$

$$\nabla^2 V = -\frac{\rho_v}{\varepsilon} \qquad (3.128)$$

where $\nabla^2 V$ is defined by Eq. (3.10).

Since the change in electric potential is equal to the work W done per unit charge q, we can obtain

$$dV = \frac{dW}{q}$$

$$= \frac{\mathbf{F}_{\text{ext}} \cdot d\mathbf{l}}{q}$$

$$= \frac{-q\mathbf{E} \cdot d\mathbf{l}}{q}$$

$$= -\mathbf{E} \cdot d\mathbf{l}. \qquad (3.129)$$

It is to be noted that Eq. (3.129) can be derived by taking the derivative of Eq. (3.124).

Electromagnetic equations for circuits

For a closed loop, the algebraic sum of the voltages for the individual branches of the loop is zero (Kirchhoff's voltage law):

$$\sum_i V_i = 0. \tag{3.130}$$

Ohm's law for a resistive material with electric resistance R (ohms [Ω]) is given by

$$V = IR. \tag{3.131}$$

The algebraic sum of current flowing out of a junction is zero (Kirchhoff's current law):

$$\sum_n I_n(t) = 0. \tag{3.132}$$

Magnetic circuit

Consider a circular ring made of a soft magnet (yoke) with an electric coil around it, Fig. 3.20(a). Upon applying an electric current I in the coil, a magnetic flux Φ (webers [Wb]) flows in the yoke. In the yoke with its cross-sectional area S, the magnetic flux is expressed by

$$\Phi = BS, \tag{3.133}$$

where B is the magnetic flux density (webers/m^2 = teslas [T]).

Similarly, one can consider a ring-shaped yoke with a permanent magnet, Fig. 3.20(b), where magnetic polarity is taken so as to realize a clockwise magnetic flux as in the electromagnet case.

Due to the analogy between the electric current flow in a conductor material with conductivity σ, and the magnetic flux in a ferromagnetic material with permeability μ, one can establish the electric resistance R and magnetic resistance R_m as

$$R = \int \frac{ds}{\sigma S}, \tag{3.134a}$$

$$R_m = \int \frac{ds}{\mu S}. \tag{3.134b}$$

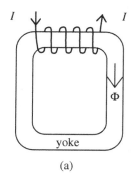 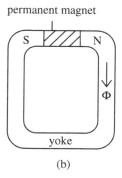

(a)   (b)

Fig 3.20   Magnetic circuit (a) in a soft magnet (yoke) with electric coil with electric current *I*, and (b) in a soft magnet with permanent magnet (Chikamizu, 1964).

Similarly to the electromotive force, the magnetomotive force (V_m) can be defined by

$$V_m = \oint H_s ds, \tag{3.135}$$

where H_s is the magnetic field along the magnetic circuit in the *s*-coordinate and $\oint$ is the closed line integral along the magnetic circuit. For the electromagnet of Fig. 3.20(a),

$$\oint H_s ds = NI, \tag{3.136}$$

where *N* is number of turns of the electric coil, and *I* (ampere (A)) is the electric current. Since the magnetic field intensity H_s is given by Eq. (3.101b) and the magnetic flux density *B* is related to magnetic flux Φ by Eqs. (3.133) and (3.134), we derive the following formula (similar to Ohm's law) for a magnetic circuit:

$$V_m = \oint H_s ds = \oint \frac{B}{\mu} ds = \oint \frac{\Phi}{\mu S} ds = \Phi \oint \frac{ds}{\mu S} = \Phi R_m, \tag{3.137a}$$

where

$$R_m = \oint \frac{ds}{\mu S}. \tag{3.137b}$$

In Eq. (3.137b), the magnetic flux Φ plays the same role as the electric current *I* in Ohm's law, i.e., Eq. (3.131).

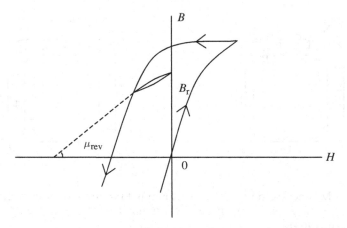

Fig. 3.21   Remanent magnetic flux density B_r and reversible magnetic permeability μ_{rev} in the second quadrant of B–H curve of a permanent magnet.

For the magnetic circuit with a permanent magnet, Fig. 3.20(b), the magneto-motive force, V_m is defined by

$$V_m = \frac{B_r l_p}{\mu_{rev}} \tag{3.138}$$

where l_p is the length of the permanent magnet, B_r is the remanent magnetic flux density and μ_{rev} is the reversible magnetic permeability defined in the second quadrant of the B–H curve of a permanent magnet, Fig. 3.21.

For the magnetic circuit of Fig. 3.20(b), Eq. (3.137a) is modified as

$$V_m = \frac{B_r l_p}{\mu_{rev}} = \Phi(R_m + R'_m), \tag{3.139}$$

where R_m is the magnetic resistance of the yoke and R'_m is the magnetic resistance of the permanent magnet. They are defined by

$$R_m = \frac{l_y}{\mu_y S_y} \tag{3.140}$$

$$R'_m = \frac{l_p}{\mu_{rev} S_p}. \tag{3.141}$$

In the above equations, l_y, μ_y, S_y are the length, magnetic permeability, and cross-sectional area of the yoke, respectively, and l_p, μ_{rev}, S_p are the length, reversible magnetic permeability (defined in Fig. 3.21), and the cross-sectional area of the permanent magnet, respectively.

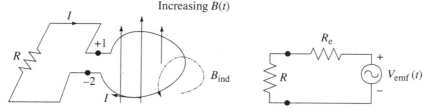

B_{ind}: induced magnetic flux due to I

Fig. 3.22 Lenz' law for time-varying fields. R is the load resistance, R_e the resistance of the coil, V_{emf} is the electromotive force.

It is noted that Kirchhoff's voltage law Eq. (3.130) and current law Eq. (3.132) are also valid for the magnetic circuit, i.e.,

$$\sum_i V_{mi} = 0, \tag{3.142}$$

$$\sum_i \Phi_i = 0. \tag{3.143}$$

3.4.3 Maxwell's equations for time-varying fields

In this subsection, we will discuss Maxwell's equations for time-varying fields.

Lenz' law

A current which creates an induced magnetic flux, B_{ind}, is always in such a direction as to oppose the change of magnetic flux $B(t)$, Fig. 3.22.

Displacement current

Let us assume that voltage is given by a time-varying cosine wave, which is applied to a capacitor with thickness d, Fig. 3.23.

Then the electric field inside the capacitor is given by

$$\mathbf{E} = \hat{\mathbf{y}} \frac{V_C}{d} = \hat{\mathbf{y}} \frac{V_0}{d} \cos \omega t. \tag{3.144}$$

We can define two types of electric current, conduction (I_c) and displacement (I_d). Current I_c is defined in a conductive line, and I_d in a capacitor. They are given by

$$I_c = \int_{S_1} \mathbf{J} \cdot d\mathbf{S}, \tag{3.145a}$$

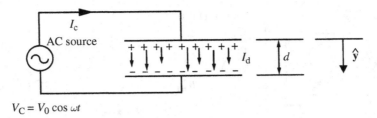

Fig. 3.23   Plate capacitor C subjected to time-varying electric field.

$$I_d = \int_{S_2} \frac{\partial \mathbf{D}}{\partial t} \cdot d\mathbf{S}. \tag{3.145b}$$

The two currents must be continuous at the interface, i.e.,

$$I_c = C\frac{dV_C}{dt} = C\frac{d}{dt}(V_0 \cos \omega t) = -CV_0\omega \sin \omega t, \tag{3.146a}$$

$$I_d = \int_S \frac{\partial \mathbf{D}}{\partial t} \cdot d\mathbf{S} = \int_S \frac{\partial}{\partial t}\left[\hat{\mathbf{y}}\frac{\varepsilon V_0}{d}\cos \omega t\right] \cdot \hat{\mathbf{y}}\, d\mathbf{S} = -\frac{\varepsilon S}{d}V_0\, \omega \sin \omega t. \tag{3.146b}$$

Since the capacitance C of an ideal plate capacitor is related to its dielectric constant ε by

$$C = \frac{\varepsilon S}{d}, \tag{3.147}$$

where S is the surface area of a capacitor, we can prove the continuity of the two types of electric current from Eqs. (3.146) and (3.147):

$$I_c = I_d. \tag{3.148}$$

Plane wave propagation

The plane wave is a continuous wave (CW) whose amplitude and phase are constant in the transverse (x- and y-) directions (perpendicular to the propagation direction z), Fig. 3.24. Since there is no variation in the transverse direction, the derivative with respect to the transverse direction becomes 0.

Rather than dealing with the time derivative, we seek solutions for the time-harmonic fields (CW case). In this case, the time derivative is replaced by $j\omega$, i.e.,

$$\frac{\partial}{\partial t} = j\omega, \quad \text{because} \quad \frac{\partial}{\partial t}\left(e^{j\omega t}\right) = j\omega e^{j\omega t}. \tag{3.149}$$

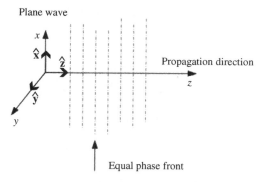

Fig. 3.24 Plane wave propagating along the *z*-direction where $\hat{\mathbf{x}}$, $\hat{\mathbf{y}}$, and $\hat{\mathbf{z}}$ are unit vectors along the *x*-, *y*-, and *z*-axes.

Therefore, Maxwell's equations are rewritten as

$$\nabla \cdot \mathbf{E} = \frac{\rho_v}{\varepsilon}, \qquad (3.150)$$

$$\nabla \times \mathbf{E} = -j\omega\mu\mathbf{H}, \qquad (3.151)$$

$$\nabla \cdot \mathbf{H} = 0, \qquad (3.152)$$

$$\nabla \times \mathbf{H} = \mathbf{J} + j\omega\varepsilon\mathbf{E}. \qquad (3.153)$$

First, we want to eliminate the current density **J** from the above equations by using the electric conduction law, Eq. (3.94b). Then we define the complex permittivity ε_c which contains the conductivity.

Equation (3.153) is rewritten as

$$\nabla \times \mathbf{H} = \mathbf{J} + j\omega\varepsilon\mathbf{E} = (\sigma + j\omega\varepsilon)\mathbf{E} = j\omega\left(\varepsilon - j\frac{\sigma}{\omega}\right)\mathbf{E} = j\omega\varepsilon_c\mathbf{E}, \qquad (3.154)$$

where

$$\varepsilon_c = \varepsilon - j\frac{\sigma}{\omega} = \varepsilon' - j\varepsilon'' = \varepsilon_r\varepsilon_0, \qquad (3.155)$$

ε_r is a relative complex dielectric constant, and ε' and ε'' are real and imaginary parts of the complex dielectric constant.

In addition, note that

$$\varepsilon_r = \varepsilon_r' - j\varepsilon_r'' \qquad (3.156)$$

where ε_r' and ε_r'' are real and imaginary parts of the relative complex dielectric constant, respectively, and $j = \sqrt{-1}$.

If there is no charge (charge-free media), we get, from Eq. (3.150),

$$\mathbf{\nabla} \cdot \mathbf{E} = \rho_v = 0. \tag{3.157}$$

Now we derive a wave equation for the source-free case ($\rho_v = 0$, $\mathbf{J} = 0$). Applying curl ($\mathbf{\nabla} \times$) to both sides of Eq. (3.151), we obtain

$$\mathbf{\nabla} \times (\mathbf{\nabla} \times \mathbf{E}) = \mathbf{\nabla} \times (-\mathrm{j}\omega\mu\mathbf{H}). \tag{3.158}$$

The left-hand side of Eq. (3.158) is replaced by the vector identity

$$\mathbf{\nabla} \times (\mathbf{\nabla} \times \mathbf{E}) = \mathbf{\nabla}(\mathbf{\nabla} \cdot \mathbf{E}) - \nabla^2\mathbf{E}. \tag{3.159}$$

The first term on the right-hand side of Eq. (3.159) vanishes from Eq. (3.157), thus Eq. (3.158) is reduced to

$$-\nabla^2\mathbf{E} = -\mathrm{j}\omega\mu(\mathbf{\nabla} \times \mathbf{H}). \tag{3.160}$$

Substituting Eq. (3.153) into the right-hand side of Eq. (3.160), and using the source-free condition ($\mathbf{J} = 0$), we obtain the following plane wave equation:

$$\nabla^2\mathbf{E} + k^2\mathbf{E} = 0, \tag{3.161}$$

where k is the propagation constant of a plane wave,

$$k = \omega\sqrt{\mu\varepsilon}, \tag{3.162a}$$

and

$$\nabla^2\mathbf{E} = \left(\frac{\partial^2}{\partial x^2} + \frac{\partial^2}{\partial y^2} + \frac{\partial^2}{\partial z^2}\right)\mathbf{E}. \tag{3.162b}$$

Hence, we seek a solution for a simple case by assuming that E- and H-fields are in the x–y plane and the wave is propagating in the z-direction. Then, there is no variation of the E-field in the x- and y-directions, and the derivatives with respect to x and y must be 0. Also, we assume that the E-field acts in the x-direction. Then Eq. (3.161) is reduced to

$$\frac{\mathrm{d}^2 E_x}{\mathrm{d}z^2} + k^2 E_x = 0. \tag{3.163}$$

Now we can obtain the one-dimensional (1D) wave equation, which is very similar to the transmission-line equations, Eqs. (3.189a)–(3.198). $E_z = 0$ and $H_z = 0$ can be obtained once we set $\partial/\partial x = 0$, $\partial/\partial y = 0$ in Maxwell's equations. To show this, we use Eq. (3.153), where we look at z-components,

$$\hat{\mathbf{z}}\left(\frac{\partial H_y}{\partial x} - \frac{\partial H_x}{\partial y}\right) = \hat{\mathbf{z}}\,\mathrm{j}\omega\varepsilon E_z$$

$$E_z = 0. \tag{3.164}$$

In general, the solution of the 1D wave equation has two terms. One is the positive-going and the other is the negative-going wave:

$$E_x(z) = E_x^+(z) + E_x^-(z) = E_{x0}^+ e^{-\mathrm{j}kz} + E_{x0}^- e^{+\mathrm{j}kz}$$

$$\begin{array}{cc} +z & -z \\ \longrightarrow & \longleftarrow \end{array} \tag{3.165}$$

Assuming the wave is propagating in the $(+z)$-direction and since the E-field has only an x-component, Eq. (3.165) is reduced to

$$\mathbf{E}(z) = \hat{\mathbf{x}} E_{x0}^+ e^{-\mathrm{j}kz}, \tag{3.166a}$$

$$E_y = 0, \tag{3.166b}$$

$$E_z = 0. \tag{3.166c}$$

Magnetic fields can be found from

$$\nabla \times \mathbf{E} = -\mathrm{j}\omega\mu\mathbf{H} = -\mathrm{j}\omega\mu(\hat{\mathbf{x}}H_x + \hat{\mathbf{y}}H_y + \hat{\mathbf{z}}H_z). \tag{3.167}$$

From Eqs. (3.166) and (3.167), we obtain the magnetic field as

$$\mathbf{H}(z) = \hat{\mathbf{y}}\frac{k}{\omega\mu}E_{x0}^+ e^{-\mathrm{j}kz}, \tag{3.168a}$$

$$H_x = 0, \tag{3.168b}$$

$$H_z = 0. \tag{3.168c}$$

From these, it is clear that both E and H are perpendicular to the wave propagation direction.

When we obtain the characteristic impedance of a transmission line (TL), we use the incident voltage and incident current. Since the E- and H-fields are related to voltage and current, we can define the impedance of a plane wave. This is known as the intrinsic impedance η and is defined as the ratio of E_x and H_y:

$$\eta = \frac{E_x}{H_y} = \frac{\omega\mu}{k} = \sqrt{\frac{\mu}{\varepsilon}}. \tag{3.169}$$

If the wave is propagating in free space, the intrinsic impedance is

$$\eta_0 = \sqrt{\frac{\mu_0}{\varepsilon_0}} = 120\,\pi\ \ \Omega$$

where the values of μ_0 and ε_0 are taken from Eq. (3.106).

Since the E and H fields are perpendicular to the direction of wave propagation, the plane wave is called a TEM (transverse electromagnetic) wave, defined by Eqs. (3.166) and (3.168).

The velocity of the plane wave is given by

$$U_P = \frac{\omega}{k} = \frac{1}{\sqrt{\mu\varepsilon}}. \tag{3.170}$$

Often we use a **k** vector to specify the direction of the wave. The vector **k** has a direction given by unit vector $\widehat{\mathbf{k}}$ and the magnitude given by $|\mathbf{k}|$, i.e.,

$$\widehat{\mathbf{k}} = \frac{\mathbf{k}}{|\mathbf{k}|}, \tag{3.171a}$$

$$|\mathbf{k}| = \omega\sqrt{\mu\varepsilon}. \tag{3.171b}$$

The general relationships among **k**, **E** and **H** are

$$\mathbf{E} = -\eta\,\widehat{\mathbf{k}} \times \mathbf{H}, \tag{3.172a}$$

$$\mathbf{H} = \frac{1}{\eta}\,\widehat{\mathbf{k}} \times \mathbf{E}. \tag{3.172b}$$

Since the wave is propagating in the $(+z)$-direction, we have

$$\widehat{\mathbf{k}} = \widehat{\mathbf{z}}, \tag{3.173a}$$

$$\mathbf{H}(z) = \frac{1}{\eta}\,\widehat{\mathbf{z}} \times (\widehat{\mathbf{x}}E_{x0}^+ e^{-jkz}) = \frac{1}{\eta}\,\widehat{\mathbf{y}}E_{x0}^+ e^{-jkz}, \tag{3.173b}$$

$$\widehat{\mathbf{y}} = \widehat{\mathbf{z}} \times \widehat{\mathbf{x}}. \tag{3.173c}$$

Wave reflection and transmission (normal incidence)

We will consider a plane wave moving through medium 1 $(\varepsilon_1, \mu_1, \eta_1)$ incident normal on a second medium $(\varepsilon_2, \mu_2, \eta_2)$, Fig.3.25(a). This situation is similar to an interface at which two transmission lines with different characteristic impedances are connected (Z_{01}, Z_{02}), Fig. 3.25(b). We need to express the incident, reflected, and transmitted waves in each medium, then apply the

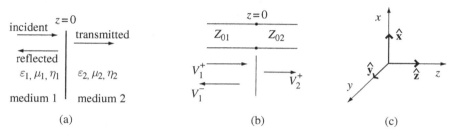

Fig. 3.25 (a) Wave reflection and transmission at interface $(z = 0)$, which is similar to (b) two transmission lines connected at $(z = 0)$. (c) Coordinates x, y, z with unit base vectors, $\hat{\mathbf{x}}$, $\hat{\mathbf{y}}$, and $\hat{\mathbf{z}}$.

boundary conditions. Because the wave is incident normal on the surface, all E- and H-fields are parallel to the surface. The boundary condition states that the tangential E- and H-fields must be continuous at the boundary.

For the incident wave:

$$\mathbf{E}^i(z) = \hat{\mathbf{x}} E_0^i e^{-jk_1 z}, \qquad (3.174a)$$

$$\mathbf{H}^i(z) = \hat{\mathbf{y}} \frac{E_0^i}{\eta_1} e^{-jk_1 z}. \qquad (3.174b)$$

For the reflected wave:

$$\mathbf{E}^r(z) = \hat{\mathbf{x}} E_0^r e^{+jk_1 z}, \qquad (3.175a)$$

$$\mathbf{H}^r(z) = -\hat{\mathbf{y}} \frac{E_0^r}{\eta_1} e^{+jk_1 z}. \qquad (3.175b)$$

For the transmitted wave, the E- and H-fields are given by

$$\mathbf{E}^t(z) = \hat{\mathbf{x}} E_0^t e^{-jk_2 z}, \qquad (3.176a)$$

$$\mathbf{H}^t(z) = \hat{\mathbf{y}} \frac{E_0^t}{\eta_2} e^{-jk_2 z}. \qquad (3.176b)$$

Moreover,

$$\eta_1 = \sqrt{\frac{\mu_1}{\varepsilon_1}}, \qquad (3.177a)$$

$$\eta_2 = \sqrt{\frac{\mu_2}{\varepsilon_2}}. \qquad (3.177b)$$

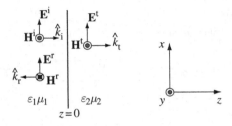

Fig. 3.26   Orientations of incident (i), transmitted (t), and reflected (r) waves.

In the above equations there are two unknowns E_0^r, E_0^t, which will be solved for by using the boundary conditions. The orientations of the E- and H-fields for the incident, transmitted, and reflected waves are illustrated in Fig. 3.26.

Boundary conditions

From Eqs. (3.108) and (3.110), the tangential E- and H-fields must be continuous at $z = 0$:

$$E_{t1} = E_{t2}, \tag{3.178a}$$

$$H_{t1} = H_{t2} \tag{3.178b}$$

where E_{ti} and H_{ti} ($i = 1, 2$) are tangential components of the E- and H-fields. It is to be noted that all the E-fields are parallel to the x-axis while all H-fields are parallel to the y-axis (cf. Figs. 3.25 and 3.26).

In medium 1:

$$\underset{\substack{\text{total}\\ \text{tangential}\\ E\text{-field}}}{E_{t1}(z = 0)} = \underset{\substack{\text{incident}\\ \text{field}}}{E^i(z = 0)} + \underset{\substack{\text{reflected}\\ \text{field}}}{E^r(z = 0)}; \tag{3.179a}$$

$$\text{TL case} - V_{\text{total}} = V_1^+ + V_1^-,$$

$$E_{t1}(z = 0) = E_0^i + E_0^r. \tag{3.179b}$$

In medium 2:

$$E_{t2}(z = 0) = E^t(z = 0) = E_0^t. \tag{3.180}$$

Therefore, at $z = 0$,

$$E_0^i + E_0^r = E_0^t. \tag{3.181}$$

Similarly, continuity of the tangential component of *H*-fields requires

$$H_{t1}(z = 0) = H^i(z = 0) + H^r(z = 0) = \frac{1}{\eta_1}\left(E_0^i - E_0^r\right), \qquad (3.182a)$$

$$H_{t2}(z = 0) = H^t(z = 0) = \frac{1}{\eta_2}E_0^t. \qquad (3.182b)$$

Therefore,

$$\frac{1}{\eta_1}\left(E_0^i - E_0^r\right) = \frac{1}{\eta_2}E_0^t. \qquad (3.183)$$

Solving for E_0^r and E_0^t in terms of E_0^i from Eqs. (3.181) and (3.183), we obtain

$$E_0^r = \frac{\eta_2 - \eta_1}{\eta_2 + \eta_1}E_0^i = \Gamma E_0^i, \qquad (3.184a)$$

$$E_0^t = \frac{2\eta_2}{\eta_2 + \eta_1}E_0^i = \tau E_0^i, \qquad (3.184b)$$

where Γ is the reflection coefficient and τ is the transmission coefficient. They are defined by

$$\Gamma = \frac{E_0^r}{E_0^i} = \frac{\eta_2 - \eta_1}{\eta_2 + \eta_1}, \qquad (3.185a)$$

$$\tau = \frac{E_0^t}{E_0^i} = \frac{2\eta_2}{\eta_2 + \eta_1}, \qquad (3.185b)$$

$$\tau = 1 + \Gamma. \qquad (3.185c)$$

For non-magnetic media, the magnetic permeabilities of media 1 and 2 are set equal to μ_0, so

$$\mu_1 = \mu_2 = \mu_0. \qquad (3.186)$$

Therefore, the reflection and transmission coefficients are reduced to

$$\Gamma = \frac{\sqrt{\varepsilon_{r1}} - \sqrt{\varepsilon_{r2}}}{\sqrt{\varepsilon_{r1}} + \sqrt{\varepsilon_{r2}}} = \frac{n_1 - n_2}{n_1 + n_2}, \qquad (3.187a)$$

$$\tau = \frac{2n_2}{n_1 + n_2}, \qquad (3.187b)$$

where n_1 and n_2, the indices of refraction, are given by

$$n_1 = \sqrt{\varepsilon_{r1}}, \quad n_2 = \sqrt{\varepsilon_{r2}}, \tag{3.188}$$

and ε_r is the relative permittivity of the medium.

Transmission lines

When we look at an arbitrary point (z) in a medium, the total voltage is composed of two waves, a wave (voltage) propagating in the $(+z)$-direction (incident), $e^{-j\beta z}$, and a wave (voltage) propagating in the $(-z)$-direction (reflected), $e^{+j\beta z}$, Fig. 3.27. Thus

$$V_{\text{total}} = V_{\text{incident}} + V_{\text{reflected}}, \tag{3.189a}$$

$$V(z) = V_0^+ e^{-j\beta z} + V_0^- e^{+j\beta z}, \tag{3.189b}$$

where β is a phase constant related to wavelength λ by

$$\beta = \frac{2\pi}{\lambda}. \tag{3.189c}$$

The currents $I(z)$ for the incident and reflected waves are obtained by dividing the corresponding voltages by Z_0, except that the reflected current has a negative sign. Thus,

$$I(z) = \frac{V_0^+}{Z_0} e^{-j\beta z} - \frac{V_0^-}{Z_0} e^{+j\beta z}, \tag{3.190}$$

where Z_0 is the characteristic impedance of the medium, given by

$$Z_0 = \frac{V^+}{I^+} = -\frac{V^-}{I^-}. \tag{3.191}$$

At $z = 0$ (load position), we get the load voltage and current

$$V(z = 0) = V_L = V_0^+ + V_0^-, \tag{3.192a}$$

$$I(z = 0) = I_L = \frac{V_0^+}{Z_0} - \frac{V_0^-}{Z_0}. \tag{3.192b}$$

The load impedance is given by

$$Z_L = \frac{V_L}{I_L} = \left(\frac{V_0^+ + V_0^-}{V_0^+ - V_0^-} \right) Z_0. \tag{3.193}$$

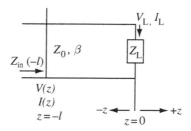

Fig. 3.27 Transmission line, where Z_0 is characteristic impedance, β is propagation constant, and Z_L is load impedance.

Using this, the reflected voltage V_0^- is written as

$$V_0^- = \Gamma_L V_0^+, \tag{3.194}$$

where Γ_L is the reflection coefficient, given by

$$\Gamma_L = \frac{V_0^-}{V_0^+} = \frac{Z_L - Z_0}{Z_L + Z_0}, \tag{3.195}$$

$$-1 \leq \Gamma_L \leq 1 \text{ if } Z_L, \ Z_0 \text{ are real}.$$

It is to be noted in Eq. (3.195) that there is no reflection ($\Gamma_L = 0$) if $Z_L = Z_0$, and total reflection ($\Gamma_L = \pm 1$) if $Z_L = 0$ or $Z_L = \infty$.

In terms of Γ_L, Eqs. (3.189) and (3.190) are written as

$$V(z) = V_0^+ \left(e^{-j\beta z} + \Gamma_L e^{j\beta z} \right), \tag{3.196a}$$

$$I(z) = \frac{V_0^+}{Z_0} \left(e^{-j\beta z} - \Gamma_L e^{j\beta z} \right). \tag{3.196b}$$

Let us consider the voltage at $z = -l$ where the reflection coefficient becomes

$$\Gamma(-l) = \frac{V^-(-l)}{V^+(-l)} = \frac{V_0^- e^{-j\beta l}}{V_0^+ e^{+j\beta l}} = \Gamma_L e^{-2j\beta l}. \tag{3.197}$$

The average power P_{ave} delivered to the load is then calculated as

$$P_{ave} = \frac{1}{T} \int_0^T P(t) \mathrm{d}t = \frac{1}{2} \mathrm{Re}[V(z) I^*(z)] = \frac{1}{2} \frac{|V_0^+|^2}{Z_0} \left(1 - |\Gamma_L|^2 \right)$$

$$= \textit{incident power} - \textit{reflected power}, \tag{3.198}$$

where T is duration or period.

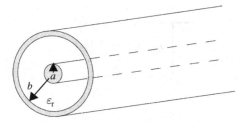

Fig. 3.28   Coaxial cable: dielectric medium with permittivity ε_r separates inner and outer conductors (shaded).

Characteristic and input impedances

The characteristic impedance Z_0 is defined by Eq. (3.191). It is to be noted that Z_0 is independent of voltage, current and position, and it is a function of material and geometry. Let us consider a coaxial cable, Fig. 3.28.

Assuming $\mu = \mu_0$, i.e., non-magnetic material, the characteristic impedance of a coaxial cable is

$$Z_0 = \frac{\eta}{2\pi}\ln(b/a) = \frac{\eta_0}{2\pi\sqrt{\varepsilon_r}}\ln\left(\frac{b}{a}\right), \qquad (3.199)$$

where b and a are the radii of the outer conducting shell and the inner conducting core, respectively, ε_r is the relative dielectric constant of the intervening medium, and η is the intrinsic impedance defined by Eq. (3.169) and given by

$$\eta = \sqrt{\frac{\mu_r}{\varepsilon_r}}\,\eta_0 = \frac{\eta_0}{\sqrt{\varepsilon_r}} = \frac{120\pi}{\sqrt{\varepsilon_r}}. \qquad (3.200)$$

The input impedance is defined at $z = -l$, Fig. 3.27:

$$Z_{in}(-l) = \frac{\text{total voltage}}{\text{total current}}.$$

By using Eqs. (3.196) $Z_{in}(-l)$ is calculated as

$$Z_{in}(-l) = \frac{V_0^+\left(e^{j\beta l} + \Gamma_L e^{-j\beta l}\right)}{(V_0^+/Z_0)(e^{j\beta l} - \Gamma_L e^{-j\beta l})} = Z_0\left[\frac{1 + \Gamma_L e^{-2j\beta l}}{1 - \Gamma_L e^{-2j\beta l}}\right]. \qquad (3.201)$$

Equation (3.201) is rewritten, when Γ_L is substituted from Eq. (3.195), as

$$Z_{in}(-l) = Z_0\left(\frac{Z_L + jZ_0\tan\beta l}{Z_0 + jZ_L\tan\beta l}\right). \qquad (3.202)$$

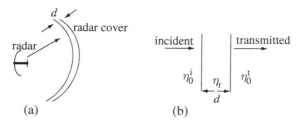

Fig. 3.29   EM waves through radome. (a) Schematic of radome structure, and (b) intrinsic impedances.

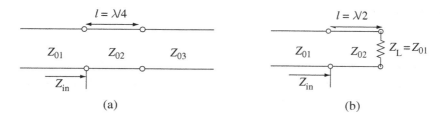

Fig. 3.30   Two cases of transmission-line design: (a) quarter wave, and (b) half wave.

Radome design

Most radar systems have covers, Fig. 3.29(a), which are called "radomes", to protect sensitive instruments from adverse weather conditions. The ideal radar cover should not have any effects at the radar operating frequency, i.e., no reflection from the radar cover ($\eta_0^i = \eta_0^t$), Fig. 3.29(b). This can be designed using transmission-line theory.

Let us consider two cases of radome (transmission line) design, Fig. 3.30.

Case 1: Quarter-wave ($\lambda/4$) transformer.

If $l = \lambda/4$, from Eq. (3.189c), $\beta l = \pi/2$. Then we obtain from Eq. (3.202)

$$Z_{in}(-l) = \frac{Z_0^2}{Z_L}. \tag{3.203}$$

It is to be noted, in comparing Fig. 3.27 with Fig. 3.30, that $Z_0 = Z_{02}$, $Z_L = Z_{03}$

If Z_{01} is set equal to Z_{03}, i.e., required by the radome design, then $Z_{02} = Z_{01}$. This prevents us from designing a radome for which

$$Z_{02} \neq Z_{01} = Z_{03}. \tag{3.204}$$

Therefore, a $\lambda/4$ transformer does not work.

Case 2: Half-wave ($\lambda/2$) TL

If $l = \lambda/2$, βl is reduced to π. From Eq. (3.203), this leads to

$$Z_{\text{in}} = Z_{\text{L}} = Z_{03}. \tag{3.205}$$

Since $Z_{03} = Z_{01}$ and the reflection coefficient Γ is given by

$$\Gamma = \frac{Z_{\text{in}} - Z_{01}}{Z_{\text{in}} + Z_{01}}, \tag{3.206}$$

it follows that the reflection coefficient Γ becomes zero from Eq. (3.206), and Z_{02} can be any value. In this case, Fig. 3.30(b), we can design a radome whose Z_{02} can have any value.

Power density and Poynting vector

The complex Poynting vector is used for showing the direction and amount of real and imaginary power flow. This is similar to the power flow of energy systems such as motors and generators.

We define the complex Poynting vector $\mathbf{S}_c$ using the phasor notation of $\mathbf{E}$ and $\mathbf{H}$.

$$\mathbf{S}_c = \frac{1}{2}(\mathbf{E} \times \mathbf{H}^*) \tag{3.207}$$

where $\mathbf{E}$ and $\mathbf{H}$ are phasors and the rms values are given by $\frac{1}{\sqrt{2}}E_0$ and $\frac{1}{\sqrt{2}}H_0$, respectively. The direction of $\mathbf{S}_c$ is the same as the propagation direction of the wave. The real part $\mathbf{S}'$ of $\mathbf{S}_c$ corresponds to the real power flow, and the imaginary part $\mathbf{S}''$ corresponds to the stored energy in space.

$$\mathbf{S}_c = \mathbf{S}' + j\mathbf{S}''. \tag{3.208}$$

Example 1: Plane wave propagating in the $(+z)$-direction.

$$\mathbf{E} = \hat{\mathbf{x}}E_0 e^{-jkz}, \tag{3.209a}$$

$$\mathbf{H} = \hat{\mathbf{y}}\frac{E_0}{\eta} e^{-jkz}, \tag{3.209b}$$

$$\mathbf{S}_c = \hat{\mathbf{z}}\frac{1}{2}\frac{E_0^2}{\eta}. \tag{3.209c}$$

The Poynting vector of the above plane wave can be expressed in terms of an explicit function of time:

$$\mathbf{S}(z, t) = \mathbf{E}(z, t) \times \mathbf{H}(z, t), \tag{3.210a}$$

$$\mathbf{E}(z, t) = \text{Re}\left[\hat{\mathbf{x}} E_0 e^{-jkz} e^{j\omega t}\right] = \hat{\mathbf{x}} E_0 \cos(\omega t - kz), \tag{3.210b}$$

$$\mathbf{H}(z, t) = \hat{\mathbf{y}} \frac{E_0}{\eta} \cos(\omega t - kz), \tag{3.210c}$$

$$\mathbf{S}(z, t) = \hat{\mathbf{z}} \frac{E_0^2}{\eta} \cos^2(\omega t - kz). \tag{3.210d}$$

The time-averaged Poynting vector $\langle \mathbf{S} \rangle$ of the plane wave can also be obtained from the real part of the complex Poynting vector:

$$\langle \mathbf{S} \rangle = \frac{1}{2\pi} \int_0^{2\pi} \mathbf{S}(z, t) dt = \hat{\mathbf{z}} \frac{E_0^2}{\eta} \frac{1}{2\pi} \int_0^{2\pi} \cos^2(\omega t - kz) \, dt = \hat{\mathbf{z}} \frac{1}{2} \frac{E_0^2}{\eta}. \tag{3.211a}$$

The time-averaged value is equal to $\mathbf{S}_c$, see Eq. (3.209c), if $\mathbf{S}_c$ is purely real.

$$\mathbf{S}_c = \frac{1}{2} \mathbf{E} \times \mathbf{H}^* = \frac{1}{2} \left(\hat{\mathbf{x}} E_0 e^{-jkz} \times \hat{\mathbf{y}} \frac{E_0}{\eta} e^{jkz} \right) = \hat{\mathbf{z}} \frac{1}{2} \frac{E_0^2}{\eta}, \tag{3.211b}$$

$$\mathbf{S}_{av} = \frac{1}{2} \text{Re}[\mathbf{S}_c] = \frac{1}{2} \text{Re}[\mathbf{E} \times \mathbf{H}^*] = \langle \mathbf{S} \rangle. \tag{3.211c}$$

Example 2: Total reflection from a conducting wall.

We have an electric field on the conductor wall $(z = 0)$, Fig. 3.31,

$$E_0^i + E_0^r = 0 \quad \text{at } z = 0. \tag{3.212a}$$

Therefore,

$$\Gamma = \frac{E_0^r}{E_0^i} = -1. \tag{3.212b}$$

The corresponding $\mathbf{E}$, $\mathbf{H}$ and $\mathbf{S}_c$ vectors are given by

$$\mathbf{E} = \hat{\mathbf{x}} E_0 e^{-jkz} - \hat{\mathbf{x}} E_0 e^{+jkz} = -\hat{\mathbf{x}} 2j E_0 \sin kz, \tag{3.213a}$$

$$\mathbf{H} = \frac{1}{\eta} \left(\hat{\mathbf{y}} E_0 e^{-jkz} + \hat{\mathbf{y}} E_0 e^{+jkz} \right) = \hat{\mathbf{y}} 2 \frac{E_0}{\eta} \cos kz, \tag{3.213b}$$

$$\mathbf{S}_c = \frac{1}{2} (\mathbf{E} \times \mathbf{H}^*) = -\hat{\mathbf{z}} \frac{1}{2} 4j \frac{|E_0|^2}{\eta} \sin kz \cos kz. \tag{3.213c}$$

It is to be noted that in Eq. (3.213c) $\mathbf{S}_c$ is purely imaginary, indicative of no energy propagation.

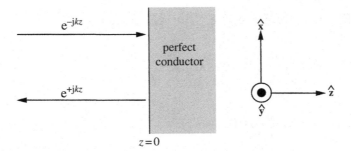

Fig. 3.31　Wave reflection from perfect conductor with boundary condition: total tangential E-field at $z = 0$ must vanish.

Power flow and reflection in lossless media

We consider the case in which E- and H-fields contain both incident and reflected terms. The wave is propagating from medium 1 (η_1) to medium 2 (η_2) (see Fig. 3.25).

An average power flow (real power flow) in medium 1 is given by the real part of the Poynting vector:

$$
\begin{aligned}
\mathbf{P}_{\text{av1}} &= \frac{1}{2}\text{Re}\left(\mathbf{E}_1 \times \mathbf{H}_1^*\right) \\
&= \frac{1}{2}\text{Re}\left\{\hat{\mathbf{x}}E_0^i\left(e^{-jk_1z} + \Gamma e^{+jk_1z}\right) \times \hat{\mathbf{y}}\frac{E_0^{i*}}{\eta_1}\left(e^{+jk_1z} - \Gamma^*e^{-jk_1z}\right)\right\} \quad (3.214a) \\
&= \hat{\mathbf{z}}\frac{\left|E_0^i\right|^2}{2\eta_1}\left(1 - |\Gamma|^2\right);
\end{aligned}
$$

$$
\mathbf{P}_{\text{av1}} = \mathbf{P}_{\text{av}}^i + \mathbf{P}_{\text{av}}^r, \tag{3.214b}
$$

$$
\mathbf{P}_{\text{av}}^i = \hat{\mathbf{x}}\frac{\left|E_0^i\right|^2}{2\eta_1}, \tag{3.214c}
$$

$$
\mathbf{P}_{\text{av}}^r = -\Gamma^2\mathbf{P}_{\text{av}}^i. \tag{3.214d}
$$

The transmitted power (real power flow) into medium 2 is

$$
\begin{aligned}
\mathbf{P}_{\text{av2}} &= \frac{1}{2}\text{Re}\left(\mathbf{E}_2 \times \mathbf{H}_2^*\right) = \frac{1}{2}\text{Re}\left\{\hat{\mathbf{x}}\tau E_0^i e^{-jk_2z} \times \hat{\mathbf{y}}\tau^*\frac{E_0^{i*}}{\eta_2}e^{+jk_2z}\right\} \\
&= \hat{\mathbf{z}}|\tau|^2\frac{\left|E_0^i\right|^2}{2\eta_2},
\end{aligned} \tag{3.215a}
$$

where

$$\Gamma = \frac{\eta_2 - \eta_1}{\eta_2 + \eta_1}, \qquad (3.215b)$$

$$\tau = \frac{2\eta_2}{\eta_2 + \eta_1}. \qquad (3.215c)$$

Since the time-averaged power flow in the $(+z)$-direction must be the same in media 1 and 2 ($\mathbf{P}_{av1} = \mathbf{P}_{av2}$), we get

$$\frac{\tau^2}{\eta_2} = \frac{1 - \Gamma^2}{\eta_1} \qquad (3.216a)$$

$$\text{or } 1 - \Gamma^2 = \left(\frac{\eta_1}{\eta_2}\right)\tau^2. \qquad (3.216b)$$

This equation corresponds to

$$\frac{incident}{power} - \frac{reflected}{power} = \frac{transmitted}{power}$$

3.5 Coupling behavior

A certain class of materials are known to exhibit coupling between their mechanical, thermal, and electromagnetic behavior. Below we discuss first linear constitutive equations of this coupling behavior – piezoelectric (mechanical/electrical), pyroelectric (electrical/thermal), and thermoelastic (mechanical/thermal). Then a brief statement will be made on non-linear coupling behavior.

3.5.1 Linear coupling behavior

Coupling between mechanical, thermal and electrical properties is inter-related as it can be given in index form as

$$\sigma_{ij} = c_{ijkl}e_{kl} + e_{nij}(-E_n) - \lambda_{ij}\theta, \qquad (3.217)$$

$$D_i = e_{imn}e_{mn} - \varepsilon_{in}(-E_n) - p_i\theta, \qquad (3.218)$$

where σ_{ij}, e_{mn}, E_n, D_i, θ are stress and strain tensors, electric field and electric flux density vectors, and temperature change, respectively, c_{ijkl}, e_{nij}, λ_{ij}, ε_{ij}, and p_i are fourth-order elastic stiffness tensor, third-order piezoelectric tensor, second-order thermal stress and dielectric tensors, and pyroelectric vector, respectively.

Equations (3.217) and (3.218) can be inverted as

$$e_{ij} = s_{ijkl}\sigma_{kl} + g_{kij}D_k + \Delta_{ij}\theta, \tag{3.219}$$

$$(-E_i) = g_{ijk}\sigma_{jk} - \beta_{ij}D_j + \gamma_i\theta, \tag{3.220}$$

where s_{ijkl}, g_{kij}, Δ_{ij}, β_{ij} and γ_i are elastic compliance tensor, piezoelectric tensor, thermal expansion tensor, dielectric compliance tensor and pyroelectric vector, respectively. The reason for the minus sign in front of the electric vector is to make the generalized property tensor R symmetric, see below.

The above equations can be expressed in symbolic form, i.e., Eqs. (3.217) and (3.218) are combined to a single algebraic equation:

$$\Sigma = R \cdot Z - \pi\theta. \tag{3.221}$$

Similarly Eqs. (3.219) and (3.220) are converted to

$$Z = F \cdot \Sigma + \Lambda\theta, \tag{3.222}$$

where Σ and Z are flux and field tensors and can be written as 9×1 column vectors:

$$\Sigma = \left\{ \begin{array}{c} \sigma \\ (6\times1) \\ D \\ (3\times1) \end{array} \right\}, \qquad Z = \left\{ \begin{array}{c} e \\ (6\times1) \\ -E \\ (3\times1) \end{array} \right\}, \tag{3.223}$$

where σ and e are defined by Eq. (3.19).

R and F are the 9×9 property tensor and compliance property tensor, defined by

$$R = \left\{ \begin{array}{cc} c & e^t \\ (6 \times 6) & (6 \times 3) \\ e & -\varepsilon \\ (3 \times 6) & (3 \times 3) \end{array} \right\}, \tag{3.224}$$

$$F = \left\{ \begin{array}{cc} s & g^t \\ (6 \times 6) & (6 \times 3) \\ g & -\beta \\ (3 \times 6) & (3 \times 3) \end{array} \right\}, \tag{3.225}$$

where e (3×6) is the piezoelectric coefficient tensor, and should be distinguished from the strain tensor (6×1).

Likewise, $\boldsymbol{\pi}$ and $\boldsymbol{\Lambda}$ are 9×1 column vectors defined by

$$\boldsymbol{\pi} = \left\{ \begin{array}{c} \boldsymbol{\lambda} \\ (6 \times 1) \\ \mathbf{p} \\ (3 \times 1) \end{array} \right\}, \qquad \boldsymbol{\Lambda} = \left\{ \begin{array}{c} \boldsymbol{\Delta} \\ (6 \times 1) \\ \boldsymbol{\gamma} \\ (3 \times 1) \end{array} \right\}. \tag{3.226}$$

Explicit expressions for all the components of the property and property compliance tensors $\boldsymbol{R}$ and $\boldsymbol{F}$ are

$$R = \begin{bmatrix} c_{11} & c_{12} & c_{13} & & & & e_{11} & e_{21} & e_{31} \\ c_{12} & c_{22} & c_{23} & & 0 & & e_{12} & e_{22} & e_{32} \\ c_{13} & c_{23} & c_{33} & & & & e_{13} & e_{23} & e_{33} \\ & & & c_{44} & & & e_{14} & e_{24} & e_{34} \\ & 0 & & & c_{55} & & e_{15} & e_{25} & e_{35} \\ & & & & & c_{66} & e_{16} & e_{26} & e_{36} \\ e_{11} & e_{12} & e_{13} & e_{14} & e_{15} & e_{16} & -\varepsilon_{11} & -\varepsilon_{12} & -\varepsilon_{13} \\ e_{21} & e_{22} & e_{23} & e_{24} & e_{25} & e_{26} & -\varepsilon_{12} & -\varepsilon_{22} & -\varepsilon_{23} \\ e_{31} & e_{32} & e_{33} & e_{34} & e_{35} & e_{36} & -\varepsilon_{13} & -\varepsilon_{23} & -\varepsilon_{33} \end{bmatrix}, \tag{3.227}$$

$$F = \begin{bmatrix} s_{11} & s_{12} & s_{13} & & & & g_{11} & g_{21} & g_{31} \\ s_{12} & s_{22} & s_{23} & & 0 & & g_{12} & g_{22} & g_{32} \\ s_{13} & s_{23} & s_{33} & & & & g_{13} & g_{23} & g_{33} \\ & & & s_{44} & & & g_{14} & g_{24} & g_{34} \\ & 0 & & & s_{55} & & g_{15} & g_{25} & g_{35} \\ & & & & & s_{66} & g_{16} & g_{26} & g_{36} \\ g_{11} & g_{12} & g_{13} & g_{14} & g_{15} & g_{16} & -\beta_{11} & -\beta_{12} & -\beta_{13} \\ g_{21} & g_{22} & g_{23} & g_{24} & g_{25} & g_{26} & -\beta_{12} & -\beta_{22} & -\beta_{23} \\ g_{31} & g_{32} & g_{33} & g_{34} & g_{35} & g_{36} & -\beta_{13} & -\beta_{23} & -\beta_{33} \end{bmatrix} \tag{3.228}$$

where piezoelectric tensors e_{nij} and g_{ijk} are shortened as e_{nm} and g_{im}, respectively. For example, by using Eq. (3.19), Eq. (3.217) in the absence of the first and third terms can be rewritten for the σ_{23} component in explicit form in the absence of ε and θ, as follows:

$$\sigma_4 = \sigma_{23} = e_{123}(-E_1) + e_{223}(-E_2) + e_{323}(-E_3)$$
$$= e_{14}(-E_1) + e_{24}(-E_2) + e_{34}(-E_3) \tag{3.229}$$

Similarly, Eq. (3.220) can provide for the $(-E_2)$ component in the absence of $\mathbf{D}$ and θ as

$$(-E_2) = g_{211}\sigma_{11} + g_{222}\sigma_{22} + g_{233}\sigma_{33} + g_{223}\sigma_{23} + g_{213}\sigma_{13} + g_{212}\sigma_{12}$$
$$= g_{21}\sigma_1 + g_{22}\sigma_2 + g_{23}\sigma_3 + g_{24}\sigma_4 + g_{25}\sigma_5 + g_{26}\sigma_6. \tag{3.230}$$

It is to be noted in the above equations that stress and strain tensors defined by σ and e are defined by T and S based on the ANSI/IEEE 1987 convention, but in this book we shall use σ and e as stress and strain, although σ can be interpreted as "electric conductivity" and e as "piezoelectric constant" in the electrical engineering community. The details of piezoelectricity are given elsewhere (Ikeda, 1990).

3.5.2 Non-linear coupling behavior

There are a number of non-linear couplings among mechanical, thermal and electromagnetic behaviors. Here, we shall examine the case of a linear piezo-electric material within the framework of thermodynamic potentials. In Sections 3.6, 3.7 and 3.8 we consider non-linear electrostriction, magnetostriction, and piezomagnetism, respectively.

We define the internal energy U of a piezoelectric body, assuming constant temperature, and its rate of change with time $\dot{U}$ by

$$U = U(e_{ij}, \ D_i),$$
$$\dot{U} = \sigma_{ij}\dot{e}_{ij} + E_i\dot{D}_i. \tag{3.231}$$

Electric enthalpy H is defined by

$$H = U - E_i D_i. \tag{3.232}$$

From Eqs. (3.231) and (3.232), the rate of change with time of H is given by

$$\dot{H} = \dot{U} - \dot{E}_i D_i - E_i \dot{D}_i = \sigma_{ij}\dot{e}_{ij} - \dot{E}_i D_i = \dot{H}(e_{ij}, E_i). \tag{3.233}$$

From (3.233)

$$\sigma_{ij} = \frac{\partial H}{\partial e_{ij}}, \tag{3.234}$$

$$D_i = -\frac{\partial H}{\partial E_i}, \tag{3.235}$$

where H is expressed as

$$H = \frac{1}{2}c^E_{ijkl}e_{ij}e_{kl} - e_{kij}E_k e_{ij} - \frac{1}{2}\varepsilon^e_{ij}E_i E_j. \tag{3.236}$$

The superscript E implies constant electric field, and superscript e constant strain.

From Eqs. (3.234)–(3.236), we obtain the constitutive equations of a piezo-electric material, Eqs. (3.217) and (3.218). When Eqs. (3.236) and (3.218) are substituted into Eq. (3.232) we get

$$
\begin{aligned}
U &= H + E_i \left\{ e_{ikl}e_{kl} + \varepsilon^e_{ij}E_j \right\} \\
&= \frac{1}{2} c^E_{ijkl}e_{ij}e_{kl} - e_{kij}E_k e_{ij} - \frac{1}{2}\varepsilon^e_{ij}E_i E_j + e_{ikl}E_i e_{kl} + \varepsilon^e_{ij}E_i E_j \\
&= \frac{1}{2} c^E_{ijkl}e_{ij}e_{kl} + \frac{1}{2}\varepsilon^e_{ij}E_i E_j \geq 0.
\end{aligned}
\tag{3.237}
$$

3.6 Electrostriction

By adding $-d_{ijmn}e_{ij}E_m E_n$ to Eq. (3.236), we obtain another function $\bar{H}$:

$$
\bar{H} = \frac{1}{2} c^E_{ijkl}e_{ij}e_{kl} - e_{kij}E_k e_{ij} - \frac{1}{2}\varepsilon^e_{ij}E_i E_j - d_{ijmn}e_{ij}E_m E_n,
\tag{3.238}
$$

$$
\sigma_{ij} = \frac{\partial \bar{H}}{\partial e_{ij}} = c^E_{ijkl}e_{kl} - e_{kij}E_k - d_{ijmn}E_m E_n,
\tag{3.239}
$$

$$
D_i = -\frac{\partial \bar{H}}{\partial E_i} = e_{imn}e_{mn} + \varepsilon^e_{ij}E_j + d_{mnij}E_j e_{mn},
\tag{3.240}
$$

where d_{mnij} is a ferro-electric coefficient. The Gibbs free energy G in the absence of a temperature term is defined by

$$
\begin{aligned}
G &= U - \sigma_{ij}e_{ij} - D_i E_i = H - \sigma_{ij}e_{ij} \\
&= -\frac{1}{2} s^E_{ijkl}\sigma_{ij}\sigma_{kl} - \frac{1}{2}\varepsilon^\sigma_{ij}E_i E_j - d_{nij}\sigma_{ij}E_n.
\end{aligned}
\tag{3.241a}
$$

$$
-\frac{\partial G}{\partial \sigma_{ij}} = e_{ij} = s^E_{ijkl}\sigma_{kl} + d_{nij}E_n
\tag{3.241b}
$$

$$
-\frac{\partial G}{\partial E_i} = D_i = d_{nij}\sigma_{ij} + \varepsilon_{ij}E_j
\tag{3.241c}
$$

where d_{nij} is the piezoelectric tensor and it is related to other piezoelectric tensors by

$$
d_{nij} = \varepsilon^\sigma_{nm}g_{mij} = e_{nkl}s^E_{klij}
\tag{3.241d}
$$

The superscript σ and E denote constant stress and constant electric field, respectively.

Let us add $-q_{ijmn}\sigma_{ij}E_mE_n$ to Eq. (3.241a) where temperature terms are ignored; then the Gibbs free energy is given by

$$\hat{G} = -\frac{1}{2}s^E_{ijkl}\sigma_{ij}\sigma_{kl} - \frac{1}{2}\varepsilon^\sigma_{mn}E_mE_n - d_{nij}\sigma_{ij}E_n - q_{ijmn}\sigma_{ij}E_mE_n, \tag{3.242}$$

$$-\frac{\partial\hat{G}}{\partial\sigma_{ij}} = e_{ij} = s^E_{ijkl}\sigma_{kl} + d_{nij}E_n + q_{ijmn}E_mE_n, \tag{3.243}$$

$$-\frac{\partial\hat{G}}{\partial E_i} = D_i = \varepsilon^\sigma_{ij}E_j + d_{imn}\sigma_{mn} + q_{ijmn}\sigma_{mn}E_j, \tag{3.244}$$

where q_{ijmn} is the piezostrictive coefficient.

3.7 Magnetostriction

Adding $-\lambda_{ijkl}\sigma_{ij}M_kM_l$ to Eq. (3.242), we obtain another energy function, $\bar{G}$:

$$\bar{G} = -\frac{1}{2}s^E_{ijkl}\sigma_{ij}\sigma_{kl} - \frac{1}{2}\varepsilon^\sigma_{mn}E_mE_n - d_{nij}\sigma_{ij}E_n - q_{ijmn}\sigma_{ij}E_mE_n$$
$$- \lambda_{ijkl}\sigma_{ij}M_kM_l, \tag{3.245}$$

$$-\frac{\partial\bar{G}}{\partial\sigma_{ij}} = e_{ij} = s^E_{ijkl}\sigma_{kl} + d_{nij}E_n + q_{ijmn}E_mE_n + \lambda_{ijkl}M_kM_l, \tag{3.246a}$$

$$-\frac{\partial\bar{G}}{\partial E_i} = D_i = \varepsilon^\sigma_{ij}E_j + d_{imn}\sigma_{mn} + q_{mnij}\sigma_{mn}E_j \tag{3.246b}$$

where λ_{ijkl} is the magnetostrictive coefficient tensor

Normally, magnetostriction occurs only in a ferromagnetic material; thus, $d_{nij}=0$, $q_{ijmn}=0$, and Eqs. (3.246) are reduced to

$$e_{ij} = s^E_{ijkl}\sigma_{kl} + \lambda_{ijkl}M_kM_l, \tag{3.247a}$$

$$D_i = \varepsilon^\sigma_{ij}E_j \tag{3.247b}$$

where M_i (A/m) is a component of the magnetization vector. From Eq. (3.102), M_i is related to magnetic field H_j by

$$M_i = \left\{\frac{\mu_{ij}}{\mu_0} - \delta_{ij}\right\}H_j. \tag{3.248}$$

Defining the magnetic susceptibility tensor χ_{ij} by

$$\chi_{ij} = \frac{1}{\mu_0}\mu_{ij} - \delta_{ij}, \tag{3.249}$$

Eq. (3.248) is written as

$$M_i = \chi_{ij}H_j. \tag{3.250}$$

3.8 Piezomagnetism

Adding $-a_{ikl}H_i\sigma_{ikl}$ to Eq. (3.245), we obtain a function $\overline{\overline{G}}$:

$$\overline{\overline{G}} = -\frac{1}{2}s^{\mathrm{E}}_{ijkl}\sigma_{ij}\sigma_{kl} - \frac{1}{2}\varepsilon^{\sigma}_{mn}E_mE_n - d_{nij}\sigma_{ij}E_n - a_{ikl}H_i\sigma_{kl} - q_{ijmn}\sigma_{ij}E_mE_n$$

$$- \lambda_{ijkl}\sigma_{ij}M_kM_l, \tag{3.251}$$

$$-\frac{\partial\overline{\overline{G}}}{\partial\sigma_{ij}} = e_{ij} = s^{\mathrm{E}}_{ijkl}\sigma_{kl} + d_{nij}E_n + q_{ijmn}E_mE_n + a_{kij}H_k + \lambda_{ijkl}M_kM_l, \tag{3.252a}$$

$$-\frac{\partial\overline{\overline{G}}}{\partial E_i} = D_i = \varepsilon^{\sigma}_{ij}E_j + d_{imn}\sigma_{mn} + q_{mnij}\sigma_{mn}E_j, \tag{3.252b}$$

where a_{ikl} is a piezomagnetic coefficient tensor

It is to be noted that Eq. (3.251) is the most general constitutive equation, accounting for coupling behavior excluding temperature effects, but including the following:

(1) piezoelectricity (d_{nij}),
(2) piezomagnetism (a_{ikl}),
(3) magnetostriction (λ_{ijkl}),
(4) electrostriction (q_{ijmn}).

The first two coupling behaviors are linear in **E** or **H** while the last two are non-linear in **E** or **H**, i.e., quadratic functions of **E** or **H**.

4

Models for electronic composites based on effective medium theory

Modeling of an electronic composite is quite diverse since the behavior of the composite involves mechanical, thermal, and electromagnetic properties. Depending on their behaviors, a number of analytical models have been developed, which can be categorized into (1) the law-of-mixtures model, (2) the effective medium model represented by the Eshelby model (1957, 1959), (3) the resistor network model, and (4) the percolation model. We shall review the fundamentals of the first two models in this chapter and the others in later chapters.

4.1 Law-of-mixtures model

The law-of-mixtures model is one-dimensional and thus is the simplest model. The model has been discussed in earlier chapters, but we shall consider it here in more detail by looking at two properties – elastic (Young's) modulus E and thermal conductivity K – of a composite which has a one-dimensional micro-structure, as shown in Fig. 4.1, and may be termed a unidirectional (continuous) fiber composite or a lamellar composite.

First, we apply a uniaxial stress σ_0 in the plane of the lamellae, or longitudinal direction. The shaded and unshaded domains in Fig. 4.1 are, respectively, fiber (or reinforcement) with Young's modulus E_f and matrix with E_m.

Assuming perfect bonding between the fiber and matrix, the condition of isostrain in the longitudinal direction can be realized, resulting in

$$e_c = e_m = e_f \tag{4.1}$$

where e_i is the longitudinal strain and subscript i denotes the ith phase; $i = m$ (matrix), f (fiber), and c (composite). For each phase, the one-dimensional Hooke's law is valid:

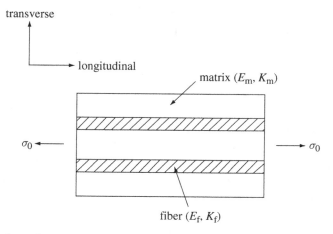

transverse

longitudinal

matrix (E_m, K_m)

σ_0 σ_0

fiber (E_f, K_f)

Fig. 4.1   One-dimensional composite model.

$$\sigma_c = \sigma_0 = E_c e_c, \tag{4.2a}$$

$$\sigma_m = E_m e_m, \tag{4.2b}$$

$$\sigma_f = E_f e_f, \tag{4.2c}$$

where σ_i and E_i are the longitudinal stress and Young's modulus of the ith phase, respectively. The composite stress σ_c along the longitudinal direction is equal to the sum of the product of the stress in each phase and its volume fraction:

$$\sigma_c = (1-f)\sigma_m + f\sigma_f, \tag{4.3}$$

where f is the volume fraction of fiber. A substitution of Eq. (4.2) into (4.3) and use of (4.1) provides the composite (longitudinal) Young's modulus E_{cL} as

$$E_{cL} = (1-f)E_m + fE_f. \tag{4.4}$$

Second, we consider the case of an applied stress σ_0 in the transverse direction, i.e., transverse to the fiber–matrix interface. In this case the transverse stress in each phase is constant throughout all the phases and equal to the applied stress:

$$\sigma_0 = \sigma_c = \sigma_m = \sigma_f. \tag{4.5}$$

The transverse strain of the composite is the sum of the transverse strain in each phase:

$$e_c = (1-f)e_m + fe_f. \tag{4.6}$$

Since Hooke's law is still valid in each phase, its use in Eqs. (4.2), (4.5), and (4.6) leads to the formula for the transverse Young's modulus E_{cT} of the composite:

$$\frac{1}{E_{cT}} = \frac{1-f}{E_m} + \frac{f}{E_f}. \tag{4.7}$$

Similarly, we can apply the law-of-mixtures model to the case of the thermal conduction behavior of the unidirectional composite of Fig. 4.1. First, an applied heat flux q_0 in the longitudinal direction is considered. In this case, the temperature gradient in the longitudinal direction in each phase is constant and equal to that of the composite:

$$\nabla\theta_c = \nabla\theta_m = \nabla\theta_f, \tag{4.8}$$

where $\nabla\theta_i$ is the temperature gradient in the ith phase in the longitudinal direction. The constitutive equation for steady-state heat conduction (Fourier's law) is valid in each phase:

$$q_0 = q_c = K_c\nabla\theta_c, \tag{4.9a}$$

$$q_m = K_m\nabla\theta_m, \tag{4.9b}$$

$$q_f = K_f\nabla\theta_f, \tag{4.9c}$$

where q_i and K_i are the heat flux and thermal conductivity of the ith phase, respectively. The heat flux q_c of the composite in the longitudinal direction is the sum of that in each phase multiplied by its volume fraction, hence

$$q_c = (1-f)q_m + fq_f. \tag{4.10}$$

A substitution of Eq. (4.9) into Eq. (4.10) and use of Eq. (4.8) yields the formula for the longitudinal thermal conductivity K_{cL} of the composite:

$$K_{cL} = (1-f)K_m + fK_f. \tag{4.11}$$

Second, for thermal conduction in the transverse direction, the transverse temperature gradient of the composite plate, i.e., the temperature difference between at the top and bottom of the composite divided by the composite thickness, is the sum of the transverse temperature gradient in each phase multiplied by its volume fraction:

$$\nabla\theta_c = (1-f)\nabla\theta_m + f\nabla\theta_f. \tag{4.12}$$

By using the constitutive equation, Eqs. (4.9), (4.11), and (4.12), one can obtain the formula for the transverse thermal conductivity of the composite, K_{cT}, as

$$\frac{1}{K_{cT}} = \frac{(1-f)}{K_m} + \frac{f}{K_f}. \tag{4.13}$$

Let us apply the above formulae based on the law-of-mixtures model to the prediction of Young's moduli (E_{cL}, E_{cT}), and thermal conductivities (K_{cL}, K_{cT}) of a unidirectional SiC Nicalon® fiber/6061 Al matrix composite. The input data for the computation are as follows:

$$6061 \text{ aluminum } E_m = 68.3 \text{ GPa}, \ K_m = 180 \text{ W}/(\text{m} \cdot \text{K});$$

$$\text{SiC Nicalon}^{®} \text{ fiber } E_f = 300 \text{ GPa}, \ K_f = 30.7 \text{ W}/(\text{m} \cdot \text{K}).$$

The results of E_c and K_c are plotted as functions of fiber volume fraction f in Fig. 4.2.

The law-of-mixtures model can be applied to other uncoupled physical behaviors, such as electric conduction (Eqs. (3.94)), electrostatics (Eq. (3.96)) and magnetostatics (Eq. (3.101b)). For such uncoupled behaviors, we note one-to-one correspondence, Table 4.1.

By using Table 4.1, one can obtain the longitudinal and transverse electromagnetic properties of a unidirectional composite based on the law-of-mixtures models:

electric conductivity σ –

$$\sigma_{cL} = (1-f)\sigma_m + f\sigma_f, \quad \frac{1}{\sigma_{cT}} = \frac{(1-f)}{\sigma_m} + \frac{f}{\sigma_f}; \tag{4.14a}$$

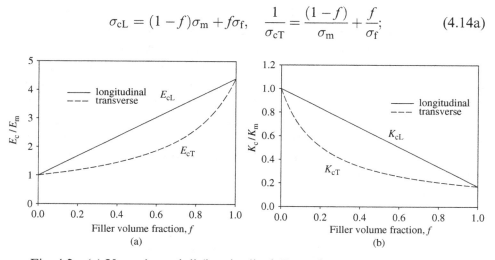

Fig. 4.2 (a) Young's moduli (longitudinal E_{cL} and transverse E_{cT}), and (b) thermal conductivities (longitudinal K_{cL} and transverse K_{cT}) of a unidirectional SiC Nicalon® fiber/6061 Al composite.

Table 4.1 *One-to-one correspondence for uncoupled behavior.*

Behavior	Constitutive equation $\Sigma = R \cdot Z$	Flux vector Σ	Field vector Z	Property tensor R
Mechanical	$\sigma = C \cdot e$	σ	e	C
Thermal	$q = K \cdot \nabla\theta$	q	$\nabla\theta$	K
Electric conduction	$J = \sigma \cdot E$	J	E	σ
Electrostatics	$D = \varepsilon \cdot E$	D	E	ε
Magnetostatics	$B = \mu \cdot H$	B	H	μ

dielectric constant ε –

$$\varepsilon_{cL} = (1-f)\varepsilon_m + f\varepsilon_f, \qquad \frac{1}{\varepsilon_{cT}} = \frac{(1-f)}{\varepsilon_m} + \frac{f}{\varepsilon_f}; \qquad (4.14b)$$

magnetic permeability μ –

$$\mu_{cL} = (1-f)\mu_m + f\mu_f, \qquad \frac{1}{\mu_{cT}} = \frac{(1-f)}{\mu_m} + \frac{f}{\mu_f}. \qquad (4.14c)$$

In the above computations, both matrix and fiber materials are assumed to be isotropic. For composites with discontinuous reinforcement (filler), the law-of-mixtures model does not provide accurate predictions of the composite properties while other models such as the shear lag model (Cox, 1952) and Eshelby's model (Eshelby, 1957) give rise to better predictions. The shear lag model was developed by Cox (1952) who applied it to the stress analysis of a unidirectional short-fiber composite. The shear lag model was extended to the analysis of the creep behavior of a unidirectional short-fiber composite (Kelly and Street, 1972). The shear lag model is, however, applicable only to the analysis of the mechanical behavior of a unidirectional short-fiber composite, thus it will not be covered in this chapter and its details have been given elsewhere (Taya and Arsenault, 1989).

4.2 Effective property tensor of a composite

The constitutive equations for uncoupled mechanical, thermal, and electro-magnetic behavior of a material can be expressed in terms of a flux vector or tensor ($\boldsymbol{\Sigma}$), a field vector or tensor ($\mathbf{Z}$) and a property tensor $\boldsymbol{R}$ as summarized in Table 4.1.

$$\boldsymbol{\Sigma} = \boldsymbol{R} \cdot \mathbf{Z}, \tag{4.15a}$$

$$\mathbf{Z} = \boldsymbol{F} \cdot \boldsymbol{\Sigma}, \tag{4.15b}$$

where $\boldsymbol{\Sigma}$ and $\mathbf{Z}$ are 6×1 column vectors for mechanical behavior with components 11, 22, 33, 23, 31, and 12, 3×1 column vectors for uncoupled thermal and electromagnetic behavior with components 1, 2, and 3, and $\boldsymbol{R}$ and $\boldsymbol{F}$ are the property and compliance tensors, respectively, whose size of matrix depends on whether we are talking about mechanical (6×6), uncoupled thermophysical (3×3) and coupled behavior (9×9). In each phase, the above constitutive equations are valid, i.e.,

$$\boldsymbol{\Sigma}_c = \boldsymbol{R}_c \cdot \mathbf{Z}_c, \tag{4.16a}$$

$$\boldsymbol{\Sigma}_m = \boldsymbol{R}_m \cdot \mathbf{Z}_m, \tag{4.16b}$$

$$\boldsymbol{\Sigma}_f = \boldsymbol{R}_f \cdot \mathbf{Z}_f, \tag{4.16c}$$

$$\mathbf{Z}_c = \boldsymbol{F}_c \cdot \boldsymbol{\Sigma}_c, \tag{4.16d}$$

$$\mathbf{Z}_m = \boldsymbol{F}_m \cdot \boldsymbol{\Sigma}_m, \tag{4.16e}$$

$$\mathbf{Z}_f = \boldsymbol{F}_f \cdot \boldsymbol{\Sigma}_f. \tag{4.16f}$$

It is to be noted that $\boldsymbol{\Sigma}_i$ and $\mathbf{Z}_i$ are the volume-averaged flux and field vectors in the ith domain with $i = $ m (matrix), f (filler), and c (composite). $\boldsymbol{\Sigma}_c$ and $\mathbf{Z}_c$ are, respectively, the sum of the volume-averaged flux and field vectors in each phase multiplied by the volume fraction of that phase. $\boldsymbol{R}_i$ and $\boldsymbol{F}_i$ are the property and compliance tensors, respectively, and $\boldsymbol{F}_i = \boldsymbol{R}_i^{-1}$.

Hence

$$\boldsymbol{\Sigma}_c = (1 - f)\boldsymbol{\Sigma}_m + f\,\boldsymbol{\Sigma}_f, \tag{4.17}$$

$$\mathbf{Z}_c = (1 - f)\mathbf{Z}_m + f\mathbf{Z}_f, \tag{4.18}$$

where f is the volume fraction of filler.

Next we shall consider two boundary conditions, field vector and flux vector boundary conditions.

4.2.1 Field vector (Z_0) boundary condition

When the field vector $\mathbf{Z}_0$ is prescribed on the boundary of a composite, Fig. 4.3 (a), the integral of the disturbance field vector $\mathbf{Z}$ over the composite domain vanishes (Dunn and Taya, 1993a), i.e.,

$$\int_{V_c} \mathbf{Z} dV = \mathbf{0}. \tag{4.19}$$

Thus, the volume-averaged field vector $\mathbf{Z}_c$ over the composite domain becomes equal to the applied field vector on the boundary, $\mathbf{Z}_0$:

$$\mathbf{Z}_c = \frac{1}{V_c} \int_{V_c} (\mathbf{Z}_0 + \mathbf{Z}) dV = \mathbf{Z}_0, \tag{4.20a}$$

since

$$\int_{V_c} \mathbf{Z} dV = \mathbf{0}. \tag{4.20b}$$

In Eqs. (4.19) and (4.20), V_c and dV are the volume of the composite and volume element, respectively, and $\mathbf{Z}$ is the disturbance of the field vector due to the existence of fillers. Eq. (4.20b) is another definition of the disturbance field vector $\mathbf{Z}$.

Suppose the volume-averaged field vector $\mathbf{Z}_f$ in the filler domain is related to the applied field vector $\mathbf{Z}_0$ on the boundary by a concentration factor tensor A:

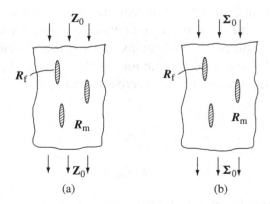

(a)   (b)

Fig. 4.3   A composite subjected to two types of boundary conditions: (a) field vector boundary condition where $\mathbf{Z}_0$ is prescribed on the boundary, and (b) flux vector boundary condition where $\mathbf{\Sigma}_0$ is prescribed on the boundary.

$$Z_f = A \cdot Z_0, \tag{4.21}$$

where the concentration factor tensor A can be obtained by a number of methods such as the Eshelby model, and the finite-element model.

A substitution of Eq. (4.16) into (4.17) and use of (4.20) and (4.21) results in

$$R_c = R_m + f(R_f - R_m) \cdot A. \tag{4.22}$$

Once the concentration factor tensor is found by an appropriate method, one can obtain the composite property tensor R_c from Eq. (4.22).

4.2.2 Flux vector (Σ_0) boundary condition

When the boundary condition is given in terms of applied flux vector Σ_0 on the composite boundary, Fig. 4.3 (b), the volume-averaged total flux vector Σ_c in the composite domain is reduced to the applied flux vector Σ_0 (Dunn and Taya, 1993a,b):

$$\Sigma_c = \frac{1}{V_c} \int_{V_c} (\Sigma_0 + \Sigma) dV = \Sigma_0, \tag{4.23}$$

where Σ is the disturbance of the flux vector due to the existence of fillers and the composite boundary.

Suppose the volume-averaged flux vector Σ_f in the filler domain is related to the applied flux vector Σ_0 by a concentration factor tensor B:

$$\Sigma_f = B \cdot \Sigma_0, \tag{4.24}$$

where again the concentration factor tensor B will be obtained by an appropriate method.

By substituting Eq. (4.16) into (4.18) and using (4.23) and (4.24), one can obtain the formula to compute the composite compliance tensor F_c:

$$F_c = F_m + f(F_f - F_m) \cdot B, \tag{4.25}$$

where F_c is the inverse of R_c.

It is to be noted that the composite property tensor R_c derived by Eq. (4.22) should coincide with that obtainable from Eq. (4.25). This requirement is called "self-consistency," which has been proven by Benveniste (1987) for the mechanical behavior of a composite and by Dunn and Taya (1993a) for coupled electrical/mechanical behavior of a piezoelectric composite.

4.2.3 Extension to hybrid composites and piezoelectric and pyroelectric composites

It is to be noted that the formulae to compute the composite property tensor R_c or compliance tensor F_c can be extended to the case of a hybrid composite containing more than one kind of embedded filler (Dunn, 1995).

$$R_c = R_m + \sum_{i=1}^{N} f_i (R_i - R_m) \cdot A_i, \tag{4.26a}$$

$$F_c = F_m + \sum_{i=1}^{N} f_i (F_i - F_m) \cdot B_i, \tag{4.26b}$$

where Eqs. (4.26a) and (4.26b) are derived assuming perfect alignment of the different kinds of filler, A_i and B_i are concentration factor tensors that relate the volume-averaged field vectors A_i and B_i in the ith filler to that in the composite, and they are related to the applied field vector (Z_0) and flux vector (Σ_0) as

$$Z_f^i = A_i \cdot Z_0, \tag{4.27a}$$

$$\Sigma_f^i = B_i \cdot \Sigma_0. \tag{4.27b}$$

The concentration factor tensors A_i and B_i satisfy

$$\sum_{i=0}^{N} f_i A_i = I, \tag{4.28a}$$

$$\sum_{i=0}^{N} f_i B_i = I \tag{4.28b}$$

where f_0 ($i = 0$) is the volume fraction of the matrix material, there are N different kinds of filler present in the composite, and I is the identity matrix.

The above formulae, Eqs. (4.22), (4.25), and (4.26), used to compute the composite compliance and property tensors, are also applicable to the case of a piezoelectric composite where the field and flux vectors are to be replaced by Eqs. (3.223) and the property and compliance tensors by Eqs. (3.224) and (3.225), respectively.

Dunn (1995) applied the above formulation to the case of a pyroelectric composite where the governing equations of a pyroelectric material are given by Eqs. (3.217) and (3.218). The pyroelectric tensor of a composite is given by

$$\pi_c = \pi_m + \sum_{i=1}^{N} f_i(\pi_i - \pi_m) + \sum_{i=1}^{N} f_i(R_i - R_m) \cdot a_i \qquad (4.29a)$$

$$\Lambda_c = \Lambda_m + \sum_{i=1}^{N} f_i(\Lambda_i - \Lambda_m) + \sum_{i=1}^{N} f_i(F_i - F_m) \cdot b_i, \qquad (4.29b)$$

where π_i and Λ_i are the vector forms of the pyroelectric tensors of the ith phase defined by Eq. (3.226) with $i=c$ (composite), m (matrix), $i=1, 2, \ldots, N$ (N different kinds of filler) and a_i and b_i are thermal concentration factor vectors defined, in the absence of electroelastic loading, by

$$\bar{Z}_i = a_i \theta, \qquad (4.30a)$$

$$\bar{\Sigma}_i = b_i \theta. \qquad (4.30b)$$

Here $\bar{Z}_i$ and $\bar{\Sigma}_i$ are the average field and flux vectors of the ith phase, and θ is the prescribed uniform temperature change in the composite, and they satisfy

$$\sum_{i=0}^{N} f_i a_i = \sum_{i=0}^{N} f_i b_i = 0, \qquad (4.31)$$

where $i=0$ denotes the matrix phase.

It is to be noted again that Eqs. (4.29) are valid only for aligned fillers, while in the case of misoriented fillers, the third term is replaced by the orientation average $\langle (R_i - R_m) \cdot a_i \rangle$.

Dunn (1995) obtained exact relations between the thermal and electroelastic property tensors for a composite composed of only one type of filler, i.e., a two-phase composite system

$$\Lambda_c = \Lambda_f + (F_c - F_f) \cdot (F_m - F_f)^{-1} \cdot (\Lambda_m - \Lambda_f) \qquad (4.32a)$$

$$\pi_c = \pi_f + (R_c - R_f) \cdot (R_m - R_f)^{-1} \cdot (\pi_m - \pi_f). \qquad (4.32b)$$

Equations (4.32) imply that, if the electroelastic property and compliance tensors are known either theoretically or experimentally, then the pyroelectric and thermal expansion coefficients are derivable.

In the above formulations, the key computational step is to obtain concentration factor tensors A, B, and vectors a and b, which will be obtained later by several methods, particularly Eshelby's method. This will be discussed in detail in Section 4.4.

4.3 Eshelby model

Eshelby (1957, 1959) proposed an algebraic method to solve for the stress and strain fields, and the strain energy, of a composite where the shape of the filler is ellipsoidal. Since the computation is reduced to a system of algebraic equations and the Eshelby method can be applicable not only to composites but also to other mechanical problems – elastic–plastic, viscoelastic, and creep behavior of a material – the Eshelby method has been quite popular among engineers and scientists in solid mechanics and materials engineering. Hatta and Taya (1985, 1986) extended the Eshelby model to the case of uncoupled thermal and electromagnetic behavior of a composite. Dunn and Taya (1993a,b) extended the Eshelby model further to the case of the coupled behavior of a composite. In the following we shall introduce first the original Eshelby model and then the extended Eshelby model for uncoupled and coupled thermophysical behavior.

4.3.1 *Inclusion problem*

Let us consider an ellipsoidal inclusion, with prescribed inelastic strain e^* (called "eigenstrain" [Mura, 1987]), which is embedded in an infinite elastic medium with its elastic stiffness tensor C_m, Fig. 4.4. Please note that the domain Ω of the ellipsoidal inclusion has the same elastic property tensor as the matrix, C_m. Examples of eigenstrain e^* are plastic strain, transformation strain, and thermal expansion strain. Eshelby found that the strain e induced by e^* becomes constant inside an ellipsoidal domain Ω for constant eigenstrain and is related linearly to the eigenstrain by

$$e = S \cdot e^*, \qquad (4.33)$$

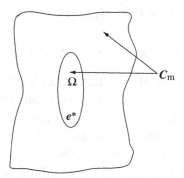

Fig. 4.4   The Eshelby inclusion model.

where S is a fourth-order tensor, called the "Eshelby tensor," which is a function of the Poisson's ratio of the matrix material and the geometry of the ellipsoid (i.e., its axial ratios). A detailed expression for S has been given elsewhere (Taya and Arsenault, 1989; Mura, 1987) as well as in Appendix A1. For constant eigenstrain e^* prescribed in the ellipsoidal domain Ω, the stress inside Ω also becomes constant and is given by Hooke's law:

$$\sigma = C_m \cdot (e - e^*). \qquad (4.34)$$

From Eqs. (4.33) and (4.34), one can obtain the stress inside Ω as

$$\sigma = C_m \cdot (S - I) \cdot e^*, \qquad (4.35)$$

where I is the identity tensor of rank four and can be expressed in 6×6 matrix form where the only non-vanishing components (i.e., 1) are the diagonal elements. The stress tensor σ and strain tensor e can be expressed as 6×1 column vectors as defined by Eqs. (3.19) where shear strain γ_{ij} is converted to engineering shear strain ($\gamma_{ij} = 2e_{ij}$) for $i \neq j$. The fourth-rank tensors C_m and S can be expressed as 6×6 square matrices

$$C_m = \begin{bmatrix} C_{11} & C_{12} & C_{13} & & & \\ C_{12} & C_{22} & C_{23} & & 0 & \\ C_{13} & C_{23} & C_{33} & & & \\ & & & C_{44} & & \\ & 0 & & & C_{55} & \\ & & & & & C_{66} \end{bmatrix}, \qquad (4.36)$$

$$S = \begin{bmatrix} S_{11} & S_{12} & S_{13} & & & \\ S_{21} & S_{22} & S_{23} & & 0 & \\ S_{31} & S_{32} & S_{33} & & & \\ & & & S_{44} & & \\ & 0 & & & S_{55} & \\ & & & & & S_{66} \end{bmatrix}. \qquad (4.37)$$

It is to be noted that C_m is symmetric, but S is not except for spherical inclusions. It is also to be noted that the inverse of the elastic stiffness tensor C_m is the elastic compliance tensor S_m. The explicit expressions for the elastic compliance tensor for orthotropic, transversely isotropic, and isotropic materials are defined in terms of engineering elastic constants (Young's and shear moduli, and Poisson's ratio) by Eqs. (3.20), (3.23), and (3.26), respectively.

Next we shall evaluate the elastic strain energy W of an infinite matrix containing an ellipsoidal inclusion with e^*

$$W = \frac{1}{2} \int_{V_c} \boldsymbol{\sigma} \cdot (e - e^*) dV, \tag{4.38}$$

where $\int_{V_c} dV$ denotes the three-dimensional volume integral over the entire domain of the infinite medium including that (Ω) of the ellipsoidal inclusion. It is to be noted that

$$\begin{aligned}
\int_{V_c} \boldsymbol{\sigma} \cdot e \, dV &= \int_{V_c} \sigma_{ij} e_{ij} dV \\
&= \int_{V_c} \sigma_{ij} \frac{1}{2} (u_{i,j} + u_{j,i}) dV \\
&= \int_{V_c} \sigma_{ij} u_{i,j} dV = \int_{V_c} (\sigma_{ij} u_i)_{,j} dV - \int_{V_c} \sigma_{ij,j} u_i dV \\
&= \int_{|V_c|} \sigma_{ij} n_j u_i ds \\
&= 0, \tag{4.39}
\end{aligned}$$

where we have used Eq. (3.13a), the symmetrical property of the stress tensor, integral by parts, Gauss' divergence theorem, Eq. (3.12) with no body force and inertia term, and the fact that $\boldsymbol{\sigma}$ is internal stress induced by the eigenstrain e^* so that there is no applied stress on the boundary, $\sigma_{ij} n_j = 0$ on $|V_c|$, and where n_j is a component of the unit normal outward vector $\mathbf{n}$ on the surface. Hence, the elastic strain energy W is reduced to

$$W = -\frac{1}{2} \int_{\Omega} \boldsymbol{\sigma} \cdot e^* dV, \tag{4.40}$$

which, for a constant e^*, is further reduced to

$$W = -\frac{1}{2} \int_{\Omega} \boldsymbol{\sigma} \cdot e^* dV = -\frac{1}{2} V_{\Omega} \boldsymbol{\sigma} \cdot e^* \tag{4.41}$$

where V_{Ω} is the volume of the ellipsoidal eigenstrain, and $\boldsymbol{\sigma}$ is obtained from Eq. (4.35).

It should be noted in Eq. (4.40) that the total elastic strain energy of the infinite body containing an ellipsoidal inclusion can be computed by knowing only the information inside Ω.

4.3.2 Inhomogeneity problem

Let us consider inhomogeneities (domain Ω) with elastic stiffness tensor C_f embedded in an infinite matrix with stiffness tensor C_m subjected to an applied stress σ_0 on the boundary, Fig. 4.5(a). We shall show that the inhomogeneity problem of Fig. 4.5(a) can be reduced to the equivalent inclusion problem of Fig. 4.5(b). First, consider a homogeneous body with C_m subjected to the applied stress σ_0. The stress σ_0 and strain e_0 are uniform over the entire domain, satisfying Hooke's law

$$\sigma_0 = C_m \cdot e_0. \tag{4.42}$$

Second, insert inhomogeneities (Ω) into the homogeneous body, which induce a disturbance field of stress σ and strain e. Focusing on a single inhomogeneity, one can write the following equation:

$$\begin{aligned}
\sigma_0 + \sigma &= C_f \cdot (e_0 + e) \\
&= C_m \cdot (e_0 + e - e^*),
\end{aligned} \tag{4.43}$$

where e^* is an unknown fictitious eigenstrain introduced into Ω while the domain of the inhomogeneities with stiffness tensor C_f is replaced by the surrounding matrix material with stiffness tensor C_m, Fig. 4.5(b). It is to be noted that Eq. (4.43) is valid for the entire domain, the inhomogeneity domain (Ω) and the matrix domain ($V_c - \Omega$). Subtracting Eq. (4.42) from (4.43), we have the stress disturbance σ given by

$$\sigma = C_m \cdot (e - e^*) \tag{4.44}$$

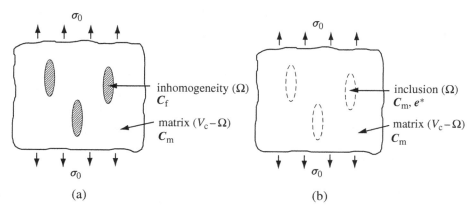

Fig. 4.5 (a) Inhomogeneity problem converted to (b) equivalent inclusion problem.

which is identical to Eq. (4.34), i.e., the Eshelby inclusion problem. Thus, as far as the disturbance stress and strain are concerned, the problem of Fig. 4.5(a) can be reduced to that of Fig. 4.5(b). Because of Eq. (4.44), the strain disturbance e is now related to e^* through the Eshelby tensor S, Eq. (4.33). Hence, our remaining computational task is to solve for e^* by using Eqs. (4.33) and (4.42). Then e^* is given by

$$e^* = D^{-1} \cdot (C_m - C_f) \cdot e_0, \tag{4.45}$$

where D^{-1} is a 6×6 matrix and is the inverse of the D matrix while D is defined by

$$D = (C_f - C_m) \cdot S + C_m. \tag{4.46}$$

Next we shall consider the case where the inhomogeneities have a prescribed eigenstrain e^p. Such an inhomogeneity is called an "inhomogeneous inclusion." The composite with inhomogeneous inclusions is subjected to an applied stress on the boundary, Fig. 4.6(a). The corresponding Eshelby equivalent inclusion problem, Fig. 4.6(b), can be set up in algebraic form by focusing on a single inhomogeneous inclusion (Ω):

$$\sigma_0 + \sigma = C_f \cdot (e_0 + e - e^p)$$
$$= C_m \cdot (e_0 + e - e^*). \tag{4.47}$$

Solving for the unknown fictitious eigenstrain e^* using Eqs. (4.47) and (4.33), we have

$$e^* = D^{-1} \cdot \{(C_m - C_f) \cdot e_0 + C_f \cdot e^p\}, \tag{4.48}$$

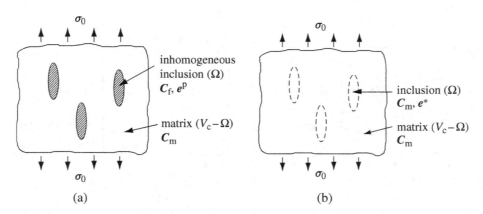

Fig. 4.6 (a) Inhomogeneous inclusion problem converted to (b) equivalent inclusion problem.

where D^{-1} is defined from Eq. (4.46). The elastic strain energy of a composite with inhomogeneous inclusions is given by (Mura, 1987)

$$W = \frac{1}{2} \int_{V_c} \sigma_0 \cdot e \, dV + \frac{1}{2} \int_{V_c} \sigma_0 \cdot e^* dV - \frac{1}{2} \int_{V_c} (\sigma_0 + \sigma) \cdot e^p dV. \qquad (4.49)$$

For $e^p = 0$, i.e., a composite with fillers (inhomogeneities) subjected to applied stress σ_0, W is reduced to

$$W = \frac{1}{2} \int_{V_c} \sigma_0 \cdot e_0 dV + \frac{1}{2} \int_{V_c} \sigma_0 \cdot e^* dV. \qquad (4.50)$$

The elastic strain energy per unit volume of a composite w is given by

$$w = \frac{1}{2} \sigma_0 \cdot e_0 + \frac{f}{2} \sigma_0 \cdot e^*, \qquad (4.51)$$

where f is the volume fraction of filler. Equation (4.51) will be used later to calculate the elastic constants of a composite.

4.3.3 Mori–Tanaka mean field theory

When the volume fraction f of fillers increases, interactions between fillers and between the fillers and the free boundary of a composite become enhanced. Thus, the preceding formulations based on the original Eshelby model are not applicable to the case of larger f. Mori and Tanaka (1973) considered the volume-averaged field to account for these interactions. Consider the case of numerous inclusions embedded in an infinite matrix, a modification of Fig. 4.4, where each ellipsoidal inclusion is given a prescribed eigenstrain e^*. Denoting the volume-averaged internal stress in the inclusion domain and in the matrix domain by $\langle \sigma \rangle_\Omega$ and $\langle \sigma \rangle_m$, respectively, these must satisfy the requirement that the integral of the internal (or disturbance of) stress field over the entire composite is zero, i.e.,

$$(1 - f) \langle \sigma \rangle_m + f \langle \sigma \rangle_f = 0, \qquad (4.52)$$

where f is the volume fraction of inclusions and its value lies between 0 and 1. Assuming that numerous inclusions with finite f exist already in an infinite matrix with stiffness tensor C_m, we now introduce another inclusion with its domain (Ω) into the matrix domain $(V_c - \Omega)$. Then, the stress inside Ω induced by this inclusion is the sum of $\langle \sigma \rangle_m$ and $\langle \sigma \rangle_\infty$ which would exist in the case of a single inclusion embedded in an infinite matrix:

$$\sigma = \langle \sigma \rangle_m + \langle \sigma \rangle_\infty, \tag{4.53}$$

where $\langle \sigma \rangle_\infty$ is given by Eq. (4.34) or (4.35). The stress defined by Eq. (4.53) is the stress inside any inclusion Ω for the case of numerous inclusions with finite f, which is equal to the averaged stress $\langle \sigma \rangle_f$ in an inclusion,

$$\langle \sigma \rangle_f = \langle \sigma \rangle_m + \langle \sigma \rangle_\infty. \tag{4.54}$$

By substituting Eq. (4.54) into (4.52), we can obtain the averaged stress in the inclusion and in the matrix in terms of an $\langle \sigma \rangle_\infty$

$$\langle \sigma \rangle_m = -f \langle \sigma \rangle_\infty \tag{4.55a}$$

$$\langle \sigma \rangle_f = (1 - f) \langle \sigma \rangle_\infty \tag{4.55b}$$

Please note that $\langle \sigma \rangle_f$ is reduced to $\langle \sigma \rangle_\infty$ for the case $f = 0$ (no fillers exist).

The preceding discussion was for the case of the inclusion problem (Mori and Tanaka, 1973), but the Mori–Tanaka theory can be applied to more complicated cases. Let us apply the Mori–Tanaka theory to a composite with inhomogeneous inclusions, Fig. 4.6(a), where now the volume fraction of fillers is large. The volume-averaged stress disturbance $\langle \sigma \rangle_m$ in the matrix domain is related to that of the strain $\bar{e}$ in the matrix by

$$\langle \sigma \rangle_m = C_m \cdot \bar{e}. \tag{4.56}$$

If the interactions are accounted for, the Eshelby equivalent inclusion equation for the inhomogeneous inclusion problem, Eq. (4.47), is modified to

$$\begin{aligned} \sigma_0 + \sigma &= C_f \cdot (e_0 + \bar{e} + e - e^p) \\ &= C_m \cdot (e_0 + \bar{e} + e - e^*). \end{aligned} \tag{4.57}$$

Subtracting the uniform field term, Eq. (4.42), from both sides of Eq. (4.57), the stress disturbance σ is obtained as

$$\sigma = C_m \cdot (\bar{e} + e - e^*). \tag{4.58}$$

Since $\int_{V_c} \sigma \, dv = 0$ (definition of the stress disturbance), we derive the following equation:

$$\bar{e} + f(e - e^*) = 0. \tag{4.59}$$

From Eqs. (4.33), (4.57) and (4.59), we can eliminate $\bar{e}$ and e to solve for e^* as

$$e^* = \bar{D}^{-1} \cdot \left[(C_m - C_f) \cdot e_0 + C_f \cdot e^p \right], \tag{4.60}$$

$$\overline{D} = (C_m - C_f) \cdot [(1-f) S + fI] + C_m. \tag{4.61}$$

By comparing Eqs. (4.48) and (4.46) with (4.60) and (4.61), respectively, one can see that the results for finite f, Eqs. (4.60) and (4.61), are reduced to those of the dilute case ($f \approx 0$), Eqs. (4.46) and (4.48).

4.3.4 Computation of elastic constants of a composite by strain energy density method

Denoting the composite stiffness tensor by C_c, the elastic strain energy density w of a composite subjected to applied stress σ_0 is

$$w = \frac{1}{2}\sigma_0 \cdot C_c^{-1} \cdot \sigma_0, \tag{4.62}$$

where the composite elastic constants are assumed to be homogeneous macroscopically with C_c. The composite elastic strain energy density can also be evaluated in terms of uniform field and disturbance field contributions, i.e., Eq. (4.51), resulting in the equivalence of the strain energy density:

$$w = \frac{1}{2}\sigma_0 \cdot C_c^{-1} \cdot \sigma_0 = \frac{1}{2}\sigma_0 \cdot C_m^{-1} \cdot \sigma_0 + \frac{f}{2}\sigma_0 \cdot e^*. \tag{4.63}$$

Suppose that a composite consists of short fibers with Young's modulus E_f and Poisson's ratio ν_f aligned in the x_3-direction and a matrix with E_m, ν_m, as shown in Fig. 4.6(a) where the axis of a short fiber is taken to lie along the x_3-axis. Thus, the composite becomes transversely isotropic with the elastic compliance tensor C_c^{-1} given by Eq. (3.23). Under an applied stress σ_0 along the x_3-axis, and using Eq. (3.20) for the composite and Eq. (3.23) for the matrix in Eq. (4.63), leads to

$$\frac{1}{2}\frac{\sigma_0^2}{E_L} = \frac{1}{2}\frac{\sigma_0^2}{E_m} + \frac{1}{2}f\sigma_0 e_{33}^*. \tag{4.64}$$

If e_{33}^* is solved from Eq. (4.57) with $e^p = 0$, and the result is rearranged as $\beta(\sigma_0/E_m)$ then the second term on the right-hand side of Eq. (4.64) becomes $(f/2)\beta\sigma_0^2$, resulting in the formula to compute the longitudinal Young's modulus E_L:

$$\frac{E_L}{E_m} = \frac{1}{1 + \beta f}, \tag{4.65}$$

where β is a function of the elastic constants of the matrix and fiber (E_m, ν_m, E_f, ν_f), the fiber volume fraction f, and the fiber aspect ratio.

4.3.5 Extended Eshelby model for uncoupled thermophysical behavior

Hatta and Taya (1985, 1986) extended the original Eshelby model to the case of uncoupled thermal and electromagnetic behavior of a composite. Here we shall derive the Eshelby formulations by taking as an example the electrostatics problem, since those of other cases can be derived automatically due to the analogy shown in Table 4.1.

The electric potential $\phi(\mathbf{x})$ due to an electric charge Q located at $\mathbf{x}'$ is given by

$$\phi(\mathbf{x}) = \frac{Q}{4\pi\varepsilon} \frac{1}{|\mathbf{x}' - \mathbf{x}|}, \tag{4.66}$$

where ε is the dielectric constant of the medium. If Q is replaced by $-Q$, and $+Q$ is added at $\mathbf{x}'$ with $|\mathbf{ds}|$ being infinitesimal, Fig. 4.7, then the resulting potential at $\mathbf{x}$, $\phi(\mathbf{x})$, is obtained by superposition of a pair of $+Q$ and $-Q$:

$$\phi(\mathbf{x}) = \frac{Q}{4\pi\varepsilon} \left\{ \frac{1}{|\mathbf{x}' + \mathbf{ds} - \mathbf{x}|} - \frac{1}{|\mathbf{x}' - \mathbf{x}|} \right\}. \tag{4.67}$$

Defining

$$f(\mathbf{x}', \mathbf{x}) = \frac{1}{|\mathbf{x}' - \mathbf{x}|}, \tag{4.68}$$

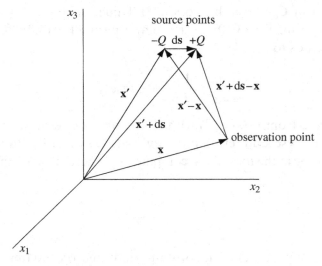

Fig. 4.7   Model for electrical potential due to a dipole.

one can rewrite Eq. (4.67) as

$$\phi(\mathbf{x}) = \frac{Q}{4\pi\varepsilon} \{f(\mathbf{x}' + d\mathbf{s}, \mathbf{x}) - f(\mathbf{x}', \mathbf{x})\}$$

$$= \frac{\mathbf{p}}{4\pi\varepsilon} \frac{\{f(\mathbf{x}' + d\mathbf{s}, \mathbf{x}) - f(\mathbf{x}', \mathbf{x})\}}{d\mathbf{s}}, \qquad (4.69)$$

where $\mathbf{p} = Q\, d\mathbf{s}$ is an electric dipole pointing in the same direction as $d\mathbf{s}$, see Fig. 4.7.

Taking a limiting sequence $|d\mathbf{s}| \to 0$ while keeping $Q\, d\mathbf{s} = \mathbf{p} = \text{constant}$, $\phi(\mathbf{x})$ can be reduced to

$$\phi(\mathbf{x}) = \frac{\mathbf{p}}{4\pi\varepsilon} \cdot \nabla \frac{1}{|\mathbf{x}' - \mathbf{x}|} \qquad (4.70)$$

or, in index form,

$$= \frac{p_j}{4\pi\varepsilon} \frac{\partial}{\partial x_j} \frac{1}{|\mathbf{x}' - \mathbf{x}|}, \qquad (4.71)$$

where $\mathbf{p}$ is the electric doublet (or dipole) and its positive direction is taken along the direction from $-Q$ to $+Q$, and ∇ is the del operator defined in section 3.1.1. The electric field $\mathbf{E}$ is derivable from the potential $\phi(\mathbf{x})$ as

$$\mathbf{E} = -\nabla\phi \qquad (4.72a)$$

or

$$E_i = -\frac{\partial}{\partial x_i} \phi(\mathbf{x}). \qquad (4.72b)$$

From Eqs. (4.71) and (4.72) we obtain

$$E_i = -\frac{p_j}{4\pi\varepsilon} \frac{\partial^2}{\partial x_i \partial x_j} \frac{1}{|\mathbf{x}' - \mathbf{x}|}. \qquad (4.73)$$

If the doublet is distributed uniformly within an ellipsoidal domain Ω, the resulting electric field E_i is obtained from

$$E_i = S_{ij} E_j^*, \qquad (4.74)$$

where E_j^* is a component of the eigen-electric-field defined by

$$E_j^* = \frac{p_j}{\varepsilon}, \qquad (4.75)$$

S_{ij} is the Eshelby tensor for uncoupled thermophysical behavior, defined by

$$S_{ij} = -\frac{1}{4\pi}\Phi_{,ij},$$ (4.76)

where

$$\Phi(\mathbf{x}) = \int_{\Omega}\frac{\mathrm{d}V'}{|\mathbf{x}' - \mathbf{x}|}$$ (4.77)

and where $\mathrm{d}V'$ is the volume element equal to $\mathrm{d}x_1'\mathrm{d}x_2'\mathrm{d}x_3'$ and $\int_{\Omega}\mathrm{d}V'$ is the volume integral in the ellipsoidal domain Ω. The details of S_{ij} are given elsewhere (Hatta and Taya, 1985, 1986; Taya and Arsenault, 1989) and in Appendix A.2. According to the potential theory (Kellog, 1953), the jump in the second derivative of Φ across the boundary of $\Omega(S)$ is given by

$$(\Phi_{,ij})_{S^+} - (\Phi_{,ij})_{S^-} = 4\pi n_i n_j,$$ (4.78)

where S^+ and S^- are surface regions encountered when approaching the surface S from outside and inside Ω, respectively, and $\mathbf{n}$ is the outward unit vector normal to S. The jump in electric field across S is then obtained from Eqs. (4.74), (4.76), and (4.78):

$$(E_i)_+ - (E_i)_- = -E_j^* n_i n_j,$$ (4.79)

where subscripts $+$ and $-$ are abbreviations of S^+ and S^-, respectively. The electric flux density vector across S must be continuous, resulting in

$$\{(D_i)_+ - (D_i)_-\}n_i = 0.$$ (4.80)

By using the constitutive equation of electrostatics for an isotropic material, Eq. (3.96), Table 4.1,

$$D_i = \varepsilon E_i,$$ (4.81)

and Eq. (4.80), one can rewrite Eq. (4.79) as

$$D_i = \varepsilon(E_i - E_i^*),$$ (4.82)

where subscript $-$ is omitted from D_i and E_i. In general tensorial form, Eq. (4.82) can be rewritten

$$D_i = \varepsilon_{ij}(E_j - E_j^*).$$ (4.83)

Equation (4.83) has the same format as the original Eshelby equation if the following one-to-one correspondence between the mechanical and electrostatic fields is recognized:

mechanical	electrostatic
stress σ	electric flux density $\mathbf{D}$
strain e	electric field $\mathbf{E}$
stiffness tensor $\mathbf{C}$	dielectric constant tensor ε
eigenstrain e^*	eigen electric field $\mathbf{E}^*$

$$\sigma = C \cdot (e - e^*) \qquad\qquad \mathbf{D} = \varepsilon \cdot (\mathbf{E} - \mathbf{E}^*) \tag{4.84a}$$

$$e = S \cdot e^* \qquad\qquad \mathbf{E} = S \cdot \mathbf{E}^* \tag{4.84b}$$

The only difference between the Eshelby model of the mechanical problem and that of the electrostatic field problem is that the mechanical problem requires higher-order tensors than the electrostatic problem. Next, consider an infinite matrix with dielectric constant ε_m, which contains an ellipsoidal inhomogeneity with ε_f (domain Ω), and is subjected to a far field on the composite boundary: an applied electric field $\mathbf{E}_0$ or an applied dielectric flux density $\mathbf{D}_0$. Before Ω was introduced, there existed a uniform field:

$$\mathbf{D}_0 = \varepsilon_m \cdot \mathbf{E}_0. \tag{4.85}$$

Upon the introduction of Ω, the total (actual) electric flux density inside Ω is given by

$$\begin{aligned} \mathbf{D}_0 + \mathbf{D} &= \varepsilon_f \cdot (\mathbf{E}_0 + \mathbf{E}) \\ &= \varepsilon_m \cdot (\mathbf{E}_0 + \mathbf{E} - \mathbf{E}^*), \end{aligned} \tag{4.86}$$

where $\mathbf{D}$ and $\mathbf{E}$ are the disturbance of the uniform field due to the existence of Ω. In Eq. (4.86), the right-hand side is replaced by the equivalent inclusion with an unknown eigen-electric field $\mathbf{E}^*$. Subtracting the uniform field, Eq. (4.85), from both sides of Eq. (4.86), we obtain

$$\mathbf{D} = \varepsilon_m \cdot (\mathbf{E} - \mathbf{E}^*) \tag{4.87}$$

which is identical to the inclusion problem, Eq. (4.83). Thus, the disturbance electric field $\mathbf{E}$ can be related to $\mathbf{E}^*$ by Eq. (4.74). In symbolic notation,

$$\mathbf{E} = S \cdot \mathbf{E}^*. \tag{4.74}'$$

From Eqs. (4.85), (4.86), and (4.74)', $\mathbf{E}^*$ can be solved. When the volume fraction f of fillers becomes finite, the interactions between fillers and those between the fillers and the free surface of the composite will be enhanced. We will consider a composite with high volume fraction of fillers where the volume-averaged disturbance electric field in the matrix domain is denoted by $\bar{\mathbf{E}}$, which is related to the volume-averaged electric flux density $\langle \mathbf{D} \rangle_m$ by

$$\langle \mathbf{D} \rangle_m = \varepsilon_m \cdot \bar{\mathbf{E}}. \tag{4.88}$$

If we insert another filler in the representative matrix domain, the local disturbance electric field $\mathbf{E}$ is induced. Then Eq. (4.86) will be changed to

$$\begin{aligned} \mathbf{D}_0 + \mathbf{D} &= \varepsilon_f \cdot (\mathbf{E}_0 + \bar{\mathbf{E}} + \mathbf{E}) \\ &= \varepsilon_m \cdot (\mathbf{E}_0 + \bar{\mathbf{E}} + \mathbf{E} - \mathbf{E}^*). \end{aligned} \tag{4.89}$$

Subtracting the uniform field, Eq. (4.85), from Eq. (4.89), the disturbance electric flux density vector $\mathbf{D}$ is obtained as

$$\mathbf{D} = \varepsilon_m \cdot (\bar{\mathbf{E}} + \mathbf{E} - \mathbf{E}^*). \tag{4.90}$$

Since the integral of $\mathbf{D}$ over the entire composite domain vanishes, and using Eq. (4.88), we arrive at

$$\bar{\mathbf{E}} = -f(\mathbf{E} - \mathbf{E}^*). \tag{4.91}$$

From Eqs. (4.85), (4.74)', (4.89) and (4.91), we can solve for $\mathbf{E}^*$.

As noted in Table 4.1, the above extended Eshelby model is valid for all uncoupled thermophysical behavior. By using this model, Hatta and Taya (1985, 1986) obtained the formulae to compute the effective thermal conductivities of a composite with various types of filler shape and orientation including hybrid composites where two different types of filler are embedded in the matrix. For a composite with a simpler microstructure the composite thermal conductivity is obtained by the extended Eshelby model in explicit form. For example, the thermal conductivity K_c of a short-fiber composite with three-dimensional random orientation is given by (Hatta and Taya, 1985)

$$\frac{K_c}{K_m} = 1 - \frac{f(K_m - K_f)\{(K_f - K_m)(2S_{33} + S_{11}) + 3K_m\}}{[3(K_f - K_m)^2(1 - f)S_{11}S_{33} + K_m(K_f - K_m)\eta + 3K_m^2]}, \tag{4.92}$$
$$\eta = 3(S_{11} + S_{33}) - f(2S_{11} + S_{33}),$$

where K_m, K_f, and f are thermal conductivity of the matrix and fiber, and volume fraction of filler, and S_{33} and S_{11} are the Eshelby tensors of a short fiber parallel and perpendicular, respectively, to the fiber axis, and are given in Appendix A.2. If a short fiber is replaced by a spherical filler, Eqs. (4.92) are reduced to

$$\frac{K_c}{K_m} = 1 + \frac{f}{(1 - f)/3 + K_m/(K_f - K_m)}. \tag{4.93}$$

Then Eq. (4.93) coincides with (i) Maxwell's solution (Maxwell, 1904), although he derived it for a composite with dielectric constant ε_c, (ii) Kerner's solution (Kerner, 1956), and (iii) the lower bound of Hashin and Shtrikman (1962). Dunn *et al.* (1993) measured the in-plane (K_{xy}) and out-of-plane (K_z) thermal conductivity of an alumina (Al_2O_3) short fiber/Kerimid® matrix composite which was developed for printed circuit boards (PCBs) and compared the measurements with the predictions based on the extended Eshelby model. Figure 4.8 shows a comparison between the measured values (symbols) and the predictions (lines). The measurements agree well with the predicted Ks if the Al_2O_3 fibers are assumed to be distributed in a two-dimensionally random manner, which is likely because the composite was processed by hot-pressing. The dotted line in Fig. 4.8 is drawn using Eqs. (4.92).

4.3.6 Extended Eshelby model for coupled behavior

The constitutive equations for coupling behavior among mechanical, thermal, and electrical properties are given in Subsection 3.5.1, i.e., piezoelectric, pyroelectric, and thermoelastic coupling behavior are all given by Eqs. (3.217)–(3.252) in index form, or in symbolic form. In order to be consistent with the formulations used in the effective property tensors of composites, Section 4.2, and also with those in Subsection 3.5.1, we shall use here symbolic formulations.

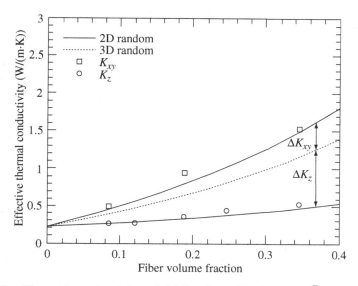

Fig. 4.8   Thermal conductivity of Al_2O_3 short fiber/Kerimid® matrix composite: measurements (symbols) and predictions by the Eshelby model (solid and dotted lines) (Dunn *et al.*, 1993, with permission of Sage Publications Ltd).

We shall apply the extended Eshelby model for coupled behavior first to piezoelectric and then to pyroelectric composites.

Piezoelectric composites

Referring to Eqs. (3.221) and (3.222) with temperature change $\theta = 0$, the constitutive equations of piezoelectric behavior can be given in symbolic form where the terms related to temperature change are dropped:

$$\Sigma = R \cdot Z, \tag{4.94a}$$

$$Z = F \cdot \Sigma, \tag{4.94b}$$

where Σ, Z are flux and field vectors defined by Eqs. (3.223), and R, F are property and compliance tensors defined by Eqs. (3.224) and (3.225), respectively. Dunn and Taya (1993a) studied the problem of an ellipsoidal inclusion with eigenfield vector Z^* embedded in an infinite homogeneous material with property tensor R_m and proved that the field vector Z induced by Z^* is linearly related to Z^* through the Eshelby tensor (electroelastic Eshelby tensor) S:

$$Z = S \cdot Z^* \tag{4.95}$$

where S is a function of (i) the property tensor R_m which includes the elastic stiffness tensor, dielectric tensor, and piezoelectric coupling tensor, and (ii) the aspect ratio of the principal axes of an ellipsoid. Detailed expressions for S are given in Appendix A3. The piezoelectric tensor S has a 9×9 matrix form. Mikata (2000, 2001) obtained explicit Eshelby tensors for cylindrical inclusions aligned or perpendicular to the symmetry axis of a transversely isotropic piezoelectric matrix. The Eshelby tensors obtained by Mikata cover more general cases, and as a special case coincide with those obtained by Dunn and Taya (1993b) and Huang and Yu (1994). Li and Dunn (1998) extended the Eshelby formulation to anisotropic coupled fields among elastic, electric, and magnetic properties and derived the tensor S explicitly for two simple geometries of inclusion – elliptical cylindrical and penny-shaped inclusions. Like the inclusion problems discussed earlier, both Z and Σ become uniform inside the ellipsoidal inclusion if the prescribed eigenfield Z^* is constant. It is to be noted that S is reduced to (i) the Eshelby tensor for an anisotropic elastic material in the absence of piezoelectric coupling and dielectric tensors, as first formulated by Kinoshita and Mura (1971), and (ii) the Eshelby tensors for uncoupled physical behaviors, as first proposed by Hatta and Taya (1985, 1986). The details of the Eshelby tensors for mechanical uncoupled thermal and physical, and coupled piezoelectric, inclusion problems are given in Appendix A. When ellipsoidal piezoelectric inhomogeneities with property tensor R_f and volume fraction f are embedded in another piezoelectric matrix with property tensor R_m subjected to either an applied flux vector Σ_0 or a field vector Z_0, this is the problem of a piezoelectric composite. The formulation based on the Eshelby model is identical to that discussed in Section 4.2. Hence the computation of the composite property tensor R_c can be obtained by either Eq. (4.22) or Eq. (4.25), depending on the boundary condition, i.e., Z_0 or Σ_0 prescribed on the boundary, respectively. The key step in the computation of R_c is to compute the concentration tensors A and B discussed in Section 4.2, the details of which will be discussed in the next section.

By using the Eshelby model extended to piezoelectric composites and the method of Section 4.2, i.e., concentration factors, Dunn and Taya (1993b) predicted the piezoelectric constants d_{ij} of PZT filler/polymer matrix composites. These are plotted as a function of filler volume fraction in Fig. 4.9 where (a) is for the case of continuous aligned fibers and (b) for random particles. In Fig. 4.9 open circles denote the experimental results and the lines are the predictions based on several methods: (1) Mori–Tanaka, (2) dilute approximation, (3) self-consistent, and (4) the differential scheme, based on McLaughlin (1990). It is to be noted that the key formulation in methods (1) to (3) is the Eshelby model extended to piezoelectric composites (Dunn and Taya, 1993a,b). The piezo-electric coupling constants d_{33} and d_{31} in Fig. 4.9 are abbreviations of d_{ijk} which relates strain ε_{ij} to electric field E_k. For example, $d_{31} = d_{311} = g_{31}/\beta_{33}$ where g_{31} and β_{33} are the components of the compliance tensor of a piezoelectric material defined by Eq. (3.228). It follows from Fig. 4.9(a) that all the predictions by the different methods coincide for the case of the unidirectionally aligned piezo-electric fiber composite and agree with experiment (Chan and Unsworth, 1989). In the case of the PZT particle/polymer composite, Fig. 4.9(b) indicates that at lower particle volume fractions most of the predictions, including the cubes model proposed by Banno (1983), appear to agree with experiment, but at high particle volume fractions, the Mori–Tanaka mean field theory based on the extended Eshelby model seems to agree well with experiment (Furukawa *et al.*, 1990). In view of the validity for a wide variety of filler geometry, and its simplicity, the Mori–Tanaka theory based on the extended Eshelby model provides accurate predictions for the overall piezoelectric properties of a composite. Formulations based on the other methods, such as the self-consistent method, are rigorous, involving heavy numerical calculations.

Li *et al.* (1999) obtained the exact thermal properties and piezoelectric properties of a polycrystalline piezoelectric material by assuming a uniform field where the properties depend on the texture orientation. Li (2000) demonstrated numerically the strong dependence on texture of the electroelastic constants of a polycrystalline piezoelectric aggregate.

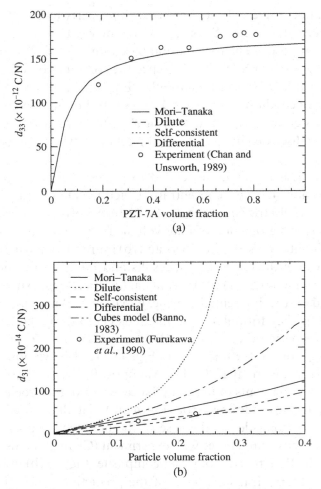

Fig. 4.9 Comparison of the predictions of composite piezoelectric constants based on several methods with experiment. (a) Aligned continuous PZT fiber/polymer matrix composite. Predictions from all models listed coincide. (b) Particle PZT/polymer matrix composite (Dunn and Taya, 1993b, with permission from Elsevier Ltd).

Pyroelectric composites

Dunn (1993) computed the pyroelectric constants of pyroelectric $BaTiO_3$/polymer matrix composites using the model developed in Section 4.2 and based on several micromechanics models: the Eshelby model extended to pyroelectric composites combined with a dilute assumption; the Mori–Tanaka mean field theory; the self-consistent model; and the differential scheme proposed by McLaughlin (1990). The predicted pyroelectric constants of p_3^σ for two cases of reinforcement – continuous fiber, and spherical $BaTiO_3$ – are

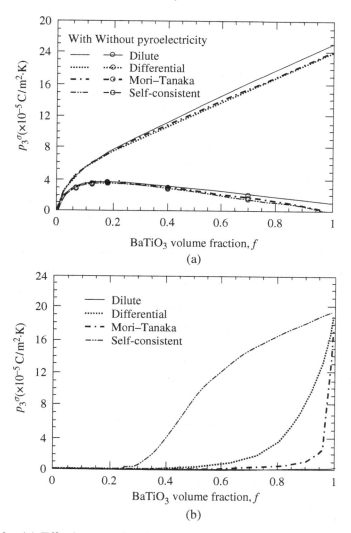

Fig. 4.10 (a) Effective pyroelectric coefficient ($p_3^\sigma = \gamma_3/\beta_{33}$) of a continuous BaTiO$_3$ fiber-reinforced/polymer matrix composite. The two sets of curves (lines, and lines with open symbols) are for computations performed when the pyro-electric effect of the BaTiO$_3$ was, and was not, considered. (b) Effective pyroelectric coefficient ($p_3^\sigma = \gamma_3/\beta_{33}$) of a spherical BaTiO$_3$ particle-reinforced/polymer matrix composite. (Dunn, 1993, with permission from American Inst. Phys.)

plotted as a function of volume fraction f of BaTiO$_3$ in Fig. 4.10(a) and (b), respectively. Superscript σ in the pyroelectric constant p_3^σ denotes the value at a constant stress while subscript 3 is the value along the x_3-axis, coincident with the direction of the BaTiO$_3$ fibers and that of applied electric flux density (D_3^0). The lines with and without open circles in Fig. 4.10(a) correspond to the cases of considering the primary and secondary pyroelectric effect, respectively. That is, the primary pyroelectric effect is due to the pyroelectric fibers of $BaTiO_3$, thus it increases with f. As a pyroelectric fiber composite is subjected to a uniform temperature change θ, a stress due to the mismatch in the thermal expansion coefficients of fiber and matrix is induced, which will in turn generate electric polarization if at least one of the constituents of the composite is piezoelectric. $BaTiO_3$ exhibits both pyroelectric and piezoelectric behavior while the polymer matrix does not. The secondary pyroelectric effect of a continuous $BaTiO_3$ fiber/polymer matrix composite is directly proportional to the thermal stress in the $BaTiO_3$ fiber, which vanishes as f reaches 1, as illustrated in Fig. 4.10(a). The primary pyroelectric effect is also visible from Fig. 4.10(a) as it increases with f. As far as the predictions by various models for the case of continuous $BaTiO_3$ fibers are concerned, they do not differ much. In the case of a spherical $BaTiO_3$/polymer matrix composite, however, the predictions by various models differ a lot and the secondary pyroelectric effect does not exist since the thermal stress in a spherical $BaTiO_3$ particle composite is neglible.

It should be noted in Fig. 4.10(b) that the predictions by the Mori–Tanaka mean field theory and the differential scheme agree for the range of low to intermediate volume fractions, but they deviate beyond that range, and the predictions by the models except for the dilute case recover the correct pyroelectric constant at $f = 1$, i.e., that of $BaTiO_3$.

4.4 Concentration factor tensors

In Section 4.2, the overall composite property tensor R_c and the compliance tensor F_c are obtained explicitly as functions of the constituents' properties, volume fractions of fillers, the Eshelby tensor, concentration factor tensors, A, B, and concentration factor vectors **a** and **b**. Here we shall state the procedure to compute the concentration factor vectors and tensors by several micromechanical models based on the Eshelby method, which covers the original and extended models.

4.4.1 Law-of-mixtures model

This model was discussed in detail in Section 4.1. Letting the concentration factor tensor $A = I$ and use of Eq. (4.22) provide the law-of-mixtures formula Eq. (4.4), which was originally put forward by Voigt (1889), while $B = I$ and use of Eq. (4.25) recovers the law-of-mixtures model Eq. (4.7), as was proposed first by Reuss (1929). The composite property tensors R_c predicted by using $A = I$, and $B = I$ coincide with the upper and lower bounds of Hashin and Shtrikman (1963). The formulae based on the law-of-mixtures model for uncoupled mechanical and thermophysical properties are explicitly given by

Eqs. (4.14). Generally, the composite property tensor $\boldsymbol{R}_c$ and compliance tensor $\boldsymbol{F}_c$ are obtained by $\boldsymbol{A} = \boldsymbol{I}$ in Eq. (4.22) and $\boldsymbol{B} = \boldsymbol{I}$ in Eq. (4.25), respectively, and given by

$$\boldsymbol{R}_c = (1 - f)\boldsymbol{R}_m + f\boldsymbol{R}_f, \tag{4.96a}$$

$$\boldsymbol{F}_c = (1 - f)\boldsymbol{F}_m + f\boldsymbol{F}_f. \tag{4.96b}$$

Equations (4.96) can be extended to hybrid composites by using Eqs. (4.26).

Dunn (1995) showed that there is identity between $\boldsymbol{A}$ and $\mathbf{a}$, and also between $\boldsymbol{B}$ and $\mathbf{b}$:

$$\mathbf{a} = (\boldsymbol{I} - \boldsymbol{A}) \cdot (\boldsymbol{R}_m - \boldsymbol{R}_f)^{-1} \cdot (\boldsymbol{\pi}_m - \boldsymbol{\pi}_f), \tag{4.97a}$$

$$\mathbf{b} = (\boldsymbol{I} - \boldsymbol{B}) \cdot (\boldsymbol{F}_m - \boldsymbol{F}_f)^{-1} \cdot (\boldsymbol{\Lambda}_m - \boldsymbol{\Lambda}_f). \tag{4.97b}$$

If the Voigt assumption $\boldsymbol{A} = \boldsymbol{I}$ is used, then $\mathbf{a} = 0$ is obtained from Eq. (4.97a). Similarly $\mathbf{b} = 0$ from Eq. (4.97b) if $\boldsymbol{B} = \boldsymbol{I}$ is used. Combined with Eqs. (4.29), we can recover the following law-of-mixtures formulae based on the Voigt and Reuss models:

$$\boldsymbol{\pi}_c = \boldsymbol{\pi}_m + f(\boldsymbol{\pi}_f - \boldsymbol{\pi}_m), \tag{4.98a}$$

$$\boldsymbol{\Lambda}_c = \boldsymbol{\Lambda}_m + f(\boldsymbol{\Lambda}_f - \boldsymbol{\Lambda}_m). \tag{4.98b}$$

4.4.2 Dilute suspension of fillers

When the volume fraction of fillers is small, the interactions between fillers can be ignored and the original Eshelby model is valid. To cover its applicability to mechanical behavior, and to uncoupled and coupled thermophysical behavior, the Eshelby model is expressed in terms of the field vector $\mathbf{Z}$ and flux vector $\boldsymbol{\Sigma}$, and the electroelastic property and compliance tensors $\boldsymbol{R}$, $\boldsymbol{F}$. The concentration factor tensors, $\boldsymbol{A}^{\text{dil}}$, $\boldsymbol{B}^{\text{dil}}$, and vectors $\mathbf{a}^{\text{dil}}$ and $\mathbf{b}^{\text{dil}}$ for the case of a dilute suspension of fillers are defined by

$$\mathbf{Z}_f = \boldsymbol{A}^{\text{dil}} \cdot \mathbf{Z}_0 + \mathbf{a}^{\text{dil}}\theta, \tag{4.99a}$$

$$\boldsymbol{\Sigma}_f = \boldsymbol{B}^{\text{dil}} \cdot \boldsymbol{\Sigma}_0 + \mathbf{b}^{\text{dil}}\theta, \tag{4.99b}$$

where $\mathbf{Z}_f$ and $\boldsymbol{\Sigma}_f$ are the average field and flux vectors in the filler domain, $\mathbf{Z}_0$ and $\boldsymbol{\Sigma}_0$ are the applied field and flux vectors on the composite boundary, and θ is the uniform temperature change imposed across the entire composite domain.

First, let us consider the case of field vector Z_0 applied to the composite boundary and a uniform temperature change θ across the composite, Fig. 4.11(a). The problem of Fig. 4.11(a) can be decomposed into two parts: the ellipsoidal filler with R_f and Λ_m embedded in a matrix with R_m and Λ_m without Z_0 and with θ; plus that with R_f and $(\Lambda_f - \Lambda_m)$ embedded in a matrix of R_m and Λ_m subjected to Z_0 and θ, Fig. 4.11(b). The first part of Fig. 4.11(b) does not produce any field and flux vectors, while the second part will provide solutions to the original problem of Fig. 4.11(a), and can be recast to the Eshelby equivalent inclusion problem, Fig. 4.11(c) which is further converted to Fig. 4.11(d), where the entire composite domain is now homogeneous (R_m, Λ_m) and the filler domain is replaced by an equivalent inclusion with unknown eigenvector field Z^*. Thus the Eshelby equivalent inclusion equation is given by

$$
\begin{aligned}
\Sigma_0 + \Sigma &= R_f \cdot (Z_0 + Z - Z^\theta) \\
&= R_m \cdot (Z_0 + Z - Z^*)
\end{aligned}
\tag{4.100}
$$

where Σ_0 is the flux vector corresponding to Z_0 and given by $\Sigma_0 = R_m \cdot Z_0$, Z is the disturbance field vector caused by the presence of the fillers and related to the eigenfield vector Z^* by

$$
Z = S \cdot Z^*,
\tag{4.101}
$$

S is the electroelastic Eshelby tensor (see Appendix A3), and Z^θ is the prescribed eigenfield (thermal expansion) vector in the filler domain, given by

$$
Z^\theta = (\Lambda_f - \Lambda_m)\theta
\tag{4.102}
$$

where Λ_i is the pyroelectric tensor of the ith phase.

The volume-averaged field vector Z_f in the filler domain is given by

$$
Z_f = Z_0 + S \cdot Z^*.
\tag{4.103}
$$

Equation (4.100) can be rewritten by using Eq. (4.103)

$$
R_f \cdot (Z_f - Z^\theta) = R_m \cdot (Z_f - Z^*).
\tag{4.104}
$$

Eliminating Z^* from Eqs. (4.103) and (4.104), we get

$$
[(R_f - R_m) + R_m \cdot S^{-1}] \cdot Z_f = R_m \cdot S^{-1} \cdot Z_0 + R_f \cdot Z^\theta.
\tag{4.105}
$$

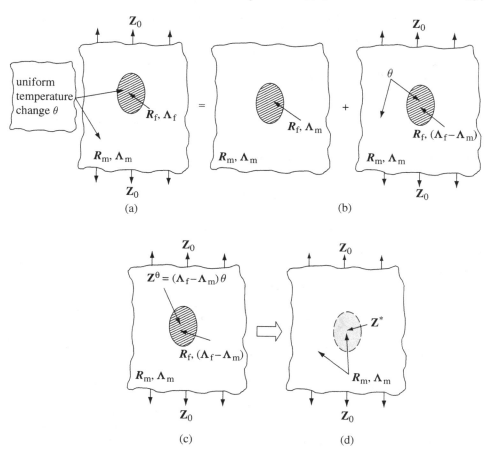

Fig. 4.11 Eshelby model applied to an electronic composite with piezo- and pyroelectric properties and field boundary condition. (a) Original problem, (b) decomposition into two parts, and (c) Eshelby equivalent inclusion problem.

Substituting Eq. (4.102) into (4.105) and rearranging, we have

$$\mathbf{Z}_f = \left[\boldsymbol{I} + \boldsymbol{S} \cdot \boldsymbol{R}_m^{-1} \cdot (\boldsymbol{R}_f - \boldsymbol{R}_m) \right]^{-1} \cdot \mathbf{Z}_0$$
$$+ \left[\boldsymbol{I} + \boldsymbol{S} \cdot \boldsymbol{R}_m^{-1} \cdot (\boldsymbol{R}_f - \boldsymbol{R}_m) \right]^{-1} \cdot \boldsymbol{S} \cdot \boldsymbol{R}_m^{-1} \cdot \boldsymbol{R}_f \cdot (\boldsymbol{\Lambda}_f - \boldsymbol{\Lambda}_m)\theta. \quad (4.106)$$

The field vector $\mathbf{Z}_f$ given by Eq. (4.106) is the solution to the problem of Fig. 4.11(c), thus the total field vector $\mathbf{Z}_f$ in the filler is the sum of Eq. (4.106) and $\boldsymbol{\Lambda}_m\theta$, i.e., the first part of Fig. 4.11(b), hence

$$\mathbf{Z}_f = \left[\boldsymbol{I} + \boldsymbol{S} \cdot \boldsymbol{R}_m^{-1} \cdot (\boldsymbol{R}_f - \boldsymbol{R}_m) \right]^{-1} \cdot \mathbf{Z}_0$$
$$+ \left[\boldsymbol{I} + \boldsymbol{S} \cdot \boldsymbol{R}_m^{-1} \cdot (\boldsymbol{R}_f - \boldsymbol{R}_m) \right]^{-1} \cdot \boldsymbol{S} \cdot \boldsymbol{R}_m^{-1} \cdot \boldsymbol{R}_f \cdot (\boldsymbol{\Lambda}_f - \boldsymbol{\Lambda}_m)\theta + \boldsymbol{\Lambda}_m\theta. \quad (4.107)$$

Comparing Eq. (4.107) with (4.99a), we obtain the concentration factor tensor A^{dil} and vector $\mathbf{a}^{\text{dil}}$ as

$$A^{\text{dil}} = \left[I + S \cdot R_{\text{m}}^{-1} \cdot (R_{\text{f}} - R_{\text{m}}) \right]^{-1}, \tag{4.108a}$$

$$\mathbf{a}^{\text{dil}} = \left[I + S \cdot R_{\text{m}}^{-1} \cdot (R_{\text{f}} - R_{\text{m}}) \right]^{-1} \cdot S \cdot R_{\text{m}}^{-1} \cdot R_{\text{f}} \cdot (\Lambda_{\text{f}} - \Lambda_{\text{m}}) + \Lambda_{\text{m}}. \tag{4.108b}$$

The equation for $\mathbf{a}^{\text{dil}}$ can be simplified as

$$\mathbf{a}^{\text{dil}} = S \cdot \left[R_{\text{m}} + (R_{\text{f}} - R_{\text{m}}) \cdot S \right]^{-1} \cdot R_{\text{f}} \cdot (\Lambda_{\text{f}} - \Lambda_{\text{m}}) + \Lambda_{\text{m}}. \tag{4.108c}$$

Second, let us consider the case of flux vector Σ_0 boundary condition with uniform temperature change θ.

Referring to Fig. 4.11(c), the flux vector Σ_{f} in a filler is given by Eq. (4.100). Solving for Z^* from Eqs. (4.100) and (4.101),

$$Z^* = S^{-1} \cdot \left[R_{\text{f}}^{-1} \cdot \Sigma_{\text{f}} - Z_0 + Z^{\theta} \right]. \tag{4.109}$$

Substituting Eq. (4.109) into (4.100) and rearranging the terms, we have

$$\begin{aligned}
\Sigma_{\text{f}} &= \left[S \cdot (R_{\text{m}}^{-1} - R_{\text{f}}^{-1}) + R_{\text{f}}^{-1} \right]^{-1} \cdot Z_0 \\
&\quad + \left[S \cdot (R_{\text{m}}^{-1} - R_{\text{f}}^{-1}) + R_{\text{f}}^{-1} \right]^{-1} \cdot (S - I) \cdot (\Lambda_{\text{f}} - \Lambda_{\text{m}}) \theta.
\end{aligned} \tag{4.110}$$

A comparison of Eq. (4.110) with (4.99b) yields the concentration factor tensor B^{dil} and vector $\mathbf{b}^{\text{dil}}$ as

$$B^{\text{dil}} = \left[S \cdot (R_{\text{m}}^{-1} - R_{\text{f}}^{-1}) + R_{\text{f}}^{-1} \right]^{-1}, \tag{4.111a}$$

$$\mathbf{b}^{\text{dil}} = \left[S \cdot (R_{\text{m}}^{-1} - R_{\text{f}}^{-1}) + R_{\text{f}}^{-1} \right]^{-1} \cdot (S - I) \cdot (\Lambda_{\text{f}} - \Lambda_{\text{m}}). \tag{4.111b}$$

4.4.3 Mori–Tanaka mean field theory

When the volume fraction of fillers becomes large, the interactions among fillers cannot be ignored. Let us first consider the field vector Z_0 prescribed on the composite boundary, Fig. 4.11(a), where numerous ellipsoidal fillers are embedded in the matrix. Now we insert an ellipsoidal filler into the volume-averaged field Z_{m} in the matrix. Then, using the solution of the dilute case, the field vector Z_{f} in any representative filler is related to Z_{m} and θ by

$$Z_{\text{f}} = A^{\text{dil}} \cdot Z_{\text{m}} + \mathbf{a}^{\text{dil}} \theta, \tag{4.112}$$

where A^{dil} and $\mathbf{a}^{\text{dil}}$ are given by Eqs. (4.108). If the Mori–Tanaka theory is valid, $\mathbf{Z}_f$ is also related to the applied field vector $\mathbf{Z}_0$ and θ, so

$$\mathbf{Z}_f = A^{\text{MT}} \cdot \mathbf{Z}_0 + \mathbf{a}^{\text{MT}}\theta, \tag{4.113}$$

where A^{MT} and $\mathbf{a}^{\text{MT}}$ are the concentration factor tensor and vector, respectively, from the Mori–Tanaka theory.

From Eqs. (4.18), (4.20) and (4.112), we obtain

$$\begin{aligned}
\mathbf{Z}_0 &= (1-f)\mathbf{Z}_m + f\,\mathbf{Z}_f \\
&= (1-f)\mathbf{Z}_m + f(A^{\text{dil}} \cdot \mathbf{Z}_m + \mathbf{a}^{\text{dil}}\theta) \\
&= [(1-f)\mathbf{I} + fA^{\text{dil}}] \cdot \mathbf{Z}_m + f\mathbf{a}^{\text{dil}}\theta,
\end{aligned} \tag{4.114}$$

where f is the volume fraction of fillers.

From Eq. (4.112)

$$\mathbf{Z}_m = A^{\text{dil}^{-1}} \cdot (\mathbf{Z}_f - \mathbf{a}^{\text{dil}}\theta). \tag{4.115}$$

Eliminating $\mathbf{Z}_m$ from Eqs. (4.114) and (4.115)

$$\mathbf{Z}_0 = \left[(1-f)A^{\text{dil}^{-1}} + f\mathbf{I}\right] \cdot \mathbf{Z}_f - (1-f)A^{\text{dil}^{-1}} \cdot \mathbf{a}^{\text{dil}}\theta. \tag{4.116}$$

Solving for $\mathbf{Z}_f$, we obtain

$$\mathbf{Z}_f = \left[(1-f)A^{\text{dil}^{-1}} + f\mathbf{I}\right]^{-1} \cdot \mathbf{Z}_0 + (1-f)\left[(1-f)A^{\text{dil}^{-1}} + f\mathbf{I}\right]^{-1} \cdot A^{\text{dil}^{-1}} \cdot \mathbf{a}^{\text{dil}}\theta. \tag{4.117}$$

A comparison between Eqs. (4.113) and (4.117) yields

$$\begin{aligned}
A^{\text{MT}} &= \left[(1-f)A^{\text{dil}^{-1}} + f\mathbf{I}\right]^{-1} \\
&= [(1-f)\mathbf{I} + fA^{\text{dil}}]^{-1} \cdot A^{\text{dil}}
\end{aligned} \tag{4.118a}$$

$$\mathbf{a}^{\text{MT}} = (1-f)[(1-f)\mathbf{I} + fA^{\text{dil}}]^{-1} \cdot \mathbf{a}^{\text{dil}}. \tag{4.118b}$$

Next we consider the boundary condition of applied flux vector Σ_0 and uniform temperature change θ. Then, the total flux vector Σ_f in the filler domain is

$$\Sigma_f = B^{\text{dil}} \cdot \Sigma_m + \mathbf{b}^{\text{dil}}\theta, \tag{4.119}$$

where $\boldsymbol{B}^{\mathrm{dil}}$ and $\mathbf{b}^{\mathrm{dil}}$ are given by Eqs. (4.111a) and (4.111b), respectively. From the Mori–Tanaka theory, $\boldsymbol{\Sigma}_{\mathrm{f}}$ is given by

$$\boldsymbol{\Sigma}_{\mathrm{f}} = \boldsymbol{B}^{\mathrm{MT}} \cdot \boldsymbol{\Sigma}_0 + \mathbf{b}^{\mathrm{MT}}\theta. \tag{4.120}$$

From Eqs. (4.17), (4.23), and (4.119)

$$\begin{aligned}
\boldsymbol{\Sigma}_0 &= (1-f)\boldsymbol{\Sigma}_{\mathrm{m}} + f\,\boldsymbol{\Sigma}_{\mathrm{f}} = (1-f)\boldsymbol{\Sigma}_{\mathrm{m}} + f\left(\boldsymbol{B}^{\mathrm{dil}} \cdot \boldsymbol{\Sigma}_{\mathrm{m}} + \mathbf{b}^{\mathrm{dil}}\theta\right) \\
&= \left[(1-f)\boldsymbol{I} + f\boldsymbol{B}^{\mathrm{dil}}\right] \cdot \boldsymbol{\Sigma}_{\mathrm{m}} + f\mathbf{b}^{\mathrm{dil}}\theta.
\end{aligned} \tag{4.121}$$

Eliminating $\boldsymbol{\Sigma}_{\mathrm{m}}$ from Eqs. (4.119) and (4.121),

$$\boldsymbol{\Sigma}_0 = \left[(1-f)\boldsymbol{B}^{\mathrm{dil}^{-1}} + f\boldsymbol{I}\right] \cdot \boldsymbol{\Sigma}_{\mathrm{f}} - (1-f)\boldsymbol{B}^{\mathrm{dil}^{-1}} \cdot \mathbf{b}^{\mathrm{dil}}\theta. \tag{4.122}$$

Solving for $\boldsymbol{\Sigma}_{\mathrm{f}}$ in Eq. (4.122),

$$\begin{aligned}
\boldsymbol{\Sigma}_{\mathrm{f}} &= \left[(1-f)\boldsymbol{B}^{\mathrm{dil}^{-1}} + f\boldsymbol{I}\right]^{-1} \cdot \boldsymbol{\Sigma}_0 \\
&+ (1-f)\left[(1-f)\boldsymbol{B}^{\mathrm{dil}^{-1}} + f\boldsymbol{I}\right]^{-1} \cdot \boldsymbol{B}^{\mathrm{dil}^{-1}} \cdot \mathbf{b}^{\mathrm{dil}}\theta.
\end{aligned} \tag{4.123}$$

A comparison between Eq. (4.120) and (4.123) provides $\boldsymbol{B}^{\mathrm{MT}}$ and $\mathbf{b}^{\mathrm{MT}}$

$$\begin{aligned}
\boldsymbol{B}^{\mathrm{MT}} &= \left[(1-f)\boldsymbol{B}^{\mathrm{dil}^{-1}} + f\boldsymbol{I}\right]^{-1} \\
&= \left[(1-f)\boldsymbol{I} + f\boldsymbol{B}^{\mathrm{dil}}\right]^{-1} \cdot \boldsymbol{B}^{\mathrm{dil}}
\end{aligned} \tag{4.124a}$$

$$\mathbf{b}^{\mathrm{MT}} = (1-f)\left[(1-f)\boldsymbol{I} + f\boldsymbol{B}^{\mathrm{dil}}\right]^{-1} \cdot \mathbf{b}^{\mathrm{dil}}. \tag{4.124b}$$

Once the above concentration factor tensors, A, B, and vectors $\mathbf{a}$, $\mathbf{b}$ are inserted into Eqs. (4.22), (4.25) for two-phase composites and into Eq. (4.26) for hybrid composites, one can obtain the property tensor $\boldsymbol{R}_{\mathrm{c}}$ and compliance tensor $\boldsymbol{F}_{\mathrm{c}}$ of a composite.

4.4.4 Eshelby model for a magnetic composite

We discussed briefly the concept of a magnetic composite within the framework of a demagnetization field in Subsection 1.2.3. Here we shall discuss the modeling of a magnetic composite within the framework of the Eshelby approach. First, consider a magnetic body of ellipsoidal shape (Ω) with a uniform magnetization vector ($\mathbf{M}$) in vacuum or air. Then the magnetic flux intensity $\mathbf{B}$ induced by this magnetization vector $\mathbf{M}$ is given by

$$\mathbf{B} = \mu_0 \left(\mathbf{M} - N_{\mathrm{d}}\mathbf{M}\right), \tag{4.125}$$

where N_d is the demagnetization factor defined by Eq. (1.36). Now, define the magnetization vector as

$$\mathbf{M} = -\mathbf{H}^*, \tag{4.126}$$

where $\mathbf{H}^*$ is the eigen-magnetic field prescribed in the domain of a magnetic body. Equation (4.125) can then be rewritten as

$$\mathbf{B} = \mu_0 \, (\mathbf{H}_d - \mathbf{H}^*). \tag{4.127}$$

Here, $\mathbf{H}_d$ is the disturbance magnetic field induced by a given applied $\mathbf{M}$(or $-\mathbf{H}^*$) in Ω, given by:

$$\mathbf{H}_d = N_d \mathbf{H}^*. \tag{4.128}$$

The above problem of a magnetic body of elliptical shape stated by Eq. (4.127) is identical to the inclusion case of the Eshelby model for an uncoupled problem, Subsection 4.3.5, if the demagnetization factor N_d is replaced by Eshelby tensor S, i.e.,

$$\mathbf{H}_d = S \cdot \mathbf{H}^*. \tag{4.129}$$

Next we consider a magnetic composite subjected to applied magnetic field $\mathbf{H}_0$ where magnetic fillers, with magnetic permeability μ_f, of ellipsoidal shape are embedded in a matrix with magnetic permeability μ_m as depicted in Fig. 1.8(a). Before introducing any magnetic fillers in the composite of Fig. 1.8(a), a uniform magnetic flux density and field existed in the homogeneous matrix, given by:

$$\mathbf{B}_0 = \mu_m \cdot \mathbf{H}_0. \tag{4.130}$$

Upon inserting the magnetic fillers, the magnetic flux density and field relations in the domain Ω take the following form:

$$\mathbf{B}_0 + \mathbf{B}_d = \mu_f \cdot (\mathbf{H}_0 + \mathbf{H}_d). \tag{4.131}$$

Let us now introduce an unknown fictitious eigen-magnetic field $\mathbf{H}^*$ in Ω (Fig. 1.8(b)) such that the following equation is valid:

$$\begin{aligned} \mathbf{B}_0 + \mathbf{B}_d &= \mu_f \cdot (\mathbf{H}_0 + \mathbf{H}_d) \\ &= \mu_m \cdot (\mathbf{H}_0 + \mathbf{H}_d - \mathbf{H}^*). \end{aligned} \tag{4.132}$$

Subtracting Eq. (4.130) from both sides of the second equation of Eq. (4.132), we get

$$\mathbf{B}_d = \mu_m \cdot (\mathbf{H}_d - \mathbf{H}^*). \tag{4.133}$$

Equation (4.133) has a form identical to Eq. (4.127) except that the magnetic permeability μ_0 is replaced by μ_m. The fields $\mathbf{H}_d$ and $\mathbf{H}^*$ in Eqs. (4.132) and (4.133) are related by Eq. (4.129) through the Eshelby tensor S. From Eqs. (4.129) and (4.132), the unknown $\mathbf{H}^*$ is solved as

$$\mathbf{H}^* = [(\mu_f - \mu_m) \cdot S + \mu_m]^{-1} \cdot (\mu_m - \mu_f) \cdot \mathbf{H}_0. \qquad (4.134)$$

It is to be noted that Eq. (4.134) has the same form as Eq. (4.45).

The computation of concentration factor tensors A and B for the dilute case of magnetic composites is the same as that in Subsection 4.4.2, and that for finite volume fractions of magnetic fillers is the same as that in Subsection 4.4.3, except that R_i is replaced by μ_i. Then, using Eqs. (4.22) and (4.25), one can compute the magnetic permeability (μ_c) of the composite.

5

Resistor network model for electrical and thermal conduction problems

The resistor network model has been applied to both electrical and thermal conduction problems for regular lattice structures, using Kirchhoff's current law and an iteration method (Kirkpatrick, 1973; Onizuka, 1975; Yuge, 1977). Based on the concept used in regular lattices, the resistor network model has been extended to a continuum system (Balberg, 1987; Balberg *et al.*, 1990). In a continuum system, analytical modeling becomes complicated since the microstructure of a composite is not a regular lattice, yielding differing resistances between filler sites, and the number of electrical (or thermal) connections between conducting filler sites is not constant.

In the following sections we describe resistor network models for both electrical and thermal conduction problems, followed by a discussion and comparison between them.

5.1 Electrical conduction

It is assumed that the ith filler site has an equipotential with voltage V_i. Considering an electrode as a filler, the top electrode has a voltage V_1 and the bottom electrode a voltage V_N. Figure 5.1 illustrates a random network of resistors (the lines) between conducting filler sites in a continuum system. If g_{ij} is the local conductance between the ith and jth filler sites, the current I_i which enters the ith filler site is the sum of currents arriving from (or leaving for) other filler sites, i.e.,

$$I_i = \sum_j^N g_{ij}(V_i - V_j), \tag{5.1}$$

where N is the number of filler sites in a composite. In the case of regular lattice structures, N is reduced to the coordination number of the lattice, which is the

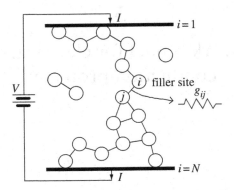

Fig. 5.1   Random network of resistors in a continuum system; g_{ij} is the local conductance of the resistor between the ith and jth filler sites (Kim, 1998).

maximum number of nearest neighboring sites (bonds). For example, N is six for a simple cubic lattice. In a continuum system, however, all filler sites must be checked to see whether they have electrical connection with each other through overlapping (penetrable conduction) or lie within tunneling distance (non-penetrable conduction, see Chapter 6). Since Eq. (5.1) is a system of linear equations, it can be rewritten as a matrix form:

$$\mathbf{I} = \boldsymbol{G} \cdot \mathbf{V}, \tag{5.2}$$

where the bold roman character represents a vector and the bold italic character a tensor. From the condition that the sum of all currents into the ith filler site vanish (Kirchhoff's current law), the current vector $\mathbf{I}$ of Eq. (5.2) is given by

$$I_i = 0, \text{ except for } i = 1 \text{ and } N, \quad \text{where } I_1 = I \text{ and } I_N = -I. \tag{5.3}$$

The conductance matrix $\boldsymbol{G}$ is a large, sparse, symmetric $N \times N$ matrix defined by

$$G_{ii} = \sum_{j=1(j \neq i)}^{N} g_{ij} \text{ for diagonal elements, otherwise } G_{ij} = G_{ji}. \tag{5.4}$$

Equation (5.2) is then described in an expanded form,

$$
\begin{Bmatrix} I \\ 0 \\ \cdot \\ \cdot \\ 0 \\ -I \end{Bmatrix}
=
\begin{bmatrix}
G_{11} & G_{12} & G_{13} & \cdot & \cdot & G_{1N} \\
 & G_{22} & G_{23} & \cdot & \cdot & G_{2N} \\
 & & G_{33} & \cdot & \cdot & \cdot \\
 & & & \cdot & \cdot & \\
 & & & & G_{(N-1)(N-1)} & G_{(N-1)N} \\
 & & & & & G_{NN}
\end{bmatrix}
\begin{Bmatrix} V_1 \\ V_2 \\ \cdot \\ \cdot \\ V_{N-1} \\ V_N \end{Bmatrix}. \tag{5.5}
$$

If the voltage difference $(V_1 - V_N)$ between the top and bottom electrodes is known, the conductance matrix G in Eq. (5.5) is reduced to an $(N-2) \times (N-2)$ matrix. The voltage V_i of each filler site is obtained by computing the inverse matrix of G, using a conjugate-gradient iteration algorithm. Finally the current I of a composite system is computed from Eq. (5.5), and the effective conductance G_c is evaluated, following Ohm's law, i.e., $I = G_c(V_1 - V_N)$.

5.2 Thermal conduction

For the thermal conduction problem, the thermal conductivity K is defined by Fourier's law as

$$Q = KA\frac{\Delta T}{\Delta x}, \tag{5.6}$$

where Q is a heat flow rate, A is the cross-sectional area through which heat is being conducted, and $\Delta T/\Delta x$ is a temperature gradient. From Eqs. (3.51) and (5.6), thermal resistance R is defined by

$$R = \frac{\Delta x}{KA}. \tag{5.7}$$

The computational procedure for the thermal conduction problem is the same as that for the electrical conduction problem. However, unlike the electrical conduction problem, the thermal conductivity ratio of filler to matrix is not very high and percolation threshold phenomena (see Section 6.4) may not take place. Therefore, it is assumed that heat flow is conducted through interfacial contact between fillers, or filler and matrix, or matrix sites, not through the tunneling mechanism mentioned earlier.

Let us look at the simple 2D problem of Fig. 5.2 which is composed of six elements including the top electrode (element 1) and bottom electrode (element 6), whose temperatures are set equal to 1 [°C or K] and 0 [°C or K], respectively. The middle layer is composed of gray material (elements 2–4), with thermal conductivity, $K_m = 0.2$ W/m·K, and white material (element 5), with $K_f = 1.0$ W/m·K. The cross-sectional area (seen as the interface line between adjacent elements) is set equal to 1.0 m². The length ΔX of each square element is set equal to 1.0 m.

The six-element model of Fig. 5.2(a) is converted to the eight-resistor model of Fig. 5.2(b). In order to assign the equivalent thermal resistance R, we use Eq. (5.7). The calculation procedure for the resistor element between two adjacent elements depends on connectivity. Figure 5.3 illustrates the method of calculation of all possible connectivities. Based on the results of thermal

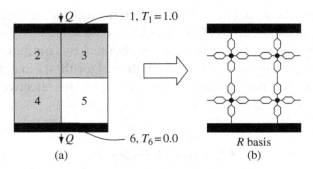

Fig. 5.2   A 2D resistor model: (a) six elements, (b) converted to resistor model.

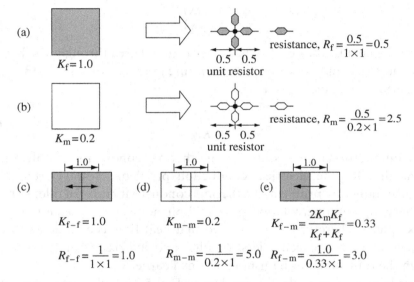

Fig. 5.3   Thermal resistance R for five different cases of connectivity: (a) between elements 1 and 2 in Fig. 5.2 with $\Delta x = 0.5$, (b) between elements 5 and 6 with $\Delta x = 0.5$, (c) between elements 2 and 3, also 2 and 4, (d) between adjacent white elements (no corresponding case in the example of Fig. 5.2), and (e) between elements 3 and 5, also 4 and 5.

resistance for all resistor elements, the resistor network equivalent to Fig. 5.2 is shown in Fig. 5.4.

If all the elements of Fig. 5.2 are considered as a composite whose thermal conductivity is K_c (or its thermal resistance R_c) then the following equations are valid:

$$\frac{Q}{A} = K_c \frac{\Delta T}{\Delta X}, \tag{5.8}$$

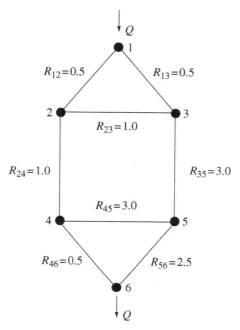

Fig. 5.4 The thermal resistor model equivalent to the problem of Fig. 5.2.

$$Q = K_c \Delta T / \Delta X = (T_1 - T_6)/R_c. \qquad (5.9)$$

Applying Kirchhoff's junction rule $\left(\sum_i Q = 0 \right)$ to the resistor model of Fig. 5.4:

Node 1: $Q + (T_2 - T_1)/0.5 + (T_3 - T_1)/0.5 = 0$ (5.10a)

Node 2: $(T_1 - T_2)/0.5 + (T_3 - T_2)/1.0 + (T_4 - T_2)/1.0 = 0$ (5.10b)

Node 3: $(T_1 - T_3)/0.5 + (T_2 - T_3)/1.0 + (T_5 - T_3)/3.0 = 0$ (5.10c)

Node 4: $(T_2 - T_4)/1.0 + (T_5 - T_4)/3.0 + (T_6 - T_4)/0.5 = 0$ (5.10d)

Node 5: $(T_3 - T_5)/3.0 + (T_4 - T_5)/3.0 + (T_6 - T_5)/2.5 = 0$ (5.10e)

Node 6: $(T_4 - T_6)/0.5 + (T_5 - T_6)/2.5 - Q = 0.$ (5.10f)

Equations (5.10) are rewritten in the form of a symmetric matrix as

$$
\begin{Bmatrix} Q \\ 0 \\ 0 \\ 0 \\ 0 \\ Q \end{Bmatrix} = \begin{Bmatrix} 4 & -2 & -2 & 0 & 0 & 0 \\ -2 & 4 & -1 & -1 & 0 & 0 \\ -2 & -1 & 3.33 & 0 & -0.33 & 0 \\ 0 & -1 & 0 & 3.33 & -0.33 & -2 \\ 0 & 0 & -0.33 & -0.33 & 1.06 & -0.4 \\ 0 & 0 & 0 & -2.0 & -0.4 & 2.4 \end{Bmatrix} \cdot \begin{Bmatrix} T_1 \\ T_2 \\ T_3 \\ T_4 \\ T_5 \\ T_6 \end{Bmatrix}. \tag{5.11}
$$

Since the top (T_1) and bottom temperature (T_6) are set equal to 1.0 and 0, respectively, Eq. (5.11) is reduced to

$$
\begin{Bmatrix} 2 \\ 2 \\ 0 \\ 0 \end{Bmatrix} = \begin{Bmatrix} 4 & -1 & -1 & 0 \\ -1 & 3.33 & 0 & -0.33 \\ -1 & 0 & 3.33 & -0.33 \\ 0 & -0.33 & -0.33 & 1.06 \end{Bmatrix} \cdot \begin{Bmatrix} T_2 \\ T_3 \\ T_4 \\ T_5 \end{Bmatrix}. \tag{5.12}
$$

Taking the inverse of Eq. (5.12), unknown temperatures (T_2, T_3, T_4 and T_5) are obtained as

$$
\begin{Bmatrix} T_2 \\ T_3 \\ T_4 \\ T_5 \end{Bmatrix} = \begin{Bmatrix} 0.7858 \\ 0.8718 \\ 0.2712 \\ 0.3559 \end{Bmatrix}. \tag{5.13}
$$

Then from Eq. (5.10a), Q is solved for as

$$
Q = 4T_1 - 2(T_2 + T_3) = 4 - 2(0.7858 + 0.8718) = 0.6848. \tag{5.14}
$$

Finally, the overall thermal resistance R of this system is solved for from Eq. (5.9):

$$
R_c = (T_1 - T_6)/Q = 1/0.6848 = 1.46. \tag{5.15}
$$

Next we shall look at the case of a 3D thermal resistor network model. In this study, the electronic composite is assumed to be a simple cubic resistor network, in which a site is considered as a unit cube and two kinds of unit cube (gray and white) are present at random, as depicted in Fig. 5.5. A gray cube represents a conductive filler phase with thermal conductivity K_f and a white cube an insulating or less conductive matrix phase with thermal conductivity K_m. If the two kinds of unit cube are assembled into an $n \times n \times n$ cube and N is the number of conductive (gray) cubes in the system, the volume fraction f of the conductive filler is defined by $f = N/n^3$. In this system there exist three kinds of connecting elements between sites, shown in Fig. 5.6: (a) conductive element with K_f between conductive filler sites, (b) less-conductive element with

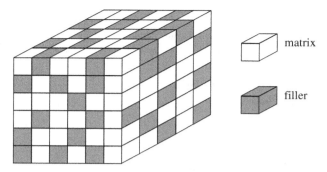

Fig. 5.5   3D resistor network model, where both gray (conductive filler) and white (insulating or less-conductive matrix) unit cubes are piled up at random into an $n \times n \times n$ cube.

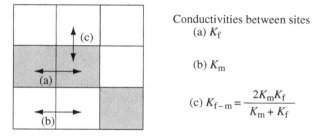

Fig. 5.6   Thermal conductivity between two adjacent sites.

K_m between matrix sites, and (c) conductive element with $2K_m K_f/(K_m + K_f)$ between conductive filler and matrix sites. Then, the temperature T_i of the ith element and the corresponding heat flow rate Q_i are computed by applying a resistor network model in terms of Kirchhoff's law:

$$Q_i = \sum_{j=1}^{N} \frac{K_{ij}A}{\Delta x}(T_i - T_j). \tag{5.16}$$

where A and Δx are the cross-sectional area and the length of a unit cube, respectively.

5.3 Comparison of model predictions with experimental data

5.3.1 Different models

Figure 5.7 shows calculated numerical results of the effective thermal conductivity of a 3D silver spherical particle/polymer matrix composite as a function of silver volume fraction, where the thermal conductivity of silver is

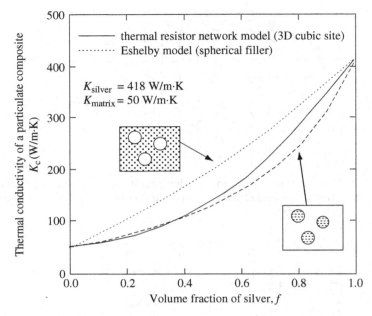

Fig. 5.7   Thermal conductivity of a 3D spherical particulate/polymer matrix composite as a function of silver volume fraction; white portion represents polymer.

418 W/m·K and that of the polymer is 50 W/m·K. The solid line represents the numerical results predicted by the 3D resistor network model discussed in the previous section while the broken lines are those predicted by the Eshelby method (Hatta and Taya, 1986), describing upper and lower bounds of the effective thermal conductivity. The upper (dotted) line was obtained by assuming that the matrix material (white) is embedded in a silver matrix while, for the lower (dashed) line, silver particles are embedded in the matrix material. It is known that in the case of a spherical particulate composite, the Eshelby method yields the same results as the Hashin–Shtrikman (1962) bound solutions. It should be observed from Fig. 5.7 that the results from the resistor network model yield a small deviation from the lower bound of the Eshelby method. Nevertheless, the results from the resistor network model agree well with those of the lower bound from the Eshelby method up to $f = 0.4$, even though in the resistor network model the spherical filler is simulated as a unit cube.

5.3.2 Comparison with experiment

Kim (1998) applied a combination of a resistor network model and a percolation model (see Chapter 6) to predict the electrical and thermal

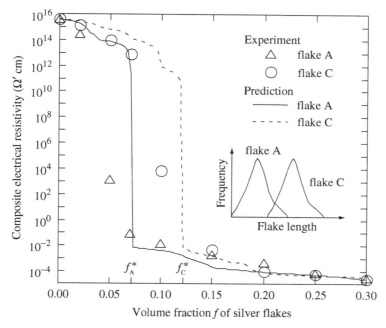

Fig. 5.8   Electrical resistivity vs. volume fraction of silver flakes. The threshold values are shown as f^* (after Kim, 1998).

conductivities of silver (flakes) filler/polymer matrix composites which are a key interconnect material between a chip and substrate or heatsink in electronic packaging applications. The predicted electrical resistivity of the silver filler/polymer matrix composite is plotted as a function of volume fraction of silver flake in Fig. 5.8, where the solid line denotes the predictions based on silver flake size distributions type A (more skewed toward smaller sizes), the dashed line denotes type C (toward larger sizes); the experimental measurements are shown as symbols. Figure 5.8 indicates a reasonably good agreement between the predictions by the percolation and resistor network models and experiment. Figure 5.9 shows results of the thermal conductivity of the same silver/polymer matrix composites. The predictions from the 3D resistor network model agree well with experiment, but those from the effective medium theory (represented by the Eshelby model) underestimate the experimental results. The reason for the underestimate by the Eshelby model (Eshelby, 1957; Hatta and Taya, 1985) is that there is a large differential of the thermal conductivities between silver and polymer matrix, and also the interconnected network structure of the silver flakes embedded in the polymer matrix makes the resistor network model better suited than the effective medium model, such as the Eshelby model, which assumes that the conductive fillers are not connected to adjacent ones.

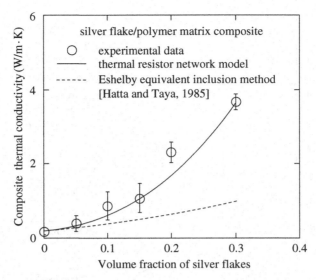

Fig. 5.9   Thermal conductivity vs. volume fraction of silver flakes, where the thermal conductivities of silver flake (K_f) and polymer matrix (K_m) are 418 and 0.19 W/m·K, respectively (Kim, 1998).

It is to be noted in Fig. 5.9 that the predicted thermal conductivity vs. silver volume fraction relation is smooth, i.e., continuous, while there exists a sharp drop in the electrical resistivity at a particular volume fraction of silver flakes, f^*, in Fig. 5.8, which is called the "threshold" volume fraction, at and above which the overall electrical behavior of the composite becomes that of a "conductor." This threshold volume fraction of conductor–insulator composites is discussed in detail in Chapter 6.

6

Percolation model

6.1 Percolation model based on base lattices

A percolation model was proposed first by Broadbent and Hammersley (1957). Since then, there have been a large number of studies of percolation problems, and two books (Stauffer and Aharony, 1991; Sahini, 1994). A percolation model (Shante and Kirkpatrick, 1971; Essam, 1972; 1980) is composed of two simple elements, a *site* and a *bond*, which are interconnected. In most cases there is a base lattice underlying sites and bonds. The key variable in percolation problems is the *probability p* of occupied bonds or sites. The probability that each bond or site is occupied or empty is entirely independent of whether its neighbors are occupied or empty. When $p = 1$, the model becomes exactly the same as the base lattice and none of the elements is empty.

There are two types of percolation problem. If the occupancy of bonds is governed by a stochastic mechanism, the problem is called a *bond percolation problem*, whereas if that of the sites is so governed, the problem is called a *site percolation problem*. Figures 6.1, 6.2, and 6.3 show examples of these problems. A two-dimensional (2D) square lattice is taken as a base lattice, as shown in Fig. 6.1. Figure 6.2 shows a bond percolation problem of this 2D base lattice, where the probability of bonds $p^b = 0.5$. Figure 6.3 show a site percolation problem of the same base lattice, where the probability of sites $p^s = 0.5$. The superscripts b and s in the above equations denote bond and site, respectively.

A *path* between two sites is said to exist if the two sites are connected by a sequence of sites and bonds, and the sites are said to be connected if there exists at least one path between them. A set of occupied sites connected to one another through occupied bonds is called a *cluster*. The flow of fluid or charge may occur between any two sites in the same cluster, but no flow occurs between sites in different clusters unless the clusters are connected. As the

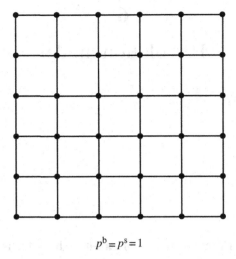

$$p^b = p^s = 1$$

Fig. 6.1   An example of a base lattice. A square lattice is taken as a base
lattice in this case.

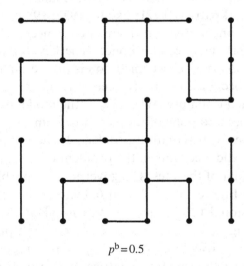

$$p^b = 0.5$$

Fig. 6.2   An example of a bond percolation problem. The base lattice is the
square lattice shown in Fig. 6.1.

probability p of occupied bonds or sites becomes larger, so do those of the
clusters. Most systems to which a percolation theory is applied contain so
many sites and bonds that the effect of the boundary (or free surface) can be
ignored and then the system is replaced by a model with an infinite number of
bonds and sites.

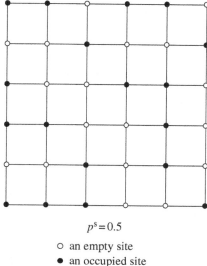

$$p^s = 0.5$$

○ an empty site
● an occupied site

Fig. 6.3   An example of a site percolation problem. The base lattice is the
square lattice shown in Fig. 6.1.

In such a system the cluster size may become infinite at some critical density
of occupancies, i.e., the critical value of p^b or p^s. At this stage the system is said
to be in a percolating state. The probability that a given site belongs to
this infinite cluster is defined as the *percolation probability* for that site. The
percolation probability is the probability that the fluid flowing through a given
site will percolate away indefinitely. The transition from a non-percolating
state to a percolating state is considered to cause various kinds of phase
transition in many applications. The percolation transition is a purely geometric phenomenon. Figure 6.2 shows a percolating cluster from the bottom to
the top, while the system in Fig. 6.3 does not have a percolating cluster. If an
electrical charge is applied to the system in Fig. 6.2, an electric current will flow
from the bottom to the top, but no current will be established in the system of
Fig. 6.3. If $p^b = 0.5$ is the critical value (which is the case with a square base
lattice), a system with $p^b < 0.5$ never has a percolating cluster. This leads to a
phase transition of electrical conduction.

   A percolation theory can be applied to a number of fields. The electrical
conduction in a mixture of conducting and insulating phases is an example of
such an application where the conductivity of the mixture exhibits a critical
behavior. A dilute ferromagnet is another example where spontaneous magnetization will occur above a critical concentration (Elliot *et al.*, 1960). Other
examples include the gelation of polymers (Flory, 1953) and the spread of

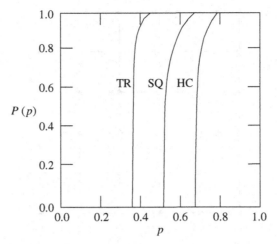

Fig. 6.4   Percolation probabilities as functions of p, where TR, SQ, and HC denote triangular, square, and honeycomb lattices, respectively (Shante and Kirkpatrick, 1971).

disease in an orchard (Broadbent and Hammerseley, 1957), where the critical probabilities mark the gel point and the point at which the disease becomes an epidemic, respectively.

6.1.1 Critical probability

In the preceding introduction the important concept of percolation theory was described. In this subsection we will define it more mathematically. To do so, one needs a concept of self-avoiding walks. A self-avoiding walk on a lattice is a random walk which visits no site more than once. In a percolation theory itself the self-avoiding walks are important because they represent the most economical way by which the fluid can flow through the medium. First, we define the percolation probability $P(p)$ by

$$P(p) = \lim_{n \to \infty} P(p, n), \qquad (6.1)$$

where $P(p, n)$ is the probability that at least one n-step self-avoiding walk exists. $P(p)$ is a monotonically increasing function of p as shown in Figure 6.4. Second, we define the critical occupancy probability p_c by

$$p_c = \sup\{p | P(p) = 0\}. \qquad (6.2)$$

It has been shown (Hammersley, 1961) that for any given medium

$$P^s(p, n) \le P^b(p, n) \qquad (6.3)$$

Table 6.1 *Critical probabilities* $(p_c^b = bond, p_c^s = site)$ *for common lattices.*
(Frisch *et al.*, 1961; Vyssotsky *et al.*, 1961; Dean, 1963; Sykes and Essam, 1964.)

Lattice	Dimensionality d	Coordination number z	Monte Carlo		Series expansion	
			p_c^s	p_c^b	p_c^s	p_c^b
honeycomb	2	3	0.668	0.640	0.698	0.6527*
square	2	4	0.581	0.498	0.593	0.5000*
Kagomé	2	4	0.655	0.435	0.6527*	–
triangular	2	6	0.493	0.349	0.5000*	0.3473*
diamond	3	4	0.436	0.390	0.428	0.388
simple cubic	3	6	0.325	0.254	0.310	0.247
body-centered cubic	3	8	–	–	0.245	0.178
face-centered cubic	3	12	0.199	0.125	0.198	0.119

Note: Results marked by * are exact results.

where the superscripts s and b denote site and bond properties, respectively. Strict inequality holds in Formula (6.3) for many cases. Equality holds if and only if $p = 0$ or 1, and if the lattice is a Bethe lattice which has a tree-like structure and no closed loops. As a consequence of Formula (6.3) we obtain

$$P^s(p) \leq P^b(p). \tag{6.4}$$

Since $P(p)$ is a monotonically increasing function of p, we have

$$P_c^s \geq p_c^b. \tag{6.5}$$

This is the reason why, with the same probability $p = 0.5$, Fig. 6.3 has no percolating cluster, whereas Fig. 6.2 has a percolating cluster. Since the critical probability is a key property in a percolation model, many studies have been reported on the critical probabilities for various types of base lattice, which are shown in Table 6.1. Although most of them are numerically obtained, there are a few exact solutions. Sykes and Essam (1964) evaluated exactly the critical probability p_c for the triangular site problem, as well as for square, triangular, and honeycomb bond problems, using the topological properties of the lattices. According to them,

$$p_c^s(\mathrm{TR}) = \frac{1}{2}, \tag{6.6}$$

$$p_c^b(\mathrm{SQ}) = \frac{1}{2}, \tag{6.7}$$

$$p_c^b(\text{TR}) = 2 \sin\left(\frac{\pi}{8}\right), \tag{6.8}$$

$$p_c^b(\text{HC}) = 1 - 2 \sin\left(\frac{\pi}{8}\right), \tag{6.9}$$

where TR, SQ, and HC denote triangular, square and honeycomb lattice, respectively.

Since the Kagomé lattice (see Fig. 6.5) is obtained as the result of bond–site topological transformation, p_c for the Kagomé site problem is also an exact result. There are two numerical methods: a Monte Carlo method, and a series expansion. We will give a brief description of the Monte Carlo method below.

Let us consider a bond problem. The argument given here is equally applicable to a site problem. Rewriting Eq. (6.1) for the bond problem, we obtain

$$\lim_{n\to\infty} P^b(p,n) = P^b(p), \tag{6.10}$$

where $P^b(p, n)$ is the probability that, starting from a given source site, at least n other sites can be reached by walking along occupied bonds, and $P^b(p)$ is the bond percolation probability. $P^b(p, n)$ can be expressed as

$$P^b(p,n) = 1 - \sum_{k=1}^{n} \langle\, n_k^0 \rangle \tag{6.11}$$

where $\langle\ \rangle$ denotes a mean value and n_k^0 is the number of clusters of k sites containing the source sites. Here, n_k^0 takes a value of either 1 or 0 determined by the existence of that cluster for each sample. In the Monte Carlo experiment (Essam, 1972; Frisch *et al.*, 1961; Vyssosky *et al.*, 1961; Hammersley and

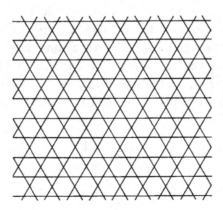

Fig. 6.5   Kagomé lattice.

Handscomb, 1964) the decision as to whether a bond is occupied or empty is made by using random numbers. The usual algorithm is to start from the source atom and keep looking for occupied bonds leading from sites already visited until n other sites that have been visited are found to be empty. After a number of trials, the fraction of trials in which n other sites were visited is equal to $P^b(p, n)$. Then $p_c^b(n)$ is calculated based on the following definition analogous to Eq. (6.2):

$$p_c^b(n) = \sup\{ p | P^b(p, n) = 0 \}. \tag{6.12}$$

Now p_c^b can be determined from a $P^b(p, n)$ vs. n curve as $n \to \infty$ (Vyssotsky *et al.*, 1961). Hammersley and Handscomb (1964) have estimated that for $n = 6000$ on a simple cubic lattice the storage required for a computer would be 36 kilobytes and that a high-speed computer in 1964 would have taken a week to obtain one point on the curve. Although the Monte Carlo experiment takes much CPU time, it is an attractive method today due to the tremendous computation speed of present-day computers. Certain features of the critical probabilities, especially of the critical bond probabilities p_c^b, have been observed to depend only on the dimensionality of the lattice and not on the specific lattice type (Vyssotsky *et al.*, 1961). As shown in Table 6.2, it was suggested that

$$p_c^b \approx \frac{2}{z}, \qquad \text{for 2D}, \tag{6.13}$$

$$p_c^b \approx \frac{1.5}{z}, \qquad \text{for 3D}, \tag{6.14}$$

Table 6.2 *Critical bond probabilities p_c (from series expansion) and dimensional invariants.* (Sources as for Table 6.1.)

Lattice	Dimensionality d	Coordination number z	p_c^b	$z p_c^b$
honeycomb	2	3	0.6527[*]	1.96
square	2	4	0.5000[*]	2.00
triangular	2	6	0.3473[*]	2.08
diamond	3	4	0.388	1.55
simple cubic	3	6	0.247	1.48
body-centered cubic	3	8	0.178	1.42
face-centered cubic	3	12	0.119	1.43

Note: Those marked by [*] are exact results.

where z is the coordination number of the lattice, i.e., the number of bonds emanating from each site. In Table 6.2, p_c^b are given by exact and by series expansion methods. Equations (6.13) and (6.14) can be represented as

$$zp_c^b \approx \frac{d}{(d-1)},$$
(6.15)

where d is the dimensionality of the lattice.

6.2 Electrical conductivity by effective medium theory

Electrical conduction in an inhomogeneous medium has been much discussed since the nineteenth century, though many problems still remain to be studied. Before percolation theory was proposed, the problem of electrical conduction in an inhomogeneous medium was solved by the effective medium theory (EMT) (Bruggeman, 1935; Landauer, 1952; Kerner, 1956).The basic assumption of EMT is the requirement of self-consistency.

Let us take Landauer's theory (Landauer, 1952) as a simple example. Suppose there is a two-component medium composed of materials 1 and 2. The shape of each element is assumed to be spherical. The main assumption in Landauer's theory is that the medium surrounding an element is considered as a homogeneous medium, which has the conductivity σ_m that characterizes the overall properties of the mixture. Then, self-consistency requires that the fluctuations of a local field about the effective value of the mixture should be averaged out. Thus, the problem is reduced to a single inclusion problem in a homogeneous medium. Let the radius of the inclusion be a, and its conductivity be σ_1 (the conductivity of material 1). From elementary electrostatics, a dipole moment associated with the volume under consideration becomes

$$\mathbf{P} = a^3 \frac{\sigma_1 - \sigma_m}{\sigma_1 + 2\sigma_m} \mathbf{E},$$
(6.16)

where bold characters stand for vector quantities, $\mathbf{P}$ is the dipole moment, and $\mathbf{E}$ is the electric field far from the inclusion. If the volume fraction of material 1 is denoted by f_1, then the polarization due to the existence of material 1 becomes

$$\mathbf{P}_1 = f_1 \frac{\sigma_1 - \sigma_m}{\sigma_1 + 2\sigma_m} \mathbf{E}.$$
(6.17)

Similarily, the existence of material 2 produces a polarization

$$\mathbf{P}_2 = f_2 \frac{\sigma_2 - \sigma_m}{\sigma_2 + 2\sigma_m} \mathbf{E}.$$
(6.18)

These polarizations produce a disturbance field from the original uniform field **E**. The space integral of the field contributed by each dipole is $-4\pi\mathbf{P}$ (Landauer, 1952). Thus, from the self-consistency requirement, the average disturbance from **E** should vanish, leading to

$$f_1 \frac{\sigma_1 - \sigma_m}{\sigma_1 + 2\sigma_m} + f_2 \frac{\sigma_2 - \sigma_m}{\sigma_2 + 2\sigma_m} = 0. \tag{6.19}$$

Equation (6.19) was derived by Bruggeman (1935). Solving for σ_m, we obtain

$$\sigma_m = \frac{1}{4}\left[\eta + (\eta^2 + 8\sigma_1\sigma_2)^{1/2}\right], \tag{6.20}$$

where

$$\eta = (3f_2 - 1)\,\sigma_2 + (3f_1 - 1)\,\sigma_1. \tag{6.21}$$

When material 2 is perfectly insulating ($\sigma_2 = 0$), Eq. (6.20) yields

$$\sigma_m = \frac{1}{2}(3f_1 - 1)\sigma_1. \tag{6.22}$$

The results of Eq. (6.22) are plotted in Fig. 6.6, where $f_1 = 0.33$ corresponds to the percolation threshold volume fraction f^*. On the other hand, a percolation theory gives $f_c = 0.145$ (Webman *et al.*, 1976) which is less than f_c based on EMT. There is some experimental evidence (Webman *et al.*, 1977) to support such a small value of f_c, i.e., $f_c = 0.15$–0.17, as a percolation threshold.

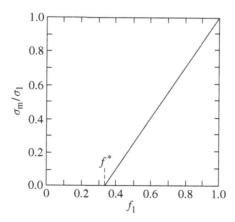

Fig. 6.6   Normalized electrical conductivity of a conductor–insulator mixture as a function of the volume fraction f_1 of the conductor as predicted by EMT.

6.3 Electrical conduction by percolation model

6.3.1 Special conduction law

Let us consider a simple example of percolation conduction. Imagine that numerous small solid circles are distributed in a two-dimensional square lattice. Each circle represents either a conductive particle (with probability p) or an insulating particle (with probability $1-p$). An electric charge flows from a conductive particle to its neighboring conductive particles only if they contact each other. Then an electric current flows through the whole lattice only if an infinite network of conductive particles is established, i.e., only for $p \geq p_c$, the critical probability. For mathematical simplicity, let us set the conductivity of the conductor equal to unity and that of the insulator to zero, and also set the length of the whole lattice equal to unity. A unit magnitude of electric potential difference is applied to two opposing faces (of the unit area) of the square lattice. Then the current flowing through the lattice gives rise to the conductivity of the lattice with $\sigma = \sigma(p)$ (Stauffer, 1979). The percolation properties of the lattice can then be described exactly by the site percolation model of a two-dimensional square lattice.

Kirkpatrick (1971, 1973) evaluated numerically the conductivity of a random resistor network produced by a Monte Carlo procedure, using a relaxation procedure based on the Kirchhoff current law. He reported that the upper conductivity far from percolation threshold was well described by a self-consistent effective medium theory but deviates from the value predicted by a percolation theory near the threshold. He assumed that the conductivity near the threshold can be described by

$$\sigma(p) \sim (p - p_c)^t, \tag{6.23}$$

where t is the critical exponent of conductivity. Such a power-law relationship also appears in various percolation phenomena. For example, the percolation probability $P(p)$, the mean cluster size $S(p)$, and the associated correlation length of the cluster $\xi(p)$ can be written as

$$P(p) \sim (p - p_c)^\beta, \quad p > p_c, \tag{6.24}$$

$$S(p) \sim |p - p_c|^{-\gamma}, \tag{6.25}$$

$$\xi(p) \sim |p - p_c|^{-\nu}, \tag{6.26}$$

where β, γ, and ν are the critical exponents of the corresponding properties (Essam, 1972, 1980; Stauffer, 1979). These power-law representations of percolation properties stem from the analogy with other phase transition phenomena, which are usually described by critical exponents (Stanley, 1971). Furthermore, the correspondence between the percolation and the thermo- dynamic functions of a ferromagnetic material leads us to expect that ideas of scaling and universality which have revolutionized our understanding of critical phenomena will also be applied to a percolation theory (Essam, 1972, 1980; Stauffer, 1979). The basic assumption of the scaling theory is that the critical behavior of percolation near a threshold region stems from the singularity of the typical cluster size which diverges at $p = p_c$ (Stauffer, 1979). According to the scaling theory (Stauffer, 1979), all the critical expo- nents of percolation properties are derived from two basic exponents, which describe the typical cluster size and the mean number of clusters. Consequently, all the critical exponents are related to each other. The current estimates of the critical exponents are $\beta = 5/36, \gamma = 43/18, \nu = 4/3$ for two- dimensional lattices and $\beta = 0.4, \gamma = 1.7$, and $\nu = 0.84$ for three-dimensional lattices (Stauffer, 1981). It should be noted that these exponents depend only on the dimensionality and not on the specific lattice character. On the other hand, the conductivity exponent t has no theoretical relationship with other exponents.

The following approximation (Levinstein *et al.*, 1976) is a rough argument but it may help us to understand the relations given by Formula (6.23). We cut a cube (or a square in the two-dimensional case) with side l of sufficiently large size out of the lattice under consideration. We apply metallic contacts to opposite faces of this cube and measure the conductance G between them. Then, the conductivity of the cube is given by

$$\sigma = \frac{Gl}{l^{d-1}}$$
$$= Gl^{2-d}, \tag{6.27}$$

where d is the dimensionality of the lattice, i.e., $d = 3$ and $d = 2$ for three- dimensional and two-dimensional problems, respectively. For a cube with side ξ, as defined by Formula (6.26), we can, in accordance with Eq. (6.27), express the conductivity of the cube as

$$\sigma = G\xi^{2-d}. \tag{6.28}$$

We should now understand how G depends on ξ. It is the percolating cluster that is responsible for the conduction in the volume of interest. Crudely speaking, we assume that photographs of clusters at different values of $(p - p_c)$ will coincide with each other if they are reduced by a factor of ξ given by Formula (6.26). Actually the percolating clusters have dead ends, crosslinks, etc., to which this scaling assumption does not apply. It seems natural to us, however, that the conductance is determined only by the large-scale properties of the clusters. We then assume that G can be expressed by

$$G = G_0 \, \xi^{-1}, \tag{6.29}$$

where G_0 is the conductivity of a bond (of unit length). Thus, we obtain the following relationships from Eqs. (6.28) and (6.29):

$$\sigma = G_0 \, \xi^{1-d} \tag{6.30}$$

$$\sim |p - p_c|^{-\nu(1-d)} \tag{6.31}$$

Comparing Eq. (6.31) with Formula (6.23), we have

$$t = (d - 1)\nu. \tag{6.32}$$

The power-law relationship of Formula (6.23) is supported by elegant table-top experiments by Last and Thouless (1971), and Watson and Leath (1974). They measured the conductivity of a piece of conducting paper as holes were punched in it at random with the aim of approximating a site percolation system.

Values of t have been estimated (Watson and Leath, 1974; Straley, 1977; Yuge and Onizuka, 1978) $t(2D)$ ranges from 1.1. to 1.38. Watson and Leath (1974) reported $t = 1.38 \pm 0.12$ for the two-dimensional site problem by means of the table-top experiment. Straley (1977) reported $t(2D \text{ bond}) = 1.10 \pm 0.05$, $t(2D \text{ site}) = 1.25 \pm 0.05$ for a 100×100 square lattice, $t(3D \text{ bond}) = 1.70 \pm 0.05$, for a $25 \times 25 \times 25$ simple cubic lattice, and $t(3D \text{ site}) = 1.75 \pm 0.1$ for $30 \times 30 \times 30$ simple cubic lattice. Yuge and Onizuka (1978) reported $t(2D \text{ site}) = 1.26 \pm 0.03$ for a 200×200 square lattice by means of an over-relaxation procedure. Lobb and Frank (1979) studied a two-dimensional random resistor network bond problem by means of a large-cell real-space renormalization group method, and reported that the exponent t was equal to the exponent ν and hence $t(2D \text{ bond}) = 1.35 \pm 0.02$. This result is consistent with Eq. (6.32).

A large-cell real-space renormalization group method was developed by Reynolds *et al.* (1977, 1978, 1980) to provide a method for accurate calculation of percolation properties, especially critical exponents. As mentioned earlier, the base-lattice size should be sufficiently large for calculation of the percolation properties. The main problem of numerical calculations is to treat the case of a large base lattice. A numerical calculation should be done on a large lattice basically by comparing it with the correlation length ξ. The physical meaning of ξ below p_c is given as follows (Deutscher, 1981). After deleting all dead ends (which do not carry a current) from the infinite cluster, one is left with "macro-bonds" connected at "nodes," with a node being the linking point of at least three macro-bonds. In the morphology of this network, macro-bonds form an irregular superlattice, with the "typical" lattice parameter or internode distance being precisely equal to ξ. This concept was already used in the approximate argument of the conduction law (Formula (6.23)). Thus, as p approaches p_c, ξ becomes larger and almost infinite. Therefore, it is impossible to construct the corresponding base lattice. The renormalization group method provides the theoretical relation between critical exponents and scaling factors and thereby can overcome the deficiency of the finite sampling calculation.

6.3.2 General conduction law

In the preceding subsection, the conductive law was obtained for a mixture with a binary conductivity distribution, which is expressed as

$$f(\sigma) = p\delta(\sigma - \sigma_0) + (1 - p)\,\delta(\sigma), \qquad (6.33)$$

where $\delta(\sigma)$ is the Dirac delta function. Eq. (6.33) can be interpreted to imply that a system is composed of elements of conductivity σ_0 and zero.

In more general cases, poorly conductive elements have finite conductances. A percolation theory also provides a conduction law for such a system (Webman *et al.*, 1975; Straley, 1977). When the conductivity distribution is given by

$$f(\sigma) = p\delta(\sigma - \sigma_0) + (1 - p)\,\delta(\sigma - \sigma_1) \qquad (6.34)$$

the macroscopic conductivity can be expressed (Cohen *et al.*, 1978; Straley, 1978) by

$$\sigma_m \sim \begin{cases} \sigma_0(p - p_c)^t, & \text{for } p > p_c, & (6.35) \\ \sigma_1(p_c - p)^{-s}, & \text{for } p < p_c, & (6.36) \\ \sigma_0(\sigma_1/\sigma_0)^u, & \text{for } p \cong p_c, & (6.37) \end{cases}$$

where

$$u = \frac{t}{t+s}. \qquad (6.38)$$

Here $\sigma_1/\sigma_0 \ll 1$ is assumed. Formulae (6.35)–(6.37) are depicted in Figure 6.7. Note that the region for Formula (6.37) is narrow. The latest estimates for these exponents are $s = 1.0$ (Cohen *et al.*, 1978), $t = 1.35$ (Lobb and Frank,

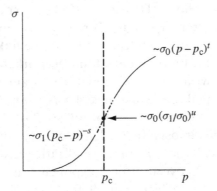

Fig. 6.7   The macroscopic conductivity of a binary mixture as a function of the probability p that a given lattice site is occupied by the conductive phase.

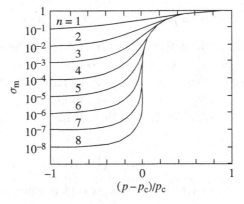

Fig. 6.8   The macroscopic conductivity σ_m of a binary mixture with various conductivity ratios σ_1/σ_0 (1 to 10^{-8}).

1979), $u = 0.57$ for two-dimensional bond percolation, and $s = 1.0$, $t = 1.6$, $u = 0.615$ for three-dimensional bond percolation (Cohen *et al.*, 1978). Figure 6.8 (Straley, 1978) shows these relations with $\sigma_0 = 1$ and $\sigma_1 = 10^{-n}$, $n = 1, 2, \ldots, 8$. In Figure 6.8, one can find that the larger n is the narrower the transition region becomes. Therefore, as σ_1/σ_0 becomes even smaller, the conduction law can be approximated by Formula (6.23). There are several experimental studies (Abeles *et al.*, 1975; Deutscher *et al.*, 1978; Kapitulnik and Deutscher, 1983) on these power-law relationships between conductivity and probability (or volume fraction).

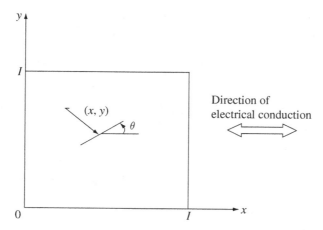

Fig. 6.9   2D fiber percolation model. (Ueda and Taya, 1986, with permission from American Inst. Phys.)

6.4 Fiber percolation model

When conductive fillers are embedded in an insulator matrix, as in a misorientated short-fiber composite (MSFC), the locations of the filler sites are not at the lattice points of any of the periodic base lattices discussed in the previous sections. Thus, one needs to develop a new percolation model aimed at these random or non-periodic microstructures.

Pike and Seager (1974) proposed a random-lattice Monte Carlo method which can account for an anisotropic percolating structure, such as one containing short fibers. Balberg and Binenbaum (1983) applied the Pike and Seager model to the case of a two-dimensionally misorientated short-fiber composite (2D-MSFC) to study the threshold morphology of a short-fiber network that yields a percolation stage. Ueda and Taya (1986) also used the Pike and Seager model, i.e., the fiber percolation model (FPM), to obtain the critical fiber length (L_c) at and above which the network of short fibers becomes percolated. Similarly, Taya and Ueda (1987) extended this 2D-FPM to a three-dimensional fiber percolation model (3D-FPM). They applied the FPM to various cases of 2D-MSFC to examine the effects of geometric anisotropy and fiber aspect ratio (L/D) on the threshold fiber volume fraction (f^*) above which the composite becomes conductive.

The FPM consists of two steps: examination of the percolation stage by the Pike and Seager method (Pike and Seager, 1974), and calculation of the effective electrical conductivity of a 2D-MSFC. A fiber is simulated by a rectangle with length L and breadth D. Every fiber has the same size, and its orientation angle θ is generated from a uniform distribution over $|\theta| \leq \alpha$, where α ($\leq 90°$) is called the in-plane limit angle, Fig. 6.9. In the second part of the FPM formulation, it is assumed that the conductivity of a percolated cluster formed of a fiber network obeys a power law given by (Deutscher, 1984)

$$\sigma = \sigma_0(p - p_c)^t \quad \text{for } p > p_c, \tag{6.39}$$

where σ_0 is the conductivity of an element of a network, p is the probability of occupied bonds, p_c is a critical probability, and t is a critical exponent of conductivity ($t = 1.35$ for a 2D lattice (Lobb and Frank, 1979)). A regular square lattice is assumed to be the base lattice for a fiber network for $\alpha = 90°$ in order to obtain p which can be given by (Ueda and Taya, 1986)

$$p = (N_f + 2N_i)/(4N_f + 2N_i), \tag{6.40}$$

where N_f and N_i are the number of fibers in the network and that of points of intersection within the network, respectively. The base lattice for $\alpha < 90°$ is assumed to be elongated along the x-axis and consists of two sets of parallel lines, one with an orientation angle of $\alpha/2$ and the other with $-\alpha/2$. Then the conductivity along the x-axis is given by

$$\sigma_x = \sigma \cot(\alpha/2), \tag{6.41}$$

where σ can be determined by Eq. (6.39). Although only σ_x is calculated in this study, the conductivity along the y-axis for $\alpha < 90°$ can be easily obtained as

$$\sigma_y = \sigma \tan(\alpha/2). \tag{6.42}$$

Balberg and Binenbaum (1983) have also examined the anisotropy of the percolating network, though they did not compute the conductivity of a composite. For smaller α, more than one percolation cluster may be established within the unit cell. Then the effective conductivity σ_e along the x-axis of a composite can be approximated by

$$\sigma_e = \sum_{i=1}^{m} b_i \sigma_{xi}, \tag{6.43}$$

where b_i and σ_{xi} are the width and conductivity of the ith cluster, respectively.

In order to calculate the effective conductivity σ_e by use of Eqs. (6.39)–(6.41) and (6.43), Ueda and Taya (1986) used the following data on the conductivities of the constituents: the conductivity of matrix $= 10^{-15}$ S/cm; that of fibers $= 10^5$ S/cm. They found that L/D also has a strong effect even on 2D systems. The results for σ_e versus fiber volume fraction (Vf) are shown in Fig. 6.10(a) for various values of L/D, with fixed $\alpha (= 90°)$. Similarly, they ran the 2D-FRM by changing $\alpha (= 15–90°)$ with fixed $L/D (= 100)$, see Fig. 6.10(b). The values

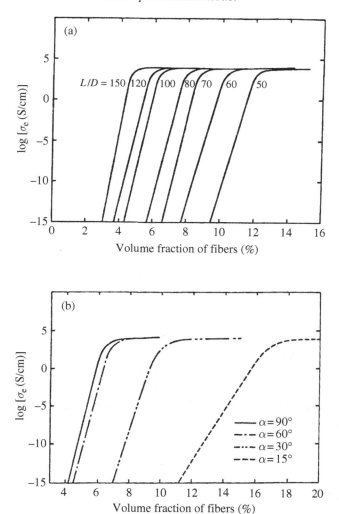

Fig. 6.10   Logarithm of σ_e for various values of (a) L/D for $\alpha = 90°$, and (b) α for $L/D = 100$. (Ueda and Taya, 1986, with permission from American Inst. Phys.)

of σ_e were obtained as a geometric mean of ten samples. Defining the threshold value of $f(f^*)$ as the mid-point of the steep slope, f^* values are plotted as a function of L/D in Fig. 6.11(a). The f^* vs. L/D curve exhibits a hyperbola shape, which was observed experimentally (Bigg, 1979). It follows from Fig. 6.11(a) that larger values of L/D are effective in enhancing the conductivity of a 2D-MSFC.

In order to examine the effect of geometric anisotropy, the results of f^* for various values of the in-plane limit angle α are shown in Fig. 6.11(b) where $L/D = 100$ was used. Figure 6.11(b) shows results similar to those obtained

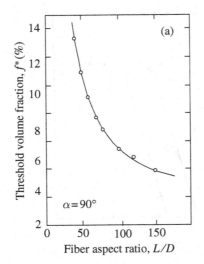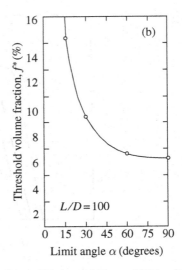

Fig. 6.11   The effect on f^* of (a) L/D, and (b) α. (Ueda and Taya, 1986, with permission from American Inst. Phys.)

by Balberg and Binenbaum (1983); namely, the more randomly short fibers are oriented, the smaller the threshold volume fraction of fibers becomes.

It should be noted that for the case of relatively isotropic geometry of the fiber network the results obtained here for a finite system represent those for an infinite system asymptotically (Balberg and Binenbaum 1983). Therefore, the results in Fig. 6.11 are considered to be valid even for an infinite system. However, for an anisotropic case, for example $\alpha = 15°$, the results for a finite system deviate from those for an infinite system, and the value of f^* along the x-axis serves as a lower bound of the exact value (Balberg and Binenbaum, 1983). Thus in the case of an anisotropic fiber network, i.e., smaller values of α, one must increase the number of short fibers to increase the accuracy of the numerical results.

3D fiber percolation model (FPM)

Short fibers are simulated by capped cylinders for the case of a 3D-MSFC (Taya and Ueda, 1987). It is assumed that they are scattered within a thin plate, and that the in-plane (θ) and out-of-plane (φ) orientation angles (defined in Fig. 6.12) are generated from a given density distribution as shown in Fig. 6.13.

In the 3D-FPM study, the distribution of out-of-plane fiber orientation angle φ is set uniform, with the out-of-plane limit angle β being $10°$ (Fig. 6.13). The effect of L/D on the threshold value of fiber volume fraction f^* was examined and the results are plotted in Fig. 6.14 where the in-plane limit

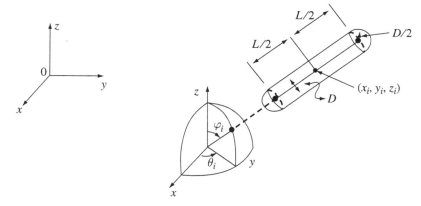

Fig. 6.12   A capped cylindrical fiber in a three-dimensional space. (Taya and Ueda, 1987, with permission from Amer. Soc. Mech. Eng.)

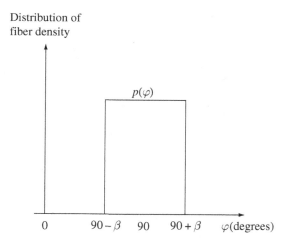

Fig. 6.13   Distribution of the out-of-plane fiber orientation angle φ of 3D-MSFC. The distribution $p(\varphi)$ is uniform over the range, $(90-\beta)° \leq \varphi \leq (90+\beta)°$ (Taya and Ueda, 1987, with permission from Amer. Soc. Mech. Eng.).

angle α is set equal to 90°. It follows from Fig. 6.14 that the effect of L/D on f^* for a 3D-MSFC is similar to that of a 2D-MSFC (see Fig. 6.11(a)) except that the f^* vs. L/D curve for the 3D-MSFC is shifted downwards. This dependency of f^* on L/D was also observed experimentally by Bigg (1979), Fig. 6.15. It is noted that the experimental results of Bigg were for the out-of-plane conductivity (σ_z) of a 3D-MSFC while the predictions of the 3D-FPM are for the in-plane conductivity (σ_x).

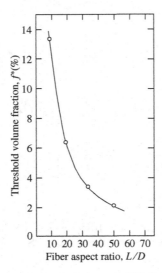

Fig. 6.14   The effect of L/D on f^* for 3D-MSFC. (Taya and Ueda, 1987, with permission from Amer. Soc. Mech. Eng.)

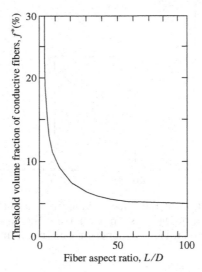

Fig. 6.15   Experimental results of f^* as a function of L/D for 3D-MSFC by Bigg (1979).

6.5 Percolation model applied to piezoresistive elastomer composites

It has been shown that an initially electrically conductive short-fiber composite becomes less conductive as straining increases, and it suddenly becomes non-conductive at and beyond a critical strain, exhibiting switching

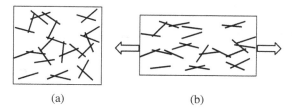

Fig. 6.16   The expected microstructure change of a composite under strain-ing: (a) percolating before straining (electrically conductive), and (b) non-percolating after straining (non-conductive). (Taya *et al.*, 1998, with permission from Elsevier Ltd.) Arrows indicate direction of e_x.

behavior (Narita, 1990; Taya *et al.*, 1998). This is mainly due to the reorientation of the conductive short fibers upon straining, as shown in Fig. 6.16, where an initially electrically percolating network degenerates to a non-percolating network. When the insulator matrix is soft, such as an elastomer, external loading causes a microstructural change of the comp-osite, resulting in a change in the overall mechanical and electrical proper-ties (Aspeden, 1986; Taya *et al.*, 1998). Electrically conductive elastomer composites, which exhibit variable conductivity in response to varying external mechanical loading, are widely used for various electronic applica-tions, touch control switches, and strain and pressure sensors for applica-tions such as robot hands or artificial limbs (Pramanik *et al.*, 1990). However, no quantitative analysis of these switchable composites has been conducted. Taya *et al.* (1998) used a modified three-dimensional fiber percolation model (3D-FPM) to predict the threshold fiber volume fraction f^* and a fiber reorientation model to predict the microstructural change of a fiber network under an incremental tensile strain Δe with the aim of obtaining piezoresistive behavior in a conductive short-fiber/elasto-mer matrix composite. The piezoresistivity is given in terms of an applied strain vs. electrical conductivity relation.

The 3D-FPM used by Taya *et al.* (1998) is shown in Fig. 6.17 and the modification is to induce the "concept of tunneling conductance" that exists between two adjacent short fibers within a critical distance $2d_c$, illustrated in Fig. 6.18. The composite conductivity σ_c is then estimated by Formula (6.23).

Fiber reorientation model

Reorientation and relocation of fibers in an elastomer composite are expected to take place under larger straining. If the microstructure of the conductive short fibers was initially percolating, then the percolating microstructure

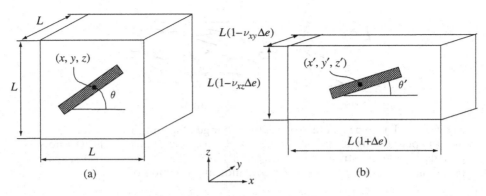

Fig. 6.17   The 3D fiber reorientation model where only angle θ is shown for
illustrative purposes: (a) before straining, and (b) after straining. (Taya *et al.*,
1998, with permission from Elsevier Ltd.) The quantities ν_{xy} and ν_{xz} are
Poisson's ratios, see Eq. (3.21).

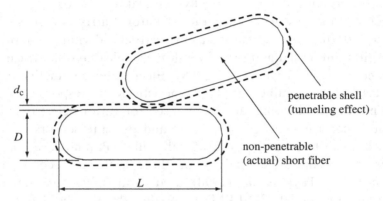

Fig. 6.18   The simulation of non-penetrable model (D = actual fiber
diameter, d_c = tunneling distance) (Taya *et al.*, 1998, with permission from
Elsevier Ltd).

degenerates to a less- or non-percolating network as the applied strain e
increases.

It is assumed that, upon applying an incremental uniaxial strain Δe along
the x-axis, the changes in the orientation and location of a short fiber are taken
into account by following an affine transformation, as depicted in Fig. 6.17,
where (a) and (b) denote, respectively, the configuration before and after the
incremental strain Δe. An affine transformation assumes that under external
loading the length components of a fiber change by the same ratio as the
corresponding dimensions of the matrix. Under the assumption of the incompressibility of an elastomer composite and an incremental strain Δe along the
x-axis, the dimensions of a rectangular specimen (L_x, L_y, L_z) and the center
coordinates (x_i, y_i, z_i) of the ith fiber are changed to

$$L'_x = L_x(1 + \Delta e), \quad L'_y = L_y(1 - \nu_{xy}\Delta e), \quad L'_z = L_z(1 - \nu_{xz}\Delta e) \quad (6.44a)$$

$$x'_i = x_i(1 + \Delta e), \quad y'_i = y_i(1 - \nu_{xy}\Delta e), \quad z'_i = z_i(1 - \nu_{xz}\Delta e) \quad (6.44b)$$

where the prime indicates the current variables after straining by Δe, and ν_{ij} is Poisson's ratio of a composite, defined as $|e_j/e_i|$. The length components in Cartesian coordinates of the ith fiber before uniaxial straining are expressed as

$$u_{xi} = u \sin \varphi_i \cos \theta_i, \quad u_{yi} = u \sin \varphi_i \sin \theta_i, \quad u_{zi} = u \cos \varphi_i, \quad (6.45a)$$

and after straining as

$$u'_{xi} = u' \sin \varphi'_i \cos \theta'_i, \quad u'_{yi} = u' \sin \varphi'_i \sin \theta'_i, \quad u'_{zi} = u' \cos \varphi'_i, \quad (6.45b)$$

where u (u') is the length of fiber before (after) straining. Applying an affine transformation with Eqs. (6.44a), (6.44b), (6.45a) and (6.45b), the following relationships are obtained:

$$1 + \Delta e = \frac{u' \sin \varphi'_i \cos \theta'_i}{u \sin \varphi_i \cos \theta_i}, \quad 1 + e_y = \frac{u' \sin \varphi'_i \sin \theta'_i}{u \sin \varphi_i \sin \theta_i}, \quad 1 + e_z = \frac{u' \cos \varphi'_i}{u \cos \varphi_i}. \quad (6.46)$$

By using Eq. (6.45), the reorientation angles of a fiber after straining are derived as

$$\theta'_i = \tan^{-1}\left[\frac{(1 - \nu_{xy}\Delta e)}{(1 + \Delta e)}\tan \theta_i\right],$$

$$\varphi'_i = \tan^{-1}\left[\frac{(1 - \nu_{xy}\Delta e)}{(1 - \nu_{xz}\Delta e)}\frac{\sin \theta_i}{\sin \varphi'_i}\tan \varphi\right]. \quad (6.47)$$

Comparison between predictions and experimental results

It is assumed that in a 3D short-fiber composite electric current flows by a tunneling mechanism. The overall fiber volume fraction (including tunneling shell d_c) and the actual fiber volume fraction (excluding the shell) at the percolation threshold for a 3D random system are plotted in Fig. 6.19 as functions of the ratio of tunneling distance to actual fiber diameter (d_c/D), where fiber length and diameter are fixed and fiber aspect ratio (L/D) is 5. The 3D fiber orientation is simulated as a completely random distribution. It is concluded from Fig. 6.19 that, as the ratio d_c/D decreases, the threshold volume fraction of fibers increases.

When d_c/D approaches zero, these two curves should converge to a single value. Thus, the d_c/D ratio in a non-penetrable model is an important

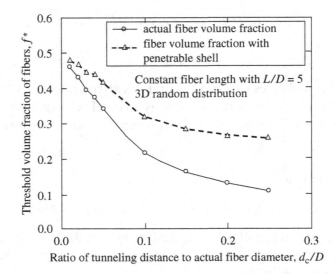

Fig. 6.19   The threshold volume fractions of actual and overall fibers as functions of the ratio of tunneling distance d_c to actual fiber diameter D (Taya *et al.*, 1998, with permission from Elsevier Ltd.)

parameter affecting the threshold volume fraction and the electrical conductivity. The tunneling distance is a characteristic of the electrical properties of the polymer matrix. However, since the electrical transport process consists of several different mechanisms, determination of the tunneling distance in a given system is difficult. Thus the tunneling distance will be approximated by matching the experimental value of threshold volume fraction f^* to the predictions by the present model.

Next, the change in the microstructure of a short-fiber/elastomer composite under straining is examined. Figure 6.20 illustrates the orientation distribution of in-plane fiber angle θ where (a) and (b) denote, respectively, the initial orientation of a 3D randomly isotropic composite with $L/D = 55$ before straining and the reorientation after the final strain of 0.5 is reached. The reason for using $L/D = 55$ is to compare the predictions with the experimental data where $L/D = 55$ was used (Narita, 1990). The final strain of 0.5 is reached by incremental strains Δe of 0.05. It is found in Fig. 6.20 that the initially uniform distribution is changed to a non-uniform distribution, becoming narrower as straining increases. For out-of-plane angle φ, although it is not shown here, similar trends of fiber reorientation distribution are observed. It is also found that the change in volume fraction f of the fibers is negligibly small since the matrix is an elastomer (Poisson's ratio $\nu \approx 0.5$).

On the basis of the analytical formulation, the computational process of the piezoresistive behavior is illustrated in Fig. 6.21. The initial input data,

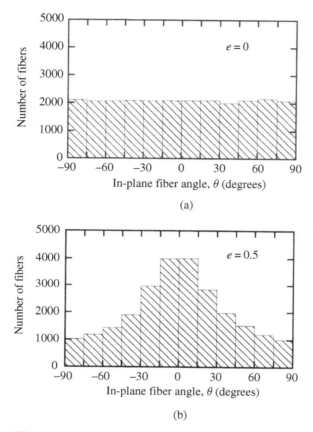

Fig. 6.20   The reorientation of fibers under straining of an isotropic composite ($L/D = 55$, total number of fibers $= 25{,}000$: (a) before straining ($e = 0$); and (b) after straining ($e = 0.5$). (Taya et al., 1998, with permission from Elsevier Ltd.)

including a seed number, fiber length L and diameter D, fiber orientations θ, φ, limit angles α, β, and tunneling distance d_c are assigned and, for a given microstructure of a composite system, the threshold volume fraction f^* of fibers is computed by applying the fiber percolation model. Then, an incremental strain along the x-axis is applied and the new microstructure of the composite with relocation and reorientation of the conductive fibers is established by applying the fiber reorientation model. The microstructure is examined to see if it forms a percolation network, and a new threshold fiber volume fraction is then computed. Finally, the effective electrical conductivity is calculated for the composite with the changed microstructure by using Eqs. (6.39) and (6.43).

The numerical results of the present model are compared with the previous experimental data for a Ni-coated graphite fiber/natural rubber composite

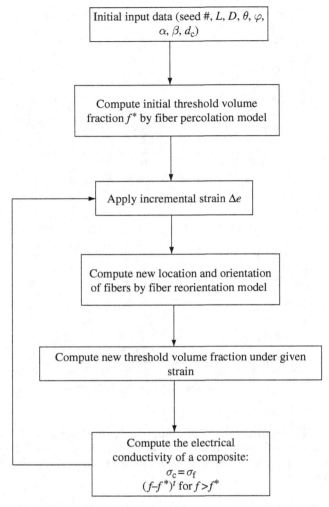

Fig. 6.21   The computational procedure for the modified 3D-FPM model
(Taya *et al.*, 1998, with permission from Elsevier Ltd).

with fiber volume fraction f of 0.05 and average fiber aspect ratio L/D of 55,
i.e., $L=426\,\mu\mathrm{m}$ and $D=7.8\,\mu\mathrm{m}$ (Narita, 1990). The specimen dimensions are
$1\times1\times1\,\mathrm{mm}^3$. It is assumed that the distribution $p(\theta, \varphi)$ of the fiber orientations of the composite is initially ($e=0$) uniform with an in-plane limit angle
$\alpha=30°$ and an out-of-plane limit angle $\beta=30°$, and the fiber length is a
normal (Gaussian) distribution with standard deviation (STD) 0.1. The ratio
of tunneling distance to fiber diameter ($d_c/D=0.05$) was approximately determined by matching the threshold volume fraction f^* predicted by the present
model with the experimental value of 0.035. The predicted effective electrical

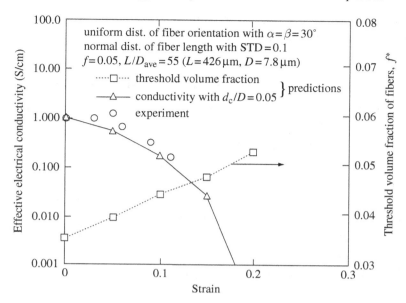

Fig. 6.22 Electrical conductivity of a composite and threshold volume fraction of conductive fibers as a function of applied strain (Taya *et al.*, 1998, with permission from Elsevier Ltd).

conductivity σ_e (triangles) along the *x*-axis (see Fig. 6.17) is plotted as a function of applied uniaxial strain *e* in Fig. 6.22, where the experimental data are also shown as circles. Figure 6.22 also illustrates the change in the threshold volume fraction (squares) under the applied strain. The threshold fiber volume fraction is seen to increase with applied strain. This explains that the initially percolated fiber network degenerates by separation of the previously existing contacts between fibers due to reorientation of the conductive fibers. This result also supported by the fact that the fiber orientation distribution becomes narrower (more aligned toward the *x*-direction) due to straining in the *x*-axis (see Fig. 6.20) and the narrower distribution gives rise to a higher threshold volume fraction of conductive fiber (Taya and Ueda, 1987). The effective electrical conductivity of the composite is calculated using Eq. (6.39) with an experimentally obtained conductivity exponent *t* of 2.0 (Narita, 1990), which is known as a universal value for a 3D system (Stauffer, 1985), and f^* is computed by the fiber percolation model, where the conductivities of coated fiber and matrix are 4.89×10^3 S/cm and 1.0×10^{-15} S/cm, respectively, and the Poisson's ratio of the composite is 0.5. Figure 6.22 clearly demonstrates a piezoresistive behavior of the composite where the composite becomes less conductive as the applied strain increases and is predicted to become non-conductive at and beyond a critical strain of 0.2, exhibiting a switching behavior. As seen in Fig. 6.22, the numerical data give a reasonably good agreement with the experimental data.

6.6 Percolation in a particle composite

For spherical-particle composites, two types of distribution of conducting particles in an insulating matrix can be considered: random, and segregated distributions (Gerteisen, 1982). In the random distribution, the conducting particles and insulating matrix grains are similar in effective size and shape, and each may occupy any site in the packing arrangement. In contrast to the random distribution, the segregated distribution is achieved through the tendency of small conducting particles to adhere to the surfaces or to segregate on grain boundaries of the much larger insulating matrix grains. With the segregated distribution, the conducting particles are in the form of long aggregate chains or links along boundaries between the insulating matrix grains, limiting penetration of the small particles into the interior of the matrix grains.

It is known that the threshold volume fraction f^* of a composite with segregated distribution is much lower than that of a composite with random distribution (Kerner, 1956; Nielsen, 1974; Bigg, 1979; Warfel, 1980; Litman and Fowler, 1981; Gerteisen, 1982). The threshold volume fraction can be estimated by considering the relative particle size ratio (the radius R_m of the insulating matrix grain to the radius R_p of the conducting particles). Gurland (1966) showed in his experimental work that, in a composite with a random distribution, the threshold volume fraction of conductor required for percolation was 35–37% ($R_m/R_p \approx 2$), whereas Kusy and Turner (1971) showed that only 6 vol.% ($R_m/R_p = 30$) was needed for a composite with segregated distribution.

There have been a few theoretical studies on the relationship between the threshold volume fraction f^* and the particle size ratio, R_m/R_p. Malliaris and Turner (1971) proposed a theoretical model for the estimate of f^* as a function of R_m/R_p on the assumption that each non-conducting matrix grain is covered with a monolayer of conducting particles at f^*. The values of f^* calculated using this assumption were found to be 2–3 times less than those observed experimentally, since many conducting particles would be isolated from percolating networks and fail to contribute to the electrical behavior of the composite. Kusy (1977) realized that the conducting particles do not have to completely cover the insulating grains in order to form a percolation path through the composite. Kusy's prediction of minimum f^* is about 0.03 and it occurs when R_m/R_p is 30 or more. Bhattacharyya and co-workers (1978, 1982) took into account the effects of the shape as well as the size of the conducting

$1\overline{\mu}m$

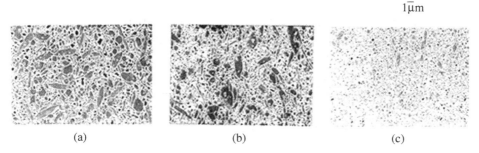

(a)   (b)   (c)

Fig. 6.23   Scanning electron micrograph of SiC particles/Si_3N_4 matrix composites: (a) 10, (b) 20, and (c) 30 wt% SiC. (Kim *et al.*, 1998b, with permission from American Inst. Phys.)

particles in estimating f^*, by assuming that the conducting particles are in the form of rectangular parallelepipeds. The previous models proposed simple relations between f^* and R_m/R_p which do not require numerical computation but assumed parameters and the critical number of conducting particles necessary to cover the surface of each matrix grain. The critical number of conducting particles is obtained from a percolation theory for a given lattice structure.

Kim *et al.* (1998b) studied the electrical resistivity of semi-conductive SiC particles/insulative Si_3N_4 matrix composites. They used a simple geometrical model based on a two-dimensional honeycomb surface to predict f^* and then used a combined effective medium theory and percolation theory (called a generalized effective medium (GEM) model) to predict accurately the relation between resistivity and volume fraction of conducting particles. In the following subsections, we will review the experimental data and models applied to this system.

Experiment

SiC powders of 0.03 μm diameter used as starting powders were mixed with Si_3N_4 powders in an ethanol solution for 72 h to form green composite solids, which were subsequently hot pressed at 1850 °C under 20 MPa in N_2 for 1 h. The hot-pressed SiC/Si_3N_4 composite plate was cut into specimens of $3 \times 4 \times 40$ mm^3. Four different fractions of SiC were prepared: 5, 10, 20, and 30 wt%. As-processed composite specimens were etched using an electron cyclotron resonance plasma for scanning electron microscopy (SEM) observations. Examples of their microstructures are shown in Fig. 6.23. The large elongated gray islands are Si_3N_4 grains, white lines are boundary layers, and the small black areas are those occupied by SiC particles but removed during the etching process. The distribution of SiC particles segregated around Si_3N_4 grain

Table 6.3 *Measured electrical resistivity of monolithic* Si_3N_4 *(0 wt% SiC) and* SiC/Si_3N_4 *composite* (Kim *et al.*, 1998b, with permission from Amer. Inst. Phys.).

SiC content (wt%)	0	5	10	20	30	100
SiC content (vol. %)	0	5.32	10.61	21.08	31.41	100
Electrical resistivity (Ω-cm)	4.2×10^8	4.3×10^7	4.1×10^6	1.2×10^3	2.8×10^1	1.0×10^{-2}

boundaries is shown in Fig. 6.24 as dark dots using energy dispersive analysis of x-rays (EDAX) in the SEM. In both Figs. 6.23 and 6.24, (a), (b), and (c) denote the cases for 10, 20, and 30 wt% SiC, respectively. The average size of SiC particles was measured as 0.1 µm, larger than 0.03 µm, the average starting size of SiC particles before processing. This is due to growth of SiC particles during the hot processing. The average starting size of the Si_3N_4 particles (α-phase Si_3N_4) was 0.2 µm, which became larger during processing in terms of β-phase Si_3N_4 grains. It was found from SEM and EDAX analyses that the majority of the SiC particles are segregated along the Si_3N_4 grain boundaries surrounding many Si_3N_4 grains and they tend to form clustered networks as the SiC content increases (Yamada and Kamiya, 1995). The electrical resistivities of monolithic Si_3N_4 and SiC/Si_3N_4 composite specimens made by the same processing route were measured using two-and four-point probe methods, the results of which are summarized in Table 6.3.

2D Percolation model

The model used by Kim *et al.* (1998b) is a 2D bond percolation model with the Monte Carlo method. Although this 2D model is not well suited to actual 3D materials, the microstructure observed by SEM and EDAX analyses (see Figs 6.23 and 6.24) is only a two-dimensional representation, so use of the 2D model is considered reasonable. The model provides the trends of percolation threshold volume fraction f^* with respect to the size ratio (R_m/R_p) of matrix grain to particle. The three-dimensional structure is first approximated to a honeycomb of insulating matrix grains surrounded by small spherical conducting particles. In the 2D model, the grains are represented by hexagons and the particles by circles, Fig. 6.25. Each of the six sides of the insulating hexagonal grain is considered as a bond, which consists of the small conducting particles. Figure 6.25(a) illustrates the bond percolation model. The thick, solid lines are the established (electrically connected) bonds while thin lines are unconnected ones. It is assumed in Fig. 6.25(b) that an (electrically conducting) bond of length L consists of n conductive particles and the size ratio of the insulating matrix grain to the conducting particle is approximated by the apparent radius ratio R_m/R_p, since the structure of the matrix grain is not a

1μm

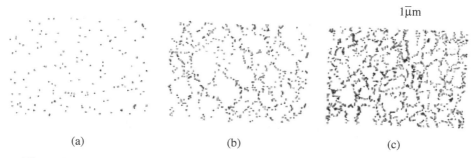

(a)   (b)   (c)

Fig. 6.24   EDAX mapping of SiC phase of SiC particles/Si_3N_4 matrix composites: (a) 10, (b) 20, and (c) 30 wt% SiC. (Kim *et al.*, 1998b, with permission from Amer. Inst. Phys.)

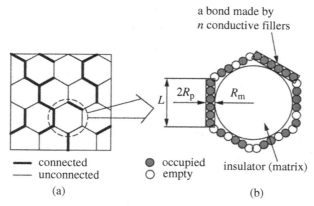

Fig. 6.25   Modified two-dimensional bond percolation model: (a) hexagonal structure, and (b) a unit of hexagonal structure. (Kim *et al.*, 1998b, with permission from Amer. Inst. Phys.)

circle but a hexagon. Assuming that all conducting particles occupy one side of the hexagonal structure, the radius of particle (R_p) is then equal to $L/2n$, hence the approximate particle size ratio R_m/R_p is described as

$$\frac{R_m}{R_p} = \sqrt{3}\,n - 1. \tag{6.48}$$

It is shown in Fig. 6.25(b) that there are overlaping portions of particles between bonds at the vertices of the hexagon, which are neglected, since they have little influence on the volume fraction of the particles when $R_m \gg R_p$, which is the case with a segregated composite.

In a Monte Carlo simulation, conducting particles are generated along the boundaries of the hexagonal structure using a pseudo-random number generator built in a computer. A bond is established when it is occupied by n

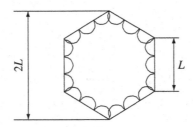

Fig. 6.26   Unit of hexagonal structure. (Kim *et al.*, 1998b, with permission from Amer. Inst. Phys.)

particles and it is not allowed to generate more particles on this established bond. After the generation of all particles, the locations and total number of established bonds are known. If two bonds were found to be connected, they are assigned a cluster identification number (CIDN). All bonds within the same cluster have the same CIDN. The CIDN is updated as each pair of bonds is being checked so that if a bond in a cluster is found to contact a new bond, which belongs to a different cluster, two clusters will be assigned the same CIDN (the smaller one of two CIDNs). If bonds in the same cluster intersect a pair of opposite boundaries of the system along a given axis, a percolating cluster is said to be established along the axis, and the number of the particles in the percolating cluster is recorded.

The volume fraction f of conductor particles is calculated by using a filling factor Φ dependent on the lattice structure, and the probability p of particles occupying the grain boundaries, which is the ratio of the number of particles present in electrically connected bonds (percolating clusters) to the total number of particles which would occupy every boundary in the computed system, and is described as

$$f = \Phi p. \tag{6.49}$$

The filling factor of a 2D hexagonal structure, as shown in Fig. 6.26, can be computed as follows. The area A_p occupied by particles is equal to the number of particles times the area of a particle inside the hexagonal unit, and is given by

$$A_p = (N_b n) \times \frac{\pi R_p^2}{2} = \frac{3\pi L^2}{4n} \tag{6.50}$$

where N_b is the number of bonds in the unit (for a hexagonal unit, $N_b = 6$) and the radius R_p of a particle is equal to $L/2n$. Since the total area A_t of the hexagonal unit is $3L^2 \cos 30°$, the filling factor $\Phi\,(= A_p/A_t)$ is given by

$$\Phi = \frac{0.9069}{n}. \tag{6.51}$$

As shown in Eq. (6.51), the filling factor is inversely proportional to the number n of particles in a bond. A substitution of Eq. (6.51) into Eq. (6.49) yields the threshold volume (area) fraction f^* of conductor particles:

$$f^* = \frac{0.9096}{n} \frac{N_{\mathrm{p}}^*}{N_{\mathrm{pt}}}, \tag{6.52}$$

where N_{p}^* and N_{pt} are the numbers of particles present when percolation occurs and the total number of particles which can occupy all the grain boundaries of the hexagonal structure, respectively. N_{p}^* is computed from the Monte Carlo simulation. Thus, the prediction of f^* as a function of particle size ratio, $R_{\mathrm{m}}/R_{\mathrm{p}}$, can be obtained from Eqs. (6.48) and (6.52).

By setting the effect of each distributed sphere on the potential outside of the spherical composite boundary equal to that obtained for the case of a composite with a single sphere, the effective resistivity ρ_{c} of a particle composite is obtained as

$$\rho_{\mathrm{c}} = \frac{2\rho_{\mathrm{p}} + \rho_{\mathrm{m}} + f(\rho_{\mathrm{p}} - \rho_{\mathrm{m}})}{2\rho_{\mathrm{p}} + \rho_{\mathrm{m}} - 2f(\rho_{\mathrm{p}} - \rho_{\mathrm{m}})}\rho_{\mathrm{m}}, \tag{6.53}$$

where ρ_{m} and ρ_{p} are the resistivities of matrix and spherical particle, respectively, and f is the volume fraction of spherical particles.

The percolation theory applies only to near the conductor–insulator transition region when the ratio of the conductivities of conductor and insulator is extremely high, i.e., $\sigma_{\mathrm{m}}/\sigma_{\mathrm{p}} \approx 0$, whereas the existing effective medium theories are applicable to the entire range of conductor volume fraction only when the conductivity ratio is moderate. In view of these problems within the framework of both percolation and effective medium theories, a generalized effective medium (GEM) equation was proposed by McLachlan *et al.* (1990). The GEM equation, which combines aspects of both percolation and Bruggeman's symmetric effective medium model, Eq. (6.19), can predict the effective electrical conductivity for a whole range of conductor volume fraction, including the conductor–insulator transition region around percolation threshold, and is expressed as

$$(1-f)\frac{\sigma_{\mathrm{m}}^{1/t} - \sigma_{\mathrm{c}}^{1/t}}{\sigma_{\mathrm{m}}^{1/t} + A\sigma_{\mathrm{c}}^{1/t}} + f\frac{\sigma_{\mathrm{p}}^{1/t} - \sigma_{\mathrm{c}}^{1/t}}{\sigma_{\mathrm{p}}^{1/t} + A\sigma_{\mathrm{c}}^{1/t}} = 0, \tag{6.54}$$

where the constant A is described by

$$A = \frac{1-f^*}{f^*} = \frac{1-L_m}{L_p},\tag{6.55}$$

where L_m and L_p are the demagnetizing factors of matrix and particle, respectively, as discussed in Subsection 1.2.3, and are the same as the Eshelby tensor.

For the limiting cases, the GEM equation is reduced to the percolation equations when $\sigma_m/\sigma_p \approx 0$ or the resistivity ratio of particle to matrix $\rho_p/\rho_m \approx 0$. Whenever $\sigma_m/\sigma_p \approx 0$ ($\sigma_m \ll \sigma_p$), the GEM equation is reduced to $\sigma_c = \sigma_p [(f-f^*)/(1-f^*)]^t$ which is valid for $f > f^*$, and conversely, whenever $\rho_p/\rho_m \approx 0$ ($\rho_m \gg \rho_p$), the equation is reduced to $\sigma_c = \sigma_m[(f^*-f)/f^*]^{-t}$, valid for $f < f^*$. Both of these equations have the mathematical form of percolation equations. The GEM equation is also reduced to Bruggeman's symmetric effective medium equation when $t = 1$ and $f^* = L_m = L_p = 1/3$.

It is known (Gurland, 1966; Kusy and Turner, 1971; Malliaris and Turner, 1971) that the size ratio of matrix grain to filler particle significantly affects the electrical resistivity of a composite. This could be demonstrated through the distribution pattern of the conducting particles in the insulating matrix grains. At higher values of the size ratio R_m/R_p, the distribution is largely segregated, thereby forming percolating networks at a very low volume fraction of particles. Figure 6.27 illustrates the correlation between the threshold conductor volume fraction f^* and the particle size ratio R_m/R_p, by applying the Kim *et al.* model, see Eqs. (6.48) and (6.52). The dramatic reduction of f^* due to an increase of R_m/R_p is clearly observed in Fig. 6.27.

From the experimental observation of the microstructure in SiC/Si_3N_4 composites (see Fig. 6.24), the size ratio of the insulating grain to the conducting particle was determined as $R_m/R_p = 10$, and the threshold volume fraction of particles was then computed as $f^* = 0.13$ by applying the 2D model (see Fig. 6.27). The value of the conductivity exponent t was obtained from the slope of the log ρ_c vs. log($f-f^*$) plot, using the experimental data (see Table 6.3). A rather high value of critical exponent ($t = 4.57$), compared with the universal value $t = 2.0$ (in three dimensions), was obtained. This high value of t has been experimentally observed in a composite with segregated distribution, yielding the range of $t = 3-7$ (Pike, 1978). The electrical resistivity of the composite was then estimated by applying several theories: the Eshelby model (Eq. (4.93)); Bruggeman's symmetric model (Eq. (6.19)); percolation theory (Formula (6.23)); and the GEM approach (Eq. (6.54)), respectively. The dependence of the electrical resistivity on the volume fraction of conducting particles is plotted in Fig. 6.28, with error bars attached to experimental data. As shown in

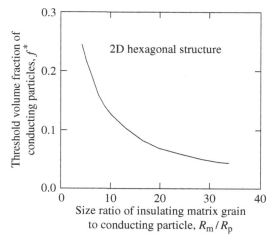

Fig. 6.27   Predicted f^* vs. R_m/R_p relation by percolation model (Kim *et al.*, 1998b, with permission from Amer. Inst. Phys.).

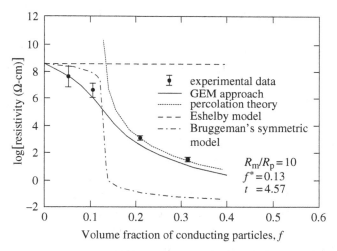

Fig. 6.28   Comparison of the predictions based on various models with the experiment in terms of composite electrical resistivity and volume fraction of conducting particles (Kim *et al.*, 1998b, with permission from Amer. Inst. Phys.).

Fig. 6.28, both the Eshelby model and Bruggeman's symmetric model fail to explain the experimental results of the composite system, while a percolation equation gives slightly better predictions beyond the threshold region than the GEM equation. For the entire range of f, however, the GEM equation gives good agreement with the experimental results.

7

Lamination model

Laminated composites have been used extensively as printed circuit board (PCB) material in electronic packaging. Determination of the macroscopic properties of a PCB made of a laminated composite is a key step in designing a new PCB or selecting a PCB tailored for a specific packaging application. A laminated composite PCB consists of a number of laminae where each lamina itself is a composite made of discontinuous filler or continuous filler (straight fibers or woven fabrics). A lamination model is used to predict the overall properties of a laminated composite for a given set of properties of the laminae. It is assumed that the properties of each lamina are already obtained by experiment or predicted by a composite model such as the Eshelby model. Here, we apply the classical lamination theory only when a *thin* plate composed of laminae is subjected to thermomechanical loading and/or an electric field with the aim of finding the deflection of the plate and stress field within the laminae. The case of thermomechanical loading is important for the analysis of PCBs while that of an electric field relates to a piezoelectric laminated plate. Similarly, the behavior of electromagnetic waves in laminated composites provides key quantitative information in designing switchable windows for radomes and surface plasmon resonance sensors. In the following, we shall discuss first the classical lamination theory focusing on (i) mechanical behavior under hygro-thermomechanical loading, then extend it to (ii) the problem of a piezoelectric laminated composite subjected to an electric field, followed by (iii) stress analysis in a thin-coating–substrate system, and (iv) electromagnetic wave propagation in a layered medium.

7.1 Classical lamination theory under hygro-thermomechanical loading

Classical lamination theory has been introduced in a number of textbooks on composites (see, for example, Christensen, 1979, and Gibson, 1994). Consider a laminate of total thickness h, consisting of n laminae lying in the x–y plane and perpendicular to the z-axis, Fig. 7.1, where the x–y plane is located at the mid-plane and the z-coordinates of the bottom and top surfaces of the ith lamina are defined by h_i and h_{i+1} ($i = 1, 2, \ldots, n$), respectively. Assume that each lamina is an orthotropic elastic material and the entire laminate is subjected to a uniform temperature change ΔT and relative moisture change m. The in-plane stress–strain relation is given by

$$
\begin{Bmatrix} e_{11} \\ e_{22} \\ 2e_{12} \end{Bmatrix} = \begin{bmatrix} S_{11} & S_{12} & 0 \\ S_{21} & S_{22} & 0 \\ 0 & 0 & S_{66} \end{bmatrix} \begin{Bmatrix} \sigma_{11} \\ \sigma_{22} \\ \sigma_{12} \end{Bmatrix} + \begin{Bmatrix} \alpha_{11} \\ \alpha_{22} \\ 0 \end{Bmatrix} \Delta T + \begin{Bmatrix} \beta_{11} \\ \beta_{22} \\ 0 \end{Bmatrix} m, \tag{7.1a}
$$

$$
\begin{Bmatrix} \sigma_{11} \\ \sigma_{22} \\ \sigma_{12} \end{Bmatrix} = \begin{bmatrix} C_{11} & C_{12} & 0 \\ C_{21} & C_{22} & 0 \\ 0 & 0 & C_{66} \end{bmatrix} \left[\begin{Bmatrix} e_{11} \\ e_{22} \\ 2e_{12} \end{Bmatrix} - \begin{Bmatrix} \alpha_{11} \\ \alpha_{22} \\ 0 \end{Bmatrix} \Delta T - \begin{Bmatrix} \beta_{11} \\ \beta_{22} \\ 0 \end{Bmatrix} m \right] \tag{7.1b}
$$

where S_{ij} is the elastic compliance tensor, whose explicit expression for an orthotropic material is given by Eq. (3.20), C_{ij} is the elastic stiffness tensor, α_i is the coefficient of thermal expansion (CTE), β_i is the coefficient of hygroscopic expansion (CHE) and m is defined by the ratio (mass of moisture in a unit volume)/(mass of dry material in a unit volume) (Gibson, 1994). It is assumed in the classical lamination theory that the out-of-plane stress components (σ_{33}, σ_{31}, and σ_{32}) are zero, i.e., a plane stress condition applies, so that C_{ij} is replaced by

$$
\left.
\begin{aligned}
C_{11} &= C_{11} - \frac{C_{13}^2}{C_{33}}, \\
C_{12} &= C_{21} = C_{12} - \frac{C_{23}C_{13}}{C_{33}}, \\
C_{22} &= C_{22} - \frac{C_{23}^2}{C_{33}}.
\end{aligned}
\right\} \tag{7.2}
$$

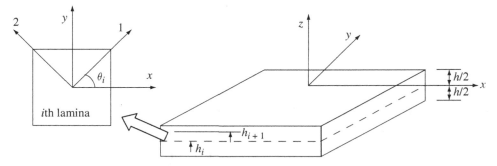

Fig 7.1 A laminated composite subjected to uniform temperature change ΔT and relative moisture change m.

Similarly, the elastic compliance S_{ij} needs to be redefined by

$$\left.\begin{array}{c} S_{11} = \dfrac{1}{E_1}, \qquad S_{22} = \dfrac{1}{E_2}, \\[2mm] S_{12} = S_{21} = -\dfrac{\nu_{21}}{E_2} = -\dfrac{\nu_{12}}{E_1}, \\[2mm] S_{66} = \dfrac{1}{G_{12}}, \end{array}\right\} \qquad (7.3)$$

where ν_{ij} is Poisson's ratio, and E_{ij}, G_{ij} are Young's and the shear modulus, respectively, see Subsection 3.2.1.

Please note that the local coordinates associated with the ith lamina are 1- and 2-axes in the x–y plane and the 3-axis coincides with the z-axis, see Fig. 7.1. Thus, the transformation matrix in transforming the x–y components in the local coordinates is defined by

$$[T] = \begin{bmatrix} \cos^2\theta_i & \sin^2\theta_i & 2\cos\theta_i\sin\theta_i \\ \sin^2\theta_i & \cos^2\theta_i & -2\cos\theta_i\sin\theta_i \\ -\sin^2\theta_i\cos\theta_i & \cos^2\theta_i & \cos^2\theta_i \end{bmatrix} \qquad (7.4)$$

where θ_i is defined as the angle between the x-axis and the 1-axis, Fig. 7.1.

Note that the stress and strain tensors in the local (1-, 2-axes) and global (x-, y-axes) coordinates are related through the transformation matrix T (do not confuse this T with temperature change ΔT)

$$\left\{\begin{array}{c} \sigma_{11} \\ \sigma_{22} \\ \sigma_{12} \end{array}\right\} = [T] \left\{\begin{array}{c} \sigma_x \\ \sigma_y \\ \sigma_{xy} \end{array}\right\}, \qquad (7.5)$$

$$\left\{\begin{array}{c} e_{11} \\ e_{22} \\ e_{12} \end{array}\right\} = [T] \left\{\begin{array}{c} e_x \\ e_y \\ e_{xy} \end{array}\right\}. \qquad (7.6)$$

Substituting Eqs. (7.5) and (7.6) into Eqs. (7.1) one can obtain the in-plane constitutive equation for the ith lamina in the global coordinates:

$$\left\{\begin{array}{c} e_x \\ e_y \\ 2e_{xy} \end{array}\right\} = \begin{bmatrix} \bar{S}_{11} & \bar{S}_{12} & \bar{S}_{16} \\ \bar{S}_{21} & \bar{S}_{22} & \bar{S}_{26} \\ \bar{S}_{16} & \bar{S}_{26} & \bar{S}_{66} \end{bmatrix}_i \left\{\begin{array}{c} \sigma_x \\ \sigma_y \\ \sigma_{xy} \end{array}\right\} + \left\{\begin{array}{c} \alpha_x \\ \alpha_y \\ \alpha_{xy} \end{array}\right\}_i \Delta T + \left\{\begin{array}{c} \beta_x \\ \beta_y \\ \beta_{xy} \end{array}\right\} m, \qquad (7.7)$$

where

$$
\begin{bmatrix}
\bar{S}_{11} & \bar{S}_{12} & \bar{S}_{16} \\
\bar{S}_{21} & \bar{S}_{22} & \bar{S}_{26} \\
\bar{S}_{16} & \bar{S}_{26} & \bar{S}_{66}
\end{bmatrix}
= [T]^{-1}
\begin{bmatrix}
S_{11} & S_{12} & 0 \\
S_{21} & S_{22} & 0 \\
0 & 0 & S_{66}
\end{bmatrix}
[T],
$$

(7.8)

$$
\left\{
\begin{array}{c}
\alpha_x \\
\alpha_y \\
\alpha_{xy}
\end{array}
\right\}
= [T]^{-1}
\left\{
\begin{array}{c}
\alpha_1 \\
\alpha_2 \\
0
\end{array}
\right\},
\quad
\left\{
\begin{array}{c}
\beta_x \\
\beta_y \\
\beta_{xy}
\end{array}
\right\}
= [T]^{-1}
\left\{
\begin{array}{c}
\beta_1 \\
\beta_2 \\
0
\end{array}
\right\}.
$$

Eq. (7.7) can be converted to the stress–strain relation in global coordinates

$$
\left\{
\begin{array}{c}
\sigma_x \\
\sigma_y \\
\sigma_{xy}
\end{array}
\right\}
=
\begin{bmatrix}
\bar{Q}_{11} & \bar{Q}_{12} & \bar{Q}_{16} \\
\bar{Q}_{21} & \bar{Q}_{22} & \bar{Q}_{26} \\
\bar{Q}_{16} & \bar{Q}_{26} & \bar{Q}_{66}
\end{bmatrix}_i
\left\{
\begin{array}{c}
e_x \\
e_y \\
2e_{xy}
\end{array}
\right\}
-
\left\{
\begin{array}{c}
\bar{\alpha}_x \\
\bar{\alpha}_y \\
\bar{\alpha}_{xy}
\end{array}
\right\}_i
\Delta T
-
\left\{
\begin{array}{c}
\bar{\beta}_x \\
\bar{\beta}_y \\
\bar{\beta}_{xy}
\end{array}
\right\}_i
m,
$$

(7.9)

where the $\bar{Q}_{ij}$ matrix is the inverse of the $\bar{S}_{ij}$ matrix, and $\bar{\alpha}_i$ and $\bar{\beta}_i$ are products of $\bar{Q}_{ij}$ with α_i and β_i, respectively.

Assuming that the Kirchhoff deformation hypothesis is valid, i.e., the plane perpendicular to the mid-plane remains normal during deformation, the displacement fields u,v, and w along the x-, y-, and z-axes are given by

$$
\left.
\begin{array}{l}
u = u^0(x, y) - z\dfrac{\partial w}{\partial x}, \\[2mm]
v = v^0(x, y) - z\dfrac{\partial w}{\partial y}, \\[2mm]
w = w^0(x, y) = w(x, y).
\end{array}
\right\}
$$

(7.10)

The two-dimensional strain components e_x, e_y, and $2e_{xy} \ (= \gamma_{xy})$ are obtained from Eq. (7.10) as

$$
\left.
\begin{array}{l}
e_x = \dfrac{\partial u}{\partial x} = e_x^0 + z\kappa_x, \\[2mm]
e_y = \dfrac{\partial v}{\partial y} = e_y^0 + z\kappa_y, \\[2mm]
\gamma_{xy} = 2e_{xy} = \dfrac{\partial u}{\partial y} + \dfrac{\partial v}{\partial x} = \gamma_x^0 + 2z\kappa_{xy},
\end{array}
\right\}
$$

(7.11)

where κ_x, κ_y, and κ_{xy} are the curvatures defined by

$$
\left.\begin{aligned}
\kappa_x &= -\frac{\partial^2 w}{\partial x^2}, \\
\kappa_y &= -\frac{\partial^2 w}{\partial y^2}, \\
\kappa_{xy} &= -\frac{\partial^2 w}{\partial x \partial y}.
\end{aligned}\right\}
\tag{7.12}
$$

In Eqs. (7.10) and (7.11), the superscript 0 denotes the field variables (displacements and strains) at the mid-plane, $z = 0$. With Eq. (7.11), Eq. (7.9) can be rewritten as

$$
\left\{ \begin{array}{c} \sigma_x \\ \sigma_y \\ \sigma_{xy} \end{array} \right\} = \left[\begin{array}{ccc} \bar{Q}_{11} & \bar{Q}_{12} & 0 \\ \bar{Q}_{21} & \bar{Q}_{22} & 0 \\ 0 & 0 & \bar{Q}_{66} \end{array} \right]_i \left\{ \begin{array}{c} e_x^0 + z\kappa_x \\ e_y^0 + z\kappa_y \\ \gamma_{xy}^0 + 2z\kappa_{xy} \end{array} \right\} - \left\{ \begin{array}{c} \bar{\alpha}_x \\ \bar{\alpha}_y \\ \bar{\alpha}_{xy} \end{array} \right\}_i \Delta T - \left\{ \begin{array}{c} \bar{\beta}_x \\ \bar{\beta}_y \\ \bar{\beta}_{xy} \end{array} \right\}_i m.
\tag{7.13}
$$

The resultant in-plane forces N_x, N_y, and N_{xy} are obtained from

$$
\left\{ \begin{array}{c} N_x \\ N_y \\ N_{xy} \end{array} \right\} = \sum_{i=1}^{n} \int_{h_i}^{h_{i+1}} \left\{ \begin{array}{c} \sigma_x \\ \sigma_y \\ \sigma_{yx} \end{array} \right\} dz.
\tag{7.14}
$$

Similarly, the resultant bending moments M_x, M_y, and M_{xy} are obtained from

$$
\left\{ \begin{array}{c} M_x \\ M_y \\ M_{xy} \end{array} \right\} = \sum_{i=1}^{n} \int_{h_i}^{h_{i+1}} \left\{ \begin{array}{c} \sigma_x \\ \sigma_y \\ \sigma_{yx} \end{array} \right\} z\,dz.
\tag{7.15}
$$

Carrying out the integration with respect to z, Eq. (7.14) is reduced to

$$
[\mathbf{N}] = [A]\{e^0\} + [\mathbf{B}]\{\kappa\} - [\mathbf{N}]_T - [\mathbf{N}]_m,
\tag{7.16}
$$

where

$$[A] = \sum_{i=1}^{n} \begin{bmatrix} \bar{Q}_{11} & \bar{Q}_{12} & \bar{Q}_{16} \\ \bar{Q}_{21} & \bar{Q}_{22} & \bar{Q}_{26} \\ \bar{Q}_{16} & \bar{Q}_{26} & \bar{Q}_{66} \end{bmatrix}_i (h_{i+1} - h_i),$$

$$[B] = \tfrac{1}{2}\sum_{i=1}^{n} \begin{bmatrix} \bar{Q}_{11} & \bar{Q}_{12} & \bar{Q}_{16} \\ \bar{Q}_{21} & \bar{Q}_{22} & \bar{Q}_{26} \\ \bar{Q}_{16} & \bar{Q}_{26} & \bar{Q}_{66} \end{bmatrix}_i (h_{i+1}^2 - h_i^2),$$

$$[\mathbf{N}]_T = \sum_{i=1}^{n} \begin{bmatrix} \bar{\alpha}_x \\ \bar{\alpha}_y \\ \bar{\alpha}_{xy} \end{bmatrix}_i (h_{i+1} - h_i)\Delta T, \qquad (7.17)$$

$$[\mathbf{N}]_m = \sum_{i=1}^{n} \begin{bmatrix} \bar{\beta}_x \\ \bar{\beta}_y \\ \bar{\beta}_{xy} \end{bmatrix}_i (h_{i+1} - h_i)m.$$

Similarly, Eq. (7.15) is reduced to

$$[\mathbf{M}] = [B]\{\mathbf{e}^0\} + [D]\{\kappa\} - [\mathbf{M}]_T - [\mathbf{M}]_m, \qquad (7.18)$$

where

$$[D] = \tfrac{1}{3}\sum_{i=1}^{n} \begin{bmatrix} \bar{Q}_{11} & \bar{Q}_{12} & \bar{Q}_{16} \\ \bar{Q}_{21} & \bar{Q}_{22} & \bar{Q}_{26} \\ \bar{Q}_{16} & \bar{Q}_{26} & \bar{Q}_{66} \end{bmatrix}_i (h_{i+1}^3 - h_i^3),$$

$$[\mathbf{M}]_T = \tfrac{1}{2}\sum_{i=1}^{n} \begin{bmatrix} \bar{\alpha}_x \\ \bar{\alpha}_y \\ \bar{\alpha}_{xy} \end{bmatrix}_i (h_{i+1}^2 - h_i^2)\Delta T, \qquad (7.19)$$

$$[\mathbf{M}]_m = \tfrac{1}{2}\sum_{i=1}^{n} \begin{bmatrix} \bar{\beta}_x \\ \bar{\beta}_y \\ \bar{\beta}_{xy} \end{bmatrix}_i (h_{i+1}^2 - h_i^2)m.$$

Combining Eqs. (7.16) and (7.19), we obtain

$$\left\{\begin{matrix} N \\ M \end{matrix}\right\} = \begin{bmatrix} A & B \\ B & D \end{bmatrix} \left\{\begin{matrix} e^0 \\ \kappa \end{matrix}\right\} - \left\{\begin{matrix} N_T \\ M_T \end{matrix}\right\} - \left\{\begin{matrix} N_m \\ M_m \end{matrix}\right\}. \qquad (7.20)$$

For a given set of loading, N, M, N_T, M_T, N_m, M_m, and material data (A, B, and D), one can solve for e^0 and κ. The strain and stress field in the *i*th lamina can be computed from Eqs. (7.11) and (7.13), respectively.

7.2 Lamination theory applied to piezoelectric laminates

A piezoelectric material exhibits coupling between its mechanical and electrical behavior and is a key actuator and sensor material. Piezoelectric plates of monomorph and bimorph types have been used extensively for bending actuators. A monomorph actuator, Fig. 7.2(a), is composed of one piezoelectric plate and one non-piezoelectric plate, while a bimorph actuator, Fig. 7.2(b), is made of two identical piezoelectric plates with the electric polarities of the two piezoelectric plates being of opposite signs. Figure 7.2(c) shows the design of a one-sided (rainbow-type) laminated composite actuator with a functionally graded microstructure (FGM) where the electroelastic properties are graded smoothly through the thickness. Recently, bimorph actuators have been modified in such a way that the top and bottom (oppositely charged) piezoelectric plates are composed of multi-layers where the electroelastic properties are graded smoothly through the plate thickness, Fig. 7.2(d). This modified bimorph is known as a "functionally graded microstructure (FGM) bimorph" (Almajid *et al.*, 2001; Almajid and Taya, 2001; Taya *et al.*, 2003a). The FGM bimorph actuator represents the most general design of piezoelectric bending actuator. A piezoelectric laminate consists of *n* laminae, each lamina being a piezoelectric material with specified electroelastic properties. The constitutive equations of a linear piezoelectric material are given by Eqs. (3.217)–(3.220) where the term related to pyroelectric coupling is also included. The in-plane constitutive equations of a piezoelectric material of crystallographic class *6mm*

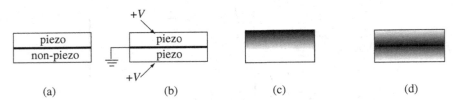

Fig. 7.2   Different types of piezoactuators: (a) monomorph, (b) bimorph, (c) one-sided FGM, and (d) FGM bimorph (see also Fig. 7.5).

under an applied electric field and a uniform temperature, and also the assumption of plane stress, can be obtained from Eqs. (7.1) as

$$
\left\{ \begin{array}{c} \sigma_x \\ \sigma_y \\ \sigma_{xy} \end{array} \right\}_i = \begin{bmatrix} \bar{Q}_{11} & \bar{Q}_{12} & 0 \\ \bar{Q}_{21} & \bar{Q}_{22} & 0 \\ 0 & 0 & \bar{Q}_{66} \end{bmatrix}_i \left\{ \begin{array}{c} e_x^0 + z\kappa_x - \alpha_{11}\Delta T \\ e_y^0 + z\kappa_y - \alpha_{11}\Delta T \\ \gamma_{xy}^0 + z\kappa_{xy} \end{array} \right\}
$$

$$
- \begin{bmatrix} 0 & 0 & \bar{e}_{31} \\ 0 & 0 & \bar{e}_{32} \\ 0 & 0 & 0 \end{bmatrix}_i \left\{ \begin{array}{c} 0 \\ 0 \\ E_z \end{array} \right\}_i , \tag{7.21}
$$

where

$$
\bar{Q}_{ij} = C_{ij} - \frac{C_{i3}}{C_{33}} C_{j3},
$$

$$
\bar{e}_{ij} = \frac{C_{i3}}{C_{33}} e_{33} - e_{ij}. \tag{7.22}
$$

It is to be noted here that $\bar{Q}_{ij}$ are the reduced stiffness constants and $\bar{e}_{ij}$ the reduced piezoelectric constants that are modified by the assumption of plane stress, e_x^0, e_y^0, and e_{xy}^0 are the in-plane strain components at mid-plane ($z = 0$), as shown in Fig. 7.1, and κ_x, κ_y, and κ_{xy} are the curvatures of the plate.

Each piezoelectric lamina is considered a capacitor whose electric field corresponds to its dielectric constant. The electric field in each layer of a multi-layer FGM piezoelectric can be obtained by considering the layers as a series of capacitors. The relationship between the electric capacitance and the voltage is given as

$$
C_t = \frac{Q}{V_t}, \tag{7.23}
$$

where C_t is the total capacitance, Q is the charge, and V_t is the total voltage. The total capacitance in a series of capacitances is found from

$$
\frac{1}{C_t} = \sum_{i=1}^{n} \frac{1}{C_i} = \sum_{i=1}^{n} \frac{t_i}{\varepsilon_i A}, \tag{7.24}
$$

where t_i and ε_i are the thickness and the dielectric constant for each lamina, and A is surface area. The electric capacitance in each layer will then be

$$
C_i = \frac{Q}{V_i} = \frac{\varepsilon_i A}{t_i}. \tag{7.25}
$$

Substituting Eqs. (7.23) and (7.24) into Eq. (7.25) results in

$$V_i = Q\frac{t_i}{A\varepsilon_i} = V_t\frac{t_i}{\varepsilon_i\sum\limits_{l=1}^{n}\frac{t_l}{\varepsilon_l}}. \tag{7.26}$$

The electric field in each layer is

$$E_i = \frac{V_i}{t_i} = V_t\frac{1}{\varepsilon_i\sum\limits_{l=1}^{n}\frac{h_{l+1}-h_l}{\varepsilon_l}} \tag{7.27}$$

where the thickness t_l of the lth lamina is given by $h_{l+1}-h_l$, see Fig. 7.1.

In-plane stress resultants (N_x, N_y, N_{xy}) and stress couples (M_x, M_y, M_{xy}) are defined by

$$\{\mathbf{N},\mathbf{M}\} = \sum_{i=1}^{n}\int_{h_i}^{h_{i+1}}\{\boldsymbol{\sigma}\}\{dz, zdz\}. \tag{7.28}$$

Substituting Eq. (7.21) into Eq. (7.28), we obtain

$$\begin{bmatrix}\mathbf{N}\\\mathbf{M}\end{bmatrix} = \begin{bmatrix}\mathbf{A} & \mathbf{B}\\\mathbf{B} & \mathbf{D}\end{bmatrix}\begin{Bmatrix}\mathbf{e}^0\\\boldsymbol{\kappa}\end{Bmatrix} - \begin{bmatrix}\mathbf{N}^E\\\mathbf{M}^E\end{bmatrix}, \tag{7.29}$$

where

$$[\mathbf{A},\mathbf{B},\mathbf{D}] = \sum_{i=1}^{n}\int_{h_i}^{h_{i+1}}[\bar{Q}]_i[dz, zdz, z^2dz], \tag{7.30}$$

$$[\mathbf{N},\mathbf{M}]^E = \sum_{i=1}^{n}\int_{h_i}^{h_{i+1}}[\bar{e}]_i\{\mathbf{E}\}_i[dz, zdz]. \tag{7.31}$$

In the above derivation, we assumed that the piezoelectric laminated plate is subjected to constant temperature and constant electric field. Under a constant applied electric field ($E_x = E_y = 0$, $E_z \neq 0$) and constant temperature ($\Delta T = 0$) in the absence of applied mechanical loading of the plate, we obtain from Eq. (7.29) the strain of the mid-plane and the curvature of the plate as

$$\begin{bmatrix}\mathbf{e}^0\\\boldsymbol{\kappa}\end{bmatrix} = \begin{bmatrix}\mathbf{a} & \mathbf{b}\\\mathbf{b} & \mathbf{d}\end{bmatrix}\begin{bmatrix}\mathbf{N}^E\\\mathbf{M}^E\end{bmatrix}, \tag{7.32}$$

where

$$\begin{bmatrix} a & b \\ b & d \end{bmatrix} = \begin{bmatrix} A & B \\ B & D \end{bmatrix}^{-1}. \tag{7.33}$$

7.3 Accurate model for cylindrical bending of piezoelectric laminate composites

The lamination model discussed in Section 7.2 is called the "classical lamination theory" (CLT). It assumes plane stress along the z-direction. The CLT cannot provide the interlaminar shear stress $\tau_{xz} = \tau_{zx}$ which is considered to be large near the free surface at the interfaces of the laminae, and hence is responsible for debonding of the interfaces of a laminated composite. To solve for τ_{xz} properly, we must use the correct plate theory based on 2D elasticity. Pagano (1969) formulated such a 2D elasticity model for a laminate plate by assuming plane strain along the y-direction. Almajid and Taya (2001) used the method of Pagano and applied it to a laminate plate composed of N layers of graded piezoelastic properties.

Consider a multi-layer laminate plate, with each layer an orthotropic piezoelectric material. The plate undergoes cylindrical bending generated by a constant electric field applied in the z-direction (Lagoudas and Zhonghe, 1994). The elasticity plate model assumes a state of "plane strain" along the y-axis, where $e_y = e_{xy} = e_{yz} = 0$, Fig. 7.1. The laminate is simply supported at both ends, $x = 0$ and $x = l$. The constitutive equations of a piezoelectric material in the absence of temperature are given by (Ikeda, 1990)

$$\begin{aligned} e_{ij} &= S_{ijkl}\sigma_{kl} + d_{mij}E_m, \\ \sigma_{ij} &= C_{ijkl}\varepsilon_{kl} - e_{mij}E_m, \end{aligned} \tag{7.34}$$

where σ_{ij}, e_{ij} are the stress and the strain tensor component, E_m is the electric field vector component, S_{ijkl} and C_{ijkl} are the elastic compliance and stiffness tensors, and e_{mij} and d_{mij} are the piezoelectric coefficients. The reduced constitutive equations of a crystallographic class $6mm$ symmetric type piezoelectric material for the case of plane strain are given as

$$\begin{Bmatrix} e_x \\ e_z \\ 2_{xz} \end{Bmatrix} = \begin{bmatrix} R_{11} & R_{13} & 0 \\ R_{13} & R_{33} & 0 \\ 0 & 0 & R_{55} \end{bmatrix} \begin{Bmatrix} \sigma_x \\ \sigma_z \\ \sigma_{xz} \end{Bmatrix} + \begin{bmatrix} 0 & 0 & \bar{d}_{31} \\ 0 & 0 & \bar{d}_{33} \\ \bar{d}_{15} & 0 & 0 \end{bmatrix} \begin{Bmatrix} E_x \\ E_y \\ E_z \end{Bmatrix}, \tag{7.35}$$

where R_{ij} and $\bar{d}_{ij}$ are the reduced compliance and piezoelectric coefficients, respectively. They are defined by

$$R_{ij} = S_{ij} - \frac{S_{i3}}{S_{33}} S_{j3},$$

$$\bar{d}_{ij} = d_{ij} - \frac{S_{i3}}{S_{33}} d_{3j},$$

(7.36)

where S_{ij} and d_{ij} are the compressed forms of S_{ijkl} and d_{mij}, see Eqs. (3.227)–(3.230). The boundary conditions of the simply supported piezoelectric laminate are given by

$$\sigma_x(x = 0, z) = \sigma_x(x = l, z) = 0,$$

$$w(x = 0, z) = w(x = l, z) = 0,$$

(7.37)

where w is the out-of-plane displacement (along the z-axis). The prescribed surface-traction-free conditions of the upper and lower surfaces of the laminate require

$$\sigma_z\left(x, -\frac{t}{2}\right) = \sigma_z\left(x, \frac{t}{2}\right) = 0,$$

$$\sigma_{xz}\left(x, -\frac{t}{2}\right) = \sigma_{xz}\left(x, \frac{t}{2}\right) = 0.$$

(7.38)

The interface continuity conditions at each layer are given by

$$\left.\begin{array}{l} \sigma_z^i\left(x, \frac{t_i}{2}\right) = \sigma_z^{i+1}\left(x, \frac{t_i}{2}\right), \\[2mm] \sigma_{xz}^i\left(x, \frac{t_i}{2}\right) = \sigma_{xz}^{i+1}\left(x, \frac{t_i}{2}\right), \\[2mm] u_i\left(x, \frac{t_i}{2}\right) = u_{i+1}\left(x, \frac{t_i}{2}\right), \\[2mm] w_i\left(x, \frac{t_i}{2}\right) = w_{i+1}\left(x, \frac{t_i}{2}\right), \end{array}\right\} (i = 1, 2, \ldots, N-1)$$

(7.39)

where u_i is the displacement along the x-axis in the ith lamina, i represents the interface ID number between the ith and the $(i+1)$th lamina and the topmost and bottommost interfaces are defined as the 1st and $(N-1)$th interfaces, respectively, while t_i represents the thickness of the ith lamina, Fig. 7.1, where the top layer corresponds to $i=1$ while the bottom layer is $i=N$. The electric field is applied in the z-direction through the thickness of the laminate. The electric field is expressed in the form of a Fourier series as

$$E_z = E_0 \sum_{n=1,3,5,\ldots}^{\infty} \frac{4}{n\pi} \sin(\lambda x),$$

$$E_x = E_y = 0.$$

(7.40)

where $\lambda = n\pi/l$, and for higher values of n a nearly constant electric field can be realized. Following Pagano (1969), the stress field in the ith layer is assumed to be

$$
\left.
\begin{aligned}
\sigma_x^i &= f_i''(z) \sum_{n=1}^{\infty} \frac{4}{n\pi} \sin\left(\frac{n\pi}{l}x\right), \\
\sigma_z^i &= -f_i(z) \sum_{n=1}^{\infty} \left(\frac{n\pi}{l}\right)^2 \frac{4}{n\pi} \sin\left(\frac{n\pi}{l}x\right), \\
\sigma_{xz}^i &= -f_i'(z) \sum_{n=1}^{\infty} \left(\frac{n\pi}{l}\right) \frac{4}{n\pi} \cos\left(\frac{n\pi}{l}x\right),
\end{aligned}
\right\}
\tag{7.41}
$$

where $f_i(z)$ is an unknown function to be determined. The equilibrium equations given by

$$
\sigma_{ij,j} = 0
\tag{7.42}
$$

are automatically satisfied if Eqs. (7.41) are substituted into Eq. (7.42). The 2D strain compatibility equation is given by

$$
2e_{xz,xz} - e_{z,xx} - e_{x,zz} = 0.
\tag{7.43}
$$

Substitution of Eq. (7.35) into Eq. (7.43) and use of Eqs. (7.40) and (7.41) leads to a fourth-order ordinary differential equation,

$$
R_{11}^i f_i''''(z) - \lambda^2 \left(2R_{13}^i + R_{55}^i\right) f_i''(z) + \lambda^4 R_{33}^i f_i(z) = \lambda^2 \left(\bar{d}_{33}\right)^i E_0.
\tag{7.44}
$$

The solutions $f_i(z)$ can be expressed by

$$
f_i(z) = \sum_{j=1}^{4} A_{ji} \exp\left(m_{ji} z_i\right) + \frac{\left(\bar{d}_{33}\right)^i E_0}{\lambda^2 R_{33}^i},
\tag{7.45}
$$

where A_{ji} are constant and the values of m_{ji} are given by

$$
\left.
\begin{aligned}
\begin{pmatrix} m_{1i} \\ m_{2i} \end{pmatrix} &= \pm\lambda\left(\frac{a_i + b_i}{c_i}\right)^{1/2}, \\
\begin{pmatrix} m_{3i} \\ m_{4i} \end{pmatrix} &= \pm\lambda\left(\frac{a_i - b_i}{c_i}\right)^{1/2},
\end{aligned}
\right\}
\tag{7.46}
$$

and

$$
\left.
\begin{aligned}
a_i &= R_{55}^i + 2R_{13}^i, \\
b_i &= \left[a_i^2 - 4R_{11}^i R_{33}^i\right]^{1/2}, \\
c_i &= 2R_{11}^i.
\end{aligned}
\right\}
\tag{7.47}
$$

The stress field solutions in the ith layer can be found using Eqs. (7.41) as

$$
\left.
\begin{aligned}
\sigma_x^i &= \sum_{n=1,3,\dots}^{\infty} \sum_{j=1}^{4} A_{ji} m_{ji}^2 \exp(m_{ji} z_i) \frac{4}{\pi} \sin(\lambda x), \\
\sigma_z^i &= \sum_{n}^{\infty} \left[\sum_{j=1}^{4} -A_{ji} \exp(m_{ji} z_i) \lambda^2 \frac{4}{\pi} \sin(\lambda x) - \frac{(\bar{d}_{33})^i E_0}{R_{33}^i} \frac{4}{\pi} \sin(\lambda x) \right], \\
\sigma_{xz}^i &= \sum_{n}^{\infty} \sum_{j=1}^{4} -A_{ji} m_{ji} \exp(m_{ji} z_i) \lambda \frac{4}{\pi} \cos(\lambda x),
\end{aligned}
\right\} \quad (7.48)
$$

while the displacement components are found from the strain–displacement relation:

$$
\left.
\begin{aligned}
u_i &= \sum_{n}^{\infty} \left\{ \sum_{j=1}^{4} A_{ji} \exp(m_{ji} z_i) \left[\lambda R_{13}^i - \frac{R_{11}^i}{\lambda} m_{ji}^2 \right] \right. \\
&\quad \left. + \left[\frac{R_{13}^i}{R_{33}^i} (\bar{d}_{33})^i - (d_{31})^i \right] \frac{E_0}{\lambda} \right\} \frac{4}{\pi} \cos(\lambda x), \\
w_i &= \sum_{n}^{\infty} \sum_{j=1}^{4} A_{ji} \exp(m_{ji} z_i) \left[R_{13}^i m_{ji} - \frac{\lambda^2 R_{33}^i}{m_{ji}} \right] \frac{4}{\pi} \sin(\lambda x).
\end{aligned}
\right\} \quad (7.49)
$$

It is to be noted that the above solutions satisfy the boundary conditions of Eqs. (7.37) while the surface-traction-free boundary conditions of Eqs. (7.38) and the interface continuity conditions of Eqs. (7.39) remain to be satisfied, resulting in $4N$ equations for $4N$ unknown constants A_{ji}.

7.4 Design of functionally graded microstructure of piezoactuators

As examples of piezoelectric laminates, we shall study two cases of piezo-electric laminate with functionally graded microstructure (FGM) where the electroelastic properties of the laminae are changed smoothly, as illustrated in Fig. 7.2(c) and (d). As a comparison we shall also examine the case of a bimorph-type piezoelectric laminate, Fig. 7.2(b).

7.4.1 Design of one-sided (rainbow-type) piezoactuators

The piezoelectric actuator with one-sided FGM (sometimes called a "rainbow" actuator) is illustrated in Fig. 7.2(c) where the electroelastic properties are graded smoothly through the thickness. The advantage of such a one-sided FGM is a reduction in the internal stress field induced by the applied electric field, but the bending displacement remains modest. Zhu and Meng (1995) designed a piezoelectric bending actuator with a one-sided FGM where the compositions are graded from high piezoelectrics mixed with low dielectrics to low piezoelectrics with high dielectrics. Wu *et al.* (1996) fabricated a piezo-electric actuator with a one-sided FGM in which the electrical resistance through the plate thickness was achieved by using diffusional control of ZnB.

Li *et al.* (2003) designed a piezoelectric ceramic bending actuator with a one-sided FGM where the grading was achieved by a smooth change in porosity through the plate thickness. They used commercially available PZT powders and laminated several layers in an Al_2O_3 mold which was cold-isostatically-pressed at 200 MPa to form a green body, then sintered in air at 1473 K for one hour. The porosity range achieved in the above processing is from 0.23 to 20.82 vol.%. The electroelastic properties of these porous PZT layers are summarized in Table 7.1.

A scanning electron microscopy (SEM) photo of the cross-section of the porous FGM actuator designed and fabricated by Li *et al.* (2003) is shown in Fig. 7.3 where pores are seen as dark elongated islands against a gray back-ground of the PZT phase. The bending displacement of the one-sided porous piezoelectric actuator was measured by two strain gauges mounted on the top (e_1) and bottom (e_2) surfaces of the actuator beam. Then the curvature κ of the cantilever beam is calculated from

$$\kappa = \frac{1}{t}\left(1 - \frac{1 + e_1}{1 + e_2}\right) \tag{7.50}$$

where t is the thickness of the beam. Figure 7.4 shows the curvature vs. applied voltage relation of the measured data and the prediction by the CLT model discussed in Section 7.2, i.e., Eq. (7.32); agreement between them is good.

Table 7.1 *Piezoelectric and elastic constants of PZT ceramics with different amounts of bulk porosity* (Li *et al.*, 2003. Reprinted with permission of the Amer. Ceram. Soc., www.ceramics.org, © 2003. All rights reserved.).

Porosity (vol.%)	Piezoelectric coeffs. (pC/N)		Coupling coeff.	Elastic constants (GPa)			Piezoelectric const. (C/m^2)
	d_{31}	d_{33}	k_{31}	C_{11}	C_{12}	C_{66}	e_{31}
0.23	−181	451	0.38	142	94.6	23.8	−5.17
4.25	−150	420	0.31	110	65.8	22	−3.49
9.47	−122	436	0.25	86.1	46.8	19.6	−0.97
14.41	−75.9	377	0.18	74.1	38.2	18	0.46
20.82	−47.2	350	0.12	58.7	27.2	15.8	1.93

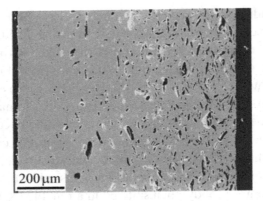

Fig. 7.3   SEM micrograph of the fabricated PZT sample with a linear porosity gradient (Li *et al.*, 2003. Reprinted with permission of the Amer. Ceram. Soc., www.ceramics.org, © 2003. All rights reserved.).

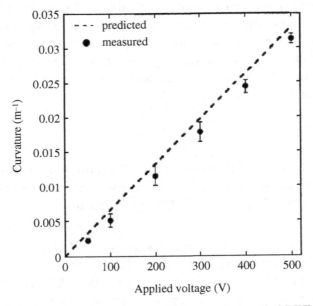

Fig. 7.4   Voltage-induced curvature of the porosity-graded PZT beam as a function of applied voltage (Li *et al.*, 2003. Reprinted with permission of the Amer. Ceram. Soc., www.ceramics.org, © 2003. All rights reserved.).

7.4.2 Design of FGM bimorph piezoactuators

The requirements of bending actuators are a large displacement and minimum stress induced in the piezoelectric layers. In this respect, the rainbow-type piezoactuators satisfy the second requirement, but not the first. In order to satisfy both requirements, Taya and his co-workers (Almajid *et al.*, 2001;

Almajid and Taya, 2001; Taya *et al.*, 2003a) designed a set of FGM-bimorph piezoactuators (see Fig. 7.2(d)).

For the purpose of modeling the behavior, the distributions of mechanical and piezoelectric properties are assumed to vary throughout the FGM plate thickness in a stepwise manner as shown in Fig. 7.5 where two cases of stepwise linear distribution are considered, type-A and type-B. For each type, six layers of FGM are considered. The electroelastic properties of the piezoelectrics used are shown in Table 7.2, where the piezoelectric (e_{ij}) and dielectric (ε) properties of PZT-C91 are higher than those of PZT-C6 while the mechanical (C_{ij}) properties are comparable. The top and bottom halves of the FGM laminate each consist of three layers. The outer and inmost layers are PZT-C91 and PZT-C6, while the intermediate layer is a composite of these two materials. Figure 7.5(a) (type-A) represents the case where the PZT-C6 material with low piezoelectric properties forms the outer layers of the top and bottom halves while PZT-C91 with high piezoelectric properties forms the inmost layers. Here the piezoelectric properties increase toward the middle. Figure 7.5(b) (type-B) represents the opposite case, where PZT-C91 forms the outer layers of the top and bottom halves, while PZT-C6 forms the inmost layers. Here the piezoelectric properties decrease towards the middle.

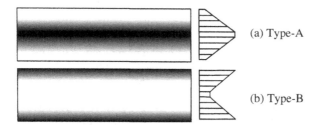

(a) Type-A

(b) Type-B

Fig. 7.5.   Electroelastic properties distribution in FGM-bimorphs (Taya *et al.*, 2003a, with permission from Elsevier Ltd).

Table 7.2. *Electroelastic properties of PZT piezoelectric materials* (Taya *et al.*, 2003a, with permission from Elsevier Ltd).

Material	Elastic constants (GPa)					Piezoelectric constants (C/m^2)			Relative dielectric constants	
	C_{11}	C_{12}	C_{13}	C_{33}	C_{44}	e_{31}	e_{33}	e_{15}	$\varepsilon_{11}/\varepsilon_0$	$\varepsilon_{33}/\varepsilon_0$
PZT-C91	120	77	77	114	24	−17.3	21.2	20.2	2557	2664
PZT-C6	123	77	80	112	19	−7.3	17.2	14.5	1039	749

$\varepsilon_0 = 8.85 \times 10^{-10} C/N \cdot m^2$

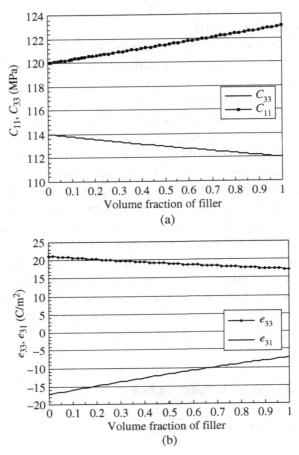

Fig. 7.6   (a) Elastic and (b) piezoelectric constants of piezoelectric composite. Filler is PZT-C6; parent material is PZT-C91 (Almajid, 2002).

The electroelastic properties of each composite lamina can be evaluated using the Eshelby model as explained in Chapter 4. The electroelastic properties of a piezoelectric composite to represent any layer (lamina) are illustrated in Fig. 7.6 where PZT-C91 is assumed to be the parent material while PZT-C6 is assumed to be the filler material. Based on the design requirements of the FGM, i.e., maximum bending displacement and minimum stresses induced in the layers, the best FGM stacking can be chosen to obtain the optimum microstructure of an actuator. FGM bimorphs type-A and type-B are investigated to develop the optimum actuator.

The electroelastic properties of each composite layer can be made to vary from pure PZT-C6 to pure PZT-C91. The volume fraction of filler required in a given composite layer (the ith layer) can be found from

$$V_{\mathrm{f}} = \left(\frac{i-1}{n-1}\right)^{1/m} \tag{7.51}$$

where m is the FGM volume fraction exponent, and n is the total number of FGM layers. For a low value of m, the volume fraction of the filler becomes very low, which represents the case where the layer is made of PZT-C91 while a high value of m represents the PZT-C6 layer, and for $m=1$ the distribution is linear.

The dimensions of the model actuator are 50 mm length by 20 mm width by 0.67 mm thickness. Each FGM layer is of 0.11 mm thickness while the middle layer is a platinum plate of 0.01 mm thickness. The average applied electric field is 100 V/mm (33 V applied voltage). The displacement and the in-plane stresses for type-A and type-B FGM bimorphs are presented in Fig. 7.7. By comparing the predicted results of Fig. 7.7(a,b)(type-A) and (c,d)(type-B), FGM bimorph type-A is found to outperform type-B in both displacement

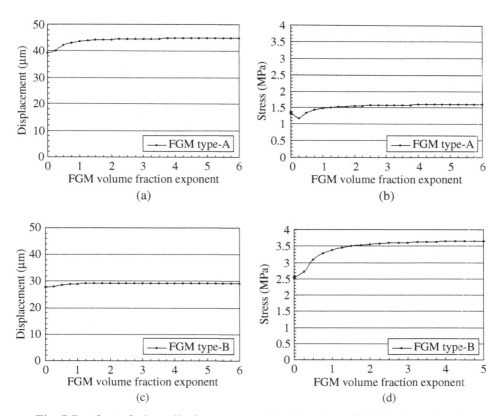

Fig. 7.7   Out-of-plane displacement and in-plane stress field for type-A and type-B FGM bimorph vs. FGM volume fraction exponent m under 100 V/mm electric field (Almajid, 2002).

and stress field. Type-A FGM bimorph is about 75% higher in displacement while about 50% lower in in-plane stresses.

The displacement and stress field in type-A increase with the increase in the FGM volume fraction exponent m. The performance of the type-A FGM bimorph stabilizes when the FGM volume fraction exponent reaches 2. The thickness of each layer of the FGM bimorph influences the performance of the actuator. The thickness of each layer is chosen according to the following equation:

$$\frac{t_i - \sum_{l=1}^{i} t_{l-1}}{h} = \left(\frac{i}{n}\right)^{1/p} \tag{7.52}$$

where t_i and h are the thickness of the ith layer and the total thickness of the laminate, respectively, n is the number of layers in the laminate, and p is the FGM thickness exponent. The influence of the thickness of each layer on the performance of the FGM bimorph actuator is illustrated in Fig. 7.8. The FGM volume fraction exponent m used here is 4.25, which represents a volume fraction of 85% of PZT-C91. The displacement and stress field increase with the increase in the FGM thickness exponent p. The displacement in Fig. 7.8(a) stabilizes around a value of $p = 1$ while the stress continues to rise with increase in p.

In the following, we will show the predicted results of the out-of-plane displacement and in-plane stress in various types of piezoelectric FGM. The FGM volume fraction exponent m is 4.25, while the FGM thickness exponent p is 1. The results predicted by the present model are given in Table 7.3.

It is to be noted from Table 7.3 that for the FGM bimorph type-A the induced in-plane stress σ_x is a minimum while the bending displacement reaches a high value comparable to that of the PZT-C91 standard bimorph. The stress profiles through the thickness for the standard and FGM bimorphs were investigated

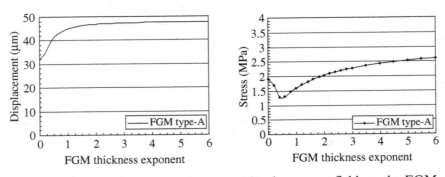

Fig. 7.8   Out-of-plane displacement and in-plane stress field vs. the FGM thickness exponent p under 100 V/mm applied electric field. FGM volume Fraction exponent m was taken to be 4.25 (85% PZT-C91) (Almajid, 2002).

Table 7.3 *Comparison of predicted out-of-plane bending displacement and in-plane stress in various types of FGM piezoelectric-composite plates* (Almajid, 2002).

Type of FGM	Standard bimorph, C91, Fig. 7.2	Standard bimorph, C6, Fig. 7.2	FGM bimorph, C91–C6 type-A, Fig. 7.5(a)	FGM bimorph, C91–C6 type-B, Fig. 7.5(b)
Applied voltage (V)	33	33	33	33
No. of layers	2	2	6	6
Thickness (mm)				
Laminate	0.67	0.67	0.67	0.67
Layer	0.33	0.33	0.11	0.11
Platinum	0.01	0.01	0.01	0.01
Electric field (V/mm)	100	100	100	100
Curvature (1/m)	0.152	0.102	0.143	0.094
Max. σ_x (MPa)	3.08	1.91	1.59	3.65
Bending displacement (μm) [for plate length 50 mm]	47.5	31.9	44.8	29.3

Table 7.4 *Comparison of experimental and predicted results for piezoactuators* (Almajid, 2002).

Bimorph system	Method	Maximum bending displacement (μm)	Max. σ_x (MPa)	Max. τ_x (MPa)
Standard bimorph PZT-C91	Experimental	49.6	—	—
	CLT	47.5	3.08	—
	2D elasticity	47.8	3.07	1.48
	ANSYS FEM	47.6	3.17	1.56
FGM bimorph type-A	Experimental	43.8	—	—
	CLT	43.6	1.48	—
	2D elasticity	44.9	1.63	0.59
	ANSYS FEM	44.8	1.65	0.63
FGM bimorph type-B	Experimental	30.4	—	—
	CLT	29.2	3.38	—
	2D elasticity	29.5	3.69	1.82
	ANSYS FEM	29.4	3.75	1.9

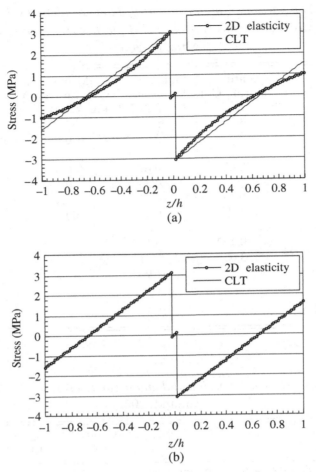

Fig. 7.9.   Normal stress σ_x through the thickness profile of standard bimorph at (a) $\ell/h = 4$, (b) $\ell/h = 75$. (Almajid, 2002.)

using CLT and 2D elasticity models where the accuracy of the CLT model for different aspect ratio, length to thickness (ℓ/h), was determined. Aspect ratios of 4 and 75 were used. The CLT predictions agreed very well with the 2D elasticity model for the high aspect ratio but there was less agreement at the low aspect ratio where the assumption of plane stress no longer applies, see Figs. 7.9 and 7.10.

To verify the above models, Taya *et al.* (2003a) processed several types of bending piezoactuators – standard bimorph, and FGM bimorphs of type-A and type-B where piezoceramics PZT-C91 and PZT-C6 are used as core material. Figure 7.11 shows the experimental data (symbols) of these three different bending actuators, which are compared with predictions made using the CLT model (lines), resulting in a good agreement. To clarify the comparison between

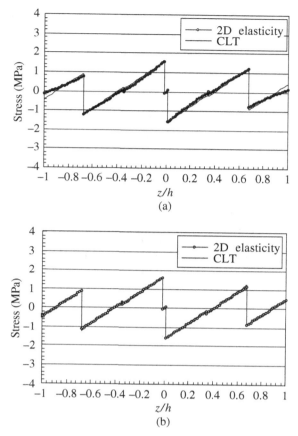

Fig. 7.10   Normal stress σ_x through the thickness profile of a six-layered FGM bimorph at (a) $\ell/h = 4$, (b) $\ell/h = 75$. (Almajid, 2002.)

predictions and measured data, they calculated the maximum bending displacements and maximum stresses (σ_x, τ_{xz}) using CLT, 2D elasticity, and ANSYS FEM. The results are summarized in Table 7.4. It is clearly seen from Table 7.4 that the experimental data are predicted with good accuracy by all three models and the type-A FGM bimorph is the best bending actuator design in maximizing the bending displacement while minimizing the induced stress field, thus enhancing its service life. It is also to be noted in Table 7.4 that the interlaminar stress (τ_{xz}), which could not be predicted by CLT, is now predicted by the 2D elasticity model whose predictions agree reasonably well with those by 3D ANSYS FEM.

A piezoelectric bending actuator of FGM bimorph type-B has also been designed by Takagi *et al.* (2002). They used PZT mixed with Pt, where three FGM layers for the top half and bottom half were used along with a central

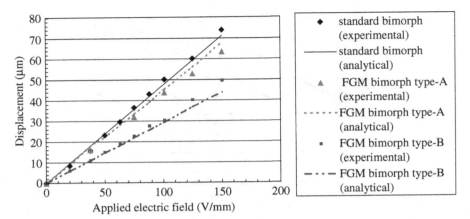

Fig. 7.11   Bending displacement vs. applied electric field for several types of piezocomposite actuators: experimental data, and predictions of CLT model (Taya *et al.*, 2003a, with permission from Elsevier Ltd.)

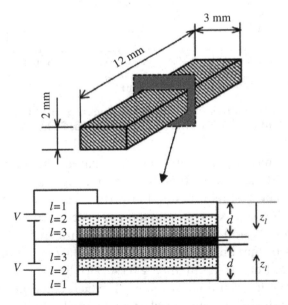

Fig. 7.12   Dimensions of the bimorph actuator and the cross-sectional lamination model (Takagi *et al.*, 2002, with permission from Elsevier Ltd.)

electrode layer composed of a PZT/30% Pt composite, Fig 7.12. The composition distribution in the upper and lower halves is given by

$$\frac{f_l^{pt} + \Delta f^{pt}}{f_{max}^{pt} + \Delta f^{pt}} = \left(\frac{z_l}{d}\right)^m \tag{7.53}$$

Where d and z_l are defined in Fig. 7.12, f_l^{pt} is the volume fraction of Pt in the lth layer, f_{max}^{pt} is the maximum volume fraction of Pt in the graded layer, i.e. for $l = 3$, and Δf^{pt} is the increment of Pt volume fraction, equal to 0.1 from one layer to the next. The electroelastic properties of the FGM layers are given in Table 7.5.

The PZT/Pt piezoelectric composite actuator of FGM biomorph type-B was fabricated by cold-isostatic-pressing of layered PZT/Pt mixtures in a steel die under a sequence of 100 MPa and 200 MPa pressure, followed by sintering at 1473 K for one hour in an excess phosphorus atmosphere. A bending actuator specimen with dimensions shown in Fig. 7.12 was cut from the as-sintered disk, and silver electrodes were attached on the top and bottom. The specimen is poled under an electric field of 1 V/mm for 30 minutes in a silicone oil bath at 393 K. The

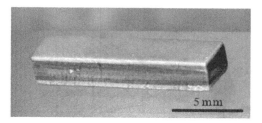

Fig. 7.13   Graded bimorph actuator with a linear composition profile ($m = 1.0$, see Eq. (7.53)) (Takagi *et al.*, 2002, with permission from Elsevier Ltd.)

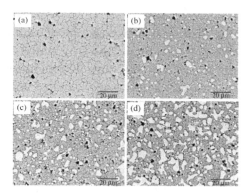

Fig. 7.14   SEM micrographs of each layer: (a) PZT/0% Pt, (b) PZT/10% Pt, (c) PZT/20% Pt, and (d) PZT/30% Pt. (Takagi *et al.*, 2002, with permission from Elsevier Ltd.)

Table 7.5 *Density, elastic and piezoelectric constants for PZT/Pt composites* (Takagi *et al.*, 2002, with permission from Elsevier Ltd).

Layer number	1	2	3	Center
Composition	PZT	PZT/10% Pt	PZT/20% Pt	PZT/30% Pt
ρ (kg/m^3)	790	920	1050	1187
C_{11} (GPa)	146	139	164	197
C_{12} (GPa)	95.4	87.2	108	74.3
C_{13} (GPa)	94.3	90.1	101	74.3
C_{33} (GPa)	128	134	145	197
C_{66} (GPa)	25.3	25.7	28.0	28.4
e_{31} (C/m^2)	-3.94	-3.08	-0.75	conductive
e_{33} (C/m^2)	17.5	17.6	17.5	conductive
$\varepsilon_{33}/\varepsilon_0$	1654	1959	2995	conductive

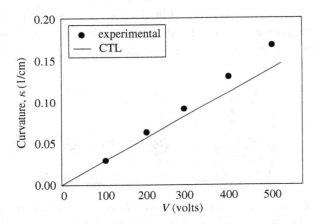

Fig. 7.15   Curvature (κ) vs. applied voltage V of PZT/Pt FGM bimorph type-B actuator (Takagi *et al.*, 2002, with permission from Elsevier Ltd).

specimen is shown in Fig. 7.13. The microstructure of each FGM layer is shown in Fig. 7.14 where the white area is Pt and the gray matrix is PZT. Black regions are caused by "plucking out"of material during preparation for the SEM.

The measured curvatures κ as a function of applied voltage V are plotted in Fig. 7.15 where the predictions by the CLT model, Eq. (7.32), are shown by the solid line, indicating good agreement for the lower voltage range, but at the higher voltage range the experimental data exceed the predictions.

7.5 Thermal stress in coating/substrate system

For electronic packaging including MEMS, a thin film is often deposited on a substrate by chemical vapour deposition (CVD) or physical vapour deposition

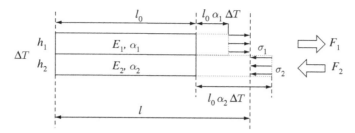

Fig. 7.16 One-dimensional model of coating–substrate system assuming α_2 greater than α_1 and complete bonding at the interface.

(PVD) which involves an intermediate- to high-temperature environment during the deposition process, followed by cooling to room temperature. In MEMS processing, one may bond the coating to the substrate using a polymer adhesive, which is then cured to secure a firm bond. This bonding process for a coating–substrate system induces unwanted thermal stress at room temperature. Here we shall review simple analytical methods to calculate thermal stress and strain fields in a coating–substrate system: first, a one-dimensional model for a laminate plate composed of two layers, then two-dimensional models and, finally, we shall examine the case of a three-layer laminate composed of coating, adhesive layer, and substrate.

7.5.1 One-dimensional model

Let us consider a one-dimensional model made of coating and substrate subjected to a uniform temperature change, Fig. 7.16, where the coating (material 1) and substrate (material 2) are initially allowed to expand freely, and then are constrained by the bond between them to attain an equilibrium state with length l; the initial length is l_0. This constraint results in internal in-plane forces, F_1 (stress σ_1) in material 1 and F_2 (stress σ_2) in material 2. If the coefficient of thermal expansion (CTE) of material 2 (α_2) exceeds that of material 1 (α_1), as in Fig. 7.16, the internal stress in material 1 (σ_1) becomes tensile while that in material 2 (σ_2) is compressive.

The in-plane one-dimensional force equilibrium requires

$$F_1 = F_2 \tag{7.54a}$$

or

$$\sigma_1 h_1 - \sigma_2 h_2 = 0. \tag{7.54b}$$

Hooke's law, $\sigma = E \cdot e$, is valid for each material, and we obtain

$$\sigma_1 = E_1 \frac{\{l - l_0(1 + \alpha_1 \Delta T)\}}{l_0} = E_1 \left\{ \frac{l}{l_0} - (1 + \alpha_1 \Delta T) \right\}, \qquad (7.55a)$$

$$\sigma_2 = E_2 \frac{\{l_0(1 + \alpha_2 \Delta T) - l\}}{l_0} = E_2 \left\{ (1 + \alpha_2 \Delta T) - \frac{l}{l_0} \right\}, \qquad (7.55b)$$

where E_1 and E_2 are Young's moduli of materials 1 and 2, respectively, and ΔT is temperature change. It is to be noted here that ΔT can be positive (as in Fig. 7.16) or negative.

Substituting Eqs. (7.55a), (7.55b) into (7.54b), we obtain the ratio of current to initial lengths of the laminated composite:

$$\left. \begin{array}{c} E_1 h_1 \left\{ \dfrac{l}{l_0} - (1 + \alpha_1 \Delta T) \right\} - E_2 h_2 \left\{ (1 + \alpha_2 \Delta T) - \dfrac{l}{l_0} \right\} = 0, \\[2mm] \dfrac{l}{l_0} = \dfrac{E_1 h_1 (1 + \alpha_1 \Delta T) + E_2 h_2 (1 + \alpha_2 \Delta T)}{E_1 h_1 + E_2 h_2}, \end{array} \right\} \qquad (7.56)$$

where h_1 and h_2 are the thicknesses of materials 1 and 2, respectively.

Substituting Eq. (7.56) into Eqs. (7.55), we obtain stresses

$$\sigma_1 = \frac{E_1 E_2 h_2 (\alpha_2 - \alpha_1) \Delta T}{E_1 h_1 + E_2 h_2}, \qquad (7.57a)$$

$$\sigma_2 = \frac{E_1 E_2 h_1 (\alpha_2 - \alpha_1) \Delta T}{E_1 h_1 + E_2 h_2}. \qquad (7.57b)$$

The effective CTE of the coating–substrate composite is given by

$$\text{CTE} = \frac{(l - l_0)}{l_0} \frac{1}{\Delta T} = \left(\frac{l}{l_0} - 1 \right) \frac{1}{\Delta T}, \qquad (7.58)$$

from which, with Eq. (7.56), CTE can be reduced to

$$\begin{aligned} \text{CTE} &= \frac{(E_1 h_1 \alpha_1 + E_2 h_2 \alpha_2) \Delta T}{(E_1 h_1 + E_2 h_2) \Delta T} = \frac{E_1 h_1 \alpha_1 + E_2 h_2 \alpha_2}{E_1 h_1 + E_2 h_2} \\[2mm] &= \frac{E_1 f_1 \alpha_1 + E_2 f_2 \alpha_2}{\bar{E}}. \end{aligned} \qquad (7.59)$$

Here $\bar{E}$ is the Young's modulus of the composite defined by the following law of mixtures:

$$\bar{E} \equiv E_1 f_1 + E_2 f_2, \qquad (7.60)$$

where f_1 and f_2 are the volume fractions of materials 1 and 2, respectively:

$$f_1 = \frac{h_1}{(h_1 + h_2)}, \tag{7.61a}$$

$$f_2 = \frac{h_2}{(h_1 + h_2)}. \tag{7.61b}$$

7.5.2 Two-dimensional model

For most electronic packaging designs, coating–substrate laminated composites are composed of rectangular or square plates that are bonded together. Therefore, at the very least the two-dimensional model is required to analyze accurately the thermal stress field in coating–substrate composites. Let us consider a square plate of thickness t and length l_0 subjected to a uniaxial tensile force F, Fig. 7.17, where the shrinkage of the square plate in the transverse direction is given by $- l_0 \nu e$, and ν is Poisson's ratio.

It is to be noted in Fig. 7.17 that, even for one-dimensional loading, a three-dimensional strain state is induced, i.e. $e_x = e$, $e_y = -\nu e$ (in-plane direction, but transverse to the loading) and $e_z = -\nu e$ (in the plate thickness direction). Then the strain e in the loading direction (x-axis) is given by one-dimensional Hooke's law

$$e = \frac{F}{E l_0 t} = \frac{\sigma_x}{E}. \tag{7.62}$$

Now let us apply an equal tensile force F in the transverse direction (y-axis) to the configuration of Fig. 7.17, thus the final configuration is a state of bi-axial tension in the plate, Fig. 7.18. Then the strains in the x- and y-directions are the same. For example e_x is obtained as

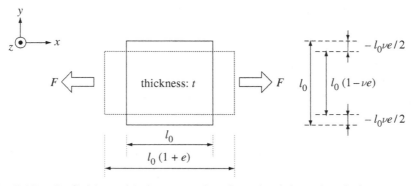

Fig. 7.17   Definition of Poisson's ratio ν in uni-axial tensional plate.

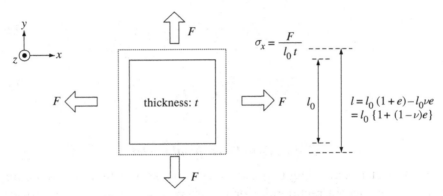

Fig. 7.18   Definition of one-dimensional (1D) effective strain in bi-axially tensioned plate.

$$\text{effective } e_x = (1 - \nu) \, e = \frac{F}{E \, l_0 \, t} (1 - \nu), \qquad (7.63)$$

where e is obtained from Eq. (7.62). Now we introduce the effective Young's modulus E_0 defined by

$$E_0 \equiv \frac{E}{1 - \nu}. \qquad (7.64)$$

Then e_x in Eq. (7.63) can be rewritten as

$$e_x = \left(\frac{F}{l_0 \, t}\right) \left(\frac{1}{E_0}\right) = \frac{\sigma_0}{E_0}. \qquad (7.65)$$

Equation (7.65) provides a one-dimensional Hooke's law for the analysis of e_x by taking into account the effect of Poisson's ratio for a two-dimensional (2D) plate. Thus, using the effective Young's modulus, one can use only one-dimensional analysis to account for 2D deformation. In the 1D model, the bending of a laminated composite plate was not considered. In reality, the bending of a laminated composite occurs due to CTE mismatch between coating (c) and substrate material (s), Fig. 7.19. To facilitate the thermal stress analysis to take into account the bending effect, we shall use a one-dimensional bending model where the effective Young's modulus defined by Eq. (7.64) is used. Figure 7.19 illustrates the one-dimensional bending model for a coating/substrate composite with width b (along y-axis) which is subjected to bending moment M (M shown in figure is defined positively) and uniform temperature change ΔT. The Young's modulus, Poisson's ratio, CTE, and thickness of the ith phase are E_i, ν_i, α_i, and h_i, with $i = c$ (coating) and s (substrate). The internal forces in the coating and substrate are defined by F_c and F_s, respectively.

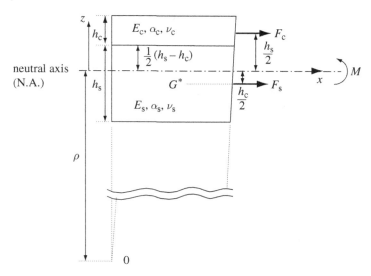

Fig. 7.19   Model for 1D bending of laminated composite.

The neutral axis of the composite beam is shown by the dash–dot line for
which the stress and strain vanish. Thus, using the effective Young's modulus,
$E_0 \equiv E/(1-\nu)$, one can use only one-dimensional analysis to account for 2D
deformation.

Force equilibrium requires

$$F_c + F_s = 0. \tag{7.66}$$

Moment equilibrium about the neutral axis (N.A) requires

$$M - \frac{F_c h_s}{2} + \frac{F_s h_c}{2} = 0. \tag{7.67}$$

Bending moment–curvature relation is given by

$$M = -\frac{1}{\rho}\left(E_s^0 I_s + E_c^0 I_c\right), \tag{7.68}$$

where E_i^0 and I_i are the effective Young's modulus, and the second moment of
inertia with respect to the N.A. of the ith material, $I_i = \int_{N.A.} z^2 \, dA$ (where A is
cross-sectional area), given by

$$I_s = \frac{b h_s^3}{12} + (b h_s) \cdot \left(\frac{h_c}{2}\right)^2, \tag{7.69a}$$

$$I_c = \frac{b h_c^3}{12} + (b h_c) \cdot \left(\frac{h_s}{2}\right)^2. \tag{7.69b}$$

The curvature ($\kappa \equiv 1/\rho$) vs. bending strain relation at location z is given by

$$e_x = -\frac{z}{\rho} = -\kappa z, \tag{7.70}$$

Where ρ is the radius of curvature defined in Fig. 7.19.

Then, the axial (in-plane, x-direction) strain due to bending in the coating is given as $-h_s/2\rho$ while the axial (in-plane, x-direction) strain due to bending in the substrate is $+h_c/2\rho$. Therefore, the total strain in the coating, e_c^t is given by

$$\frac{F_c}{E_c^0 h_c b} - \frac{h_s}{2\rho} + \Delta T \alpha_c = e_c^t. \tag{7.71}$$

         thermal expansion

      by bending

     by F_c

Similarly, the total strain in the substrate e_s^t is given by

$$\frac{F_s}{E_s^0 h_s b} + \frac{h_c}{2\rho} + \Delta T \alpha_s = e_s^t. \tag{7.72}$$

Compatibility of strain (i.e., no debonding at the interface) requires that

$$e_c^t = e_s^t. \tag{7.73}$$

From Eqs. (7.71) through (7.73), we obtain

$$\frac{F_c}{E_c^0 h_c b} - \frac{F_s}{E_s^0 h_s b} - \frac{1}{2\rho}(h_s + h_c) = \Delta T(\alpha_s - \alpha_c). \tag{7.74}$$

By using Eq. (7.66), we can eliminate F_c to arrive at

$$-\frac{F_s}{b}\left(\frac{1}{E_c^0 h_c} + \frac{1}{E_s^0 h_s}\right) - \frac{1}{2\rho}(h_s + h_c) = \Delta T(\alpha_s - \alpha_c). \tag{7.75}$$

From Eqs. (7.67), (7.68) we eliminate M to obtain

$$\frac{F_c h_s}{2} - \frac{F_s h_c}{2} = -\frac{1}{\rho}\left(E_s^0 I_s + E_c^0 I_c\right). \tag{7.76}$$

By using Eqs. (7.66) and (7.76), we can solve for F_s:

$$F_s = \frac{2\left(E_s^0 I_s + E_c^0 I_c\right)}{\rho(h_s + h_c)}. \tag{7.77}$$

Upon substitution of Eq. (7.77) into Eq. (7.75), we can solve for curvature $1/\rho$ as

$$\frac{1}{\rho} = \frac{-\Delta T (\alpha_s - \alpha_c)}{\left\{ 2 \dfrac{(E_s^0 I_s + E_c^0 I_c)}{(h_s + h_c)\, b} \left(\dfrac{1}{E_c^0 h_c} + \dfrac{1}{E_s^0 h_s} \right) + \dfrac{(h_s + h_c)}{2} \right\}} . \tag{7.78}$$

By substituting Eq. (7.78) into Eq. (7.77) and using Eq. (7.66), we can solve for F_s and F_c as

$$F_s = \frac{-\Delta T (\alpha_s - \alpha_c)}{\dfrac{1}{b} \left(\dfrac{1}{E_c^0 h_c} + \dfrac{1}{E_s^0 h_s} \right) + \dfrac{(h_s + h_c)^2}{4 \,(E_s^0 I_s + E_c^0 I_c)}} , \tag{7.79}$$

$$F_c = \frac{\Delta T (\alpha_s - \alpha_c)}{\dfrac{1}{b} \left(\dfrac{1}{E_c^0 h_c} + \dfrac{1}{E_s^0 h_s} \right) + \dfrac{(h_s + h_c)^2}{4 \,(E_s^0 I_s + E_c^0 I_c)}} . \tag{7.80}$$

Let us consider the case of a thin coating on a thick substrate, Fig. 7.20; then

$$h_c \ll h_s. \tag{7.81}$$

The second moments of inertia of the thin coating and thick substrate are approximated from Eq. (7.69) as

$$I_c = (bh_c) \left(\frac{h_s}{2} \right)^2 = \frac{bh_c h_s^2}{4}, \tag{7.82a}$$

$$I_s = \frac{bh_s^3}{12}, \tag{7.82b}$$

where the following approximation is used:

$$h_s + h_c \approx h_s. \tag{7.83}$$

Applying this to Eq. (7.80), we obtain the following reductions:

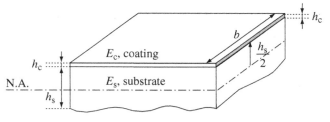

Fig. 7.20 Model for thin-coating–thick-substrate system.

$$\frac{(h_s + h_c)^2}{(E_s^0 I_s + E_c^0 I_c)} \approx \frac{h_s^2}{E_s^0 \frac{bh_s^3}{12} + E_c^0 \frac{bh_c h_s^2}{4}} = \frac{1}{E_s^0 \frac{bh_s}{12} + E_c^0 \frac{bh_c}{4}}$$

$$= \frac{12}{b(E_s^0 h_s + 3 E_c^0 h_c)} \approx \frac{12}{b E_s^0 h_s}, \tag{7.84a}$$

$$\frac{1}{E_c^0 h_c} + \frac{1}{E_s^0 h_s} \approx \frac{1}{E_c^0 h_c}. \tag{7.84b}$$

With Eqs. (7.84a) and (7.84b), F_c is simplified as

$$F_c \approx \frac{\Delta T (\alpha_s - \alpha_c)}{\frac{1}{b} \frac{1}{E_c^0 h_c} + \frac{3}{b E_s^0 h_s}} = \frac{\Delta T (\alpha_s - \alpha_c)\, b}{\frac{1}{E_c^0 h_c} + \frac{3}{E_s^0 h_s}}. \tag{7.85}$$

Thus, the stress in the thin coating is given by

$$\sigma_c = \frac{F_c}{h_c b} = \frac{\Delta T (\alpha_s - \alpha_c)}{\frac{1}{E_c^0} + \frac{3h_c}{E_s^0 h_s}} = \frac{\Delta T (\alpha_s - \alpha_c)\, E_c^0}{1 + \left(\dfrac{E_c^0}{E_s^0}\right) \dfrac{3h_c}{h_s}}. \tag{7.86}$$

If we ignore the high-order term in the denominator, σ_c is further approximated as

$$\sigma_c \approx E_c^0 \Delta T (\alpha_s - \alpha_c). \tag{7.87a}$$

With Eq. (7.64), Eq. (7.87a) is rewritten as

$$\sigma_c \approx \frac{E_c}{1 - \nu_c} \Delta T (\alpha_s - \alpha_c). \tag{7.87b}$$

It is to be noted that, for $\Delta T < 0$, and $\alpha_c > \alpha_s$ (for example, a metal coating on a ceramic substrate), the thermal stress in the coating is tensile, $\sigma_c < 0$. This is the case for a tensile stress developed in a metal coating after cooling down to room temperature, which sometimes results in the cracking of the coating.

7.5.3 *Alternative method of solving for the in-plane stress in a thin coating*

Consider the two-dimensional thermal stress problem of a thin metal coating on a thick ceramic substrate, which undergoes a temperature change ΔT, Fig. 7.21(a).

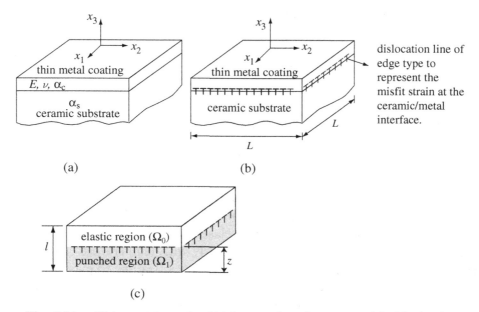

Fig. 7.21   Thin metal coating/thick ceramic substrate model: (a) elastic–plastic analysis, (b) dislocation model, and (c) punched region in metal coating.

We have

$$\varepsilon^* = \left(\varepsilon^*_{11}, \, \varepsilon^*_{22}, \, 0\right) = \left(\varepsilon^*, \, \varepsilon^*, \, 0\right), \tag{7.88}$$

where the CTE misfit strain ε^* is defined by

$$\varepsilon^* = (\alpha_c - \alpha_s)\,\Delta T. \tag{7.89}$$

Assuming plane strain along the x_1-, x_2-directions, the total strains e^t in the x_1- and x_2-directions vanish:

$$\left.\begin{aligned} e^t_{11} &= e_{11} + \varepsilon^*_{11} = 0, \\ e^t_{22} &= e_{22} + \varepsilon^*_{22} = 0 \end{aligned}\right\} \quad\Rightarrow\quad \left\{\begin{aligned} e_{11} &= -\varepsilon^*_{11}, \\ e_{22} &= -\varepsilon^*_{22}, \\ e_{11} &= e_{22} = -\varepsilon^*. \end{aligned}\right\} \tag{7.90}$$

The 3D Hooke's law (of an isotropic elastic material) for normal stress and strains is given by

$$\left.\begin{aligned} e_{11} &= \frac{\sigma_{11}}{E} - \frac{\nu}{E}(\sigma_{22} + \sigma_{33}), \\ e_{22} &= \frac{\sigma_{22}}{E} - \frac{\nu}{E}(\sigma_{11} + \sigma_{33}), \\ e_{33} &= \frac{\sigma_{33}}{E} - \frac{\nu}{E}(\sigma_{11} + \sigma_{22}). \end{aligned}\right\} \tag{7.91}$$

Assuming plane stress in the x_3-direction, $\sigma_{33} = 0$, we obtain from the 3D Hooke's law

$$\sigma_{11} - \nu\sigma_{22} = E e_{11},\tag{7.92a}$$

$$\sigma_{22} - \nu\sigma_{11} = E e_{22}.\tag{7.92b}$$

From Eqs. (7.89), (7.90), and (7.92), we can solve for σ_{11} and σ_{22}:

$$\sigma_{11} = \sigma_{22} = \frac{E_c\Delta T\,(\alpha_s - \alpha_c)}{(1 - \nu_c)}.\tag{7.93}$$

It is to be noted that the in-plane stress in a thin coating given by Eq. (7.93) coincides with the solution obtained from the previous 2D model, Eq. (7.87b). The above calculation is based on pure elastic analysis. The thermal stress in the thin coating, Eq. (7.93), increases linearly with temperature change ΔT. Hence for larger ΔT the stress field becomes larger, leading to plastic yielding in the metal coating. We are interested in the critical temperature change $(\Delta T)_{\text{crit}}$ at and above which the metal coating yields.

We introduce the effective (von Mises') stress $\bar{\sigma}$ defined by, see Eq. (3.34),

$$\bar{\sigma}^2 = \frac{1}{2}\left\{ (\sigma_{11} - \sigma_{22})^2 + (\sigma_{22} - \sigma_{33})^2 + (\sigma_{33} - \sigma_{11})^2 \right\}.\tag{7.94}$$

Since $\sigma_{11} = \sigma_{22}$, and $\sigma_{33} = 0$ in the coating, the effective stress of Eq. (7.94) is reduced to

$$\bar{\sigma} = \sigma_{11}.\tag{7.95}$$

When yielding occurs, $\bar{\sigma} = \sigma_y = \sigma_{11}$, then we can impose the yield condition for the metal coating using Eqs. (7.93) and (7.95),

$$\sigma_y \leq \frac{E_c\Delta T\,(\alpha_s - \alpha_c)}{(1 - \nu_c)} = \sigma_{11}.\tag{7.96}$$

We can solve for $(\Delta T)_{\text{crit}}$ from Formula (7.96)

$$\Delta T \geq \frac{\sigma_y(1 - \nu_c)}{E_c\,(\alpha_s - \alpha_c)} \equiv (\Delta T)_{\text{crit}},\tag{7.97}$$

where σ_y is the yield stress of the coating. Formula (7.97) is the condition under which yielding of the metal coating takes place.

The above problem was solved by using simple elastic–plastic analysis based on the assumption that the state of stress and strain in a thin metal coating is plane strain in the in-plane direction and plane stress in the thickness direction.

The key driving source for thermal stress field is the CTE mismatch strain ε^* given by Eq. (7.89). As far as the elastic–plastic analysis of the above problem is concerned, Taya and Mori (1994) solved the problem using a dislocation punching model, Fig. 7.21, which illustrates the configuration (b) before and (c) after the array of dislocations is punched over the distance z.

Assuming that the dislocation punching took place with its frontal boundary located a distance z from the interface, Fig. 7.21(b), we shall calculate the stress and strain field within the elastic region (domain Ω_0) and the punched region (plastic domain Ω_1), see Fig. 7.21(c), and also the condition for punching.

In elastic region (Ω_0)

Stress σ_{ij} and strain e_{ij} in the elastic region (white domain in Fig. 7.21(c)) are the same as those in the coating before punching, i.e., given by Eq. (7.93).

In plastic region (Ω_1)

Denoting plastic strain in the punched region (shaded in Fig. 7.21(c)) by e_{ij}^p and using the plane strain condition in the x_1–x_2 plane and incompressibility of plastic strain, one arrives at

$$\left.\begin{array}{l} e_{11}^p = e_{22}^p = -\varepsilon^*, \\ e_{33}^p = 2\varepsilon^*, \\ \sigma_{ij} = 0 \end{array}\right\} \tag{7.98}$$

where ε^* is again the CTE misfit strain, defined by Eq. (7.89). It is assumed that stresses in Ω_1 are all zero due to complete relaxation.

Computation of elastic energy U and plastic work W

The elastic energy per unit width along the x_2-axis is given by

$$U = \frac{1}{2}\sigma_{ij}e_{ij}(l - z), \tag{7.99}$$

where l is the thickness of the metal coating (see Fig. 7.21(b)).

A substitution of Eqs. (7.90) and (7.93) into Eq. (7.99) yields

$$U_c = \frac{E_c\varepsilon^*(l - z)}{(1 - \nu_c)}. \tag{7.100}$$

The plastic work done in the coating per unit width W_c is given by

$$W_c = 2kNbz, \tag{7.101}$$

where k is the friction stress (shear stress at yield), N is the number of dislocation lines along the x_2-axis (or x_1-axis) and b is the Burgers vector. Noting that $Nb = L\varepsilon^*$, the above equation is reduced to

$$W = 2kL\varepsilon^* z. \tag{7.102}$$

Then the plastic work per unit width is given by

$$W = 2k\varepsilon^* z. \tag{7.103}$$

The punching criterion is given by

$$-\frac{\partial U}{\partial z} \geq \frac{\partial W}{\partial z}, \tag{7.104}$$

where the left-hand term represents the driving force for punching while the right is the retarding force against punching.

By substituting Eqs. (7.100) and (7.103) into (7.104) we obtain the condition for punching

$$\varepsilon^* \geq \frac{2k(1-\nu)}{E}. \tag{7.105}$$

Inequality (7.105) can be derived by invoking the yield criterion across the frontal boundary of the punched region, i.e., the interface between Ω_0 and Ω_1, Fig.7.21(c). The average stresses across the elastic–plastic boundary are calculated by using Eqs. (7.93) and (7.98):

$$\bar{\sigma}_{11} = \bar{\sigma}_{22} = -\frac{E\varepsilon^*}{2(1-\nu)}, \tag{7.106}$$

which is substituted into Hill's yield criterion

$$|\bar{\sigma}_{33} - \bar{\sigma}_{11}| \geq 2k. \tag{7.107}$$

A substitution of Eq. (7.106) and $\sigma_{33} = 0$ into (7.107) leads to Inequality (7.105).

A more detailed analysis of the dislocation theory and induced stress field is summarized in a book by Freund and Suresh (2003).

7.5.4 Shear stress in the adhesive between chip and substrate

In electronic packaging, a bonding material is often used, for example bonding of a chip to a substrate using a thin adhesive polymer or a solder. Here, we consider a simple model for a chip bonded to a substrate by a thin adhesive, Fig. 7.22.

First, we shall examine the case of elastic deformation of the chip/adhesive/ substrate problem, then elastic–plastic deformation of the adhesive layer.

Elastic solutions

Consider the free-body diagram in Fig.7.23, in which the upper layer (material 1) is a chip;

$$-F_1 + F_1 + \frac{\partial F_1}{\partial x}dx - \tau\,dx = 0.$$

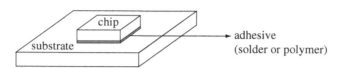

Fig. 7.22 Chip mounted on a substrate and bonded with adhesive.

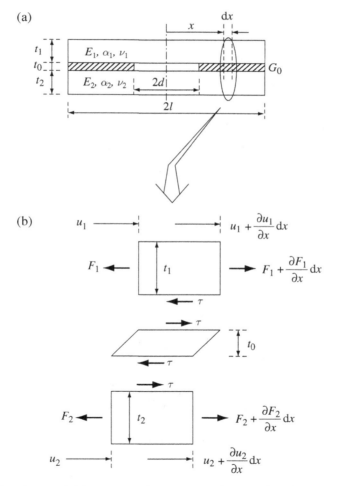

Fig. 7.23 1D model for two laminae bonded by interfacial adhesive layer.

Similarly, for the substrate (material 2)

$$- F_2 + F_2 + \frac{\partial F_2}{\partial x} dx + \tau \, dx = 0.$$

Therefore, we have

$$\frac{\partial F_1}{\partial x} - \tau = 0, \qquad (7.108a)$$

$$\frac{\partial F_2}{\partial x} + \tau = 0, \qquad (7.108b)$$

where F_i is force per unit width, and τ is shear stress per unit width.
The total strains in materials 1 and 2 are given by

$$e_1 = \frac{\partial u_1}{\partial x} = \frac{F_1}{E_1^0 \, t_1} + \alpha_1 \Delta T, \qquad (7.109a)$$

$$e_2 = \frac{\partial u_2}{\partial x} = \frac{F_2}{E_2^0 \, t_2} + \alpha_2 \Delta T, \qquad (7.109b)$$

where E_1^0 and E_2^0 are the effective Young moduli of materials 1 and 2, respectively, see Eq. (7.64), and they are given by

$$E_1^0 = \frac{E_1}{1 - \nu_1}, \qquad (7.110a)$$

$$E_2^0 = \frac{E_2}{1 - \nu_2}. \qquad (7.110b)$$

For the adhesive, Hooke's law gives

$$\frac{\tau}{G_0} = \frac{u_1 - u_2}{t_0} \qquad (7.111)$$

where G_0 is the shear modulus of the adhesive.
Applying d/dx to Eq. (7.111), we obtain

$$\frac{1}{G_0} \frac{d\tau}{dx} = \frac{1}{t_0} \left(\frac{du_1}{dx} - \frac{du_2}{dx} \right). \qquad (7.112)$$

A substitution of Eqs. (7.109) into (7.112) results in

$$\frac{t_0}{G_0} \frac{d\tau}{dx} = \left(\frac{F_1}{E_1^0 \, t_1} - \frac{F_2}{E_2^0 \, t_2} \right) + (\alpha_1 - \alpha_2) \, \Delta T. \qquad (7.113)$$

Applying d/dx to Eq. (7.113), we obtain

$$\frac{t_0}{G_0} \frac{d^2\tau}{dx^2} = \left(\frac{1}{E_1^0 \, t_1} \frac{dF_1}{dx} - \frac{1}{E_2^0 \, t_2} \frac{dF_2}{dx} \right). \tag{7.114}$$

Substituting Eqs. (7.108) into Eq. (7.114), we have the second-order differential equation for shear stress τ in the adhesive layer,

$$\frac{d^2\tau}{dx^2} - \beta^2 \tau = 0, \tag{7.115}$$

where

$$\beta^2 = \frac{G_0}{t_0} \left(\frac{1}{E_1^0 \, t_1} + \frac{1}{E_2^0 \, t_2} \right). \tag{7.116}$$

Equation (7.115) has the following general solution:

$$\tau(x) = A \, \sinh \, \beta(x - d) + B \, \cosh \, \beta(x - d) \tag{7.117}$$

for $d \le |x| \le l$.

Equation (7.117) contains two unknown constants, A and B, which will be solved for by using boundary conditions. To this end, we consider the strain and stress at $x = d$ (innermost boundary of the adhesive layer), Fig.7.23(a). The displacements u_1, u_2 (Fig.7.23(b)) in materials 1 and 2 are given from Eqs. (7.109) as

$$u_1 = \left(\frac{F_1}{E_1^0 \, t_1} + \alpha_1 \Delta T \right) d, \tag{7.118a}$$

$$u_2 = \left(\frac{F_2}{E_2^0 \, t_2} + \alpha_2 \Delta T \right) d. \tag{7.118b}$$

Taking the difference of the displacements at $x = d$, we have

$$\frac{u_1 - u_2}{d} = \left(\frac{F_1}{E_1^0 \, t_1} - \frac{F_1}{E_1^0 \, t_1} \right) + (\alpha_1 - \alpha_2) \, \Delta T. \tag{7.119}$$

From Eq (7.111), the left-hand side of Eq. (7.119) is equal to $(t_0/G_0)(\tau/d)$ while, from Eq (7.113), the right-hand side of Eq. (7.119) is equal to $(t_0/G_0)(d\tau/dx)$. Therefore, at $x = d$ we have

$$\left. \frac{d\tau}{dx} \right|_{x=d} = \left. \frac{\tau}{d} \right|_{x=d}. \tag{7.120}$$

From Eq. (7.117) we obtain

$$\frac{d\tau}{dx} = A\beta \cosh \beta(x-d) + B\beta \sinh \beta(x-d). \tag{7.121}$$

From Eqs. (7.117), (7.120) and (7.121), for $x = d$, we obtain

$$\left.\begin{array}{l} \left.\dfrac{d\tau}{dx}\right|_{x=d} = A\beta, \\[2mm] \left.\dfrac{\tau}{d}\right|_{x=d} = \dfrac{B}{d}, \end{array}\right\} \quad A\beta = \frac{B}{d}. \tag{7.122}$$

Next we consider the second boundary condition at $x = l$ where $F_1 = F_2 = 0$. Using Eq. (7.113) with $F_1 = F_2 = 0$, we obtain

$$\left.\frac{t_0}{G_0}\frac{d\tau}{dx}\right|_{x=l} = (\alpha_1 - \alpha_2)\,\Delta T = \frac{t_0}{G_0}\,\{A\beta\cosh\beta(l-d) + B\beta\sinh\beta(l-d)\}. \tag{7.123}$$

From Eqs. (7.122) and (7.123), we can solve for unknown coefficients A, B as

$$A = \frac{G_0(\alpha_1 - \alpha_2)\,\Delta T}{t_0\,\beta\,\{\cosh\beta(l-d) + \beta d\sinh\beta(l-d)\}}, \tag{7.124a}$$

$$B = \frac{G_0\,d\,(\alpha_1 - \alpha_2)\,\Delta T}{t_0\,\{\cosh\beta(l-d) + \beta d\sinh\beta(l-d)\}}. \tag{7.124b}$$

Substituting Eqs. (7.124) into Eq. (7.117), the shear stress in the adhesive layer is obtained as

$$\tau(x) = \frac{(\alpha_1 - \alpha_2)\,\Delta T\,G_0\,\{\sinh\beta(x-d) + \beta d\cosh\beta(x-d)\}}{t_0\,\beta\,\{\cosh\beta(l-d) + \beta d\sinh\beta(l-d)\}}. \tag{7.125}$$

The maximum shear stress occurs at $x = l$:

$$\tau(l) = \frac{(\alpha_1 - \alpha_2)\,\Delta T\,G_0\,\{\sinh\beta(l-d) + \beta d\cosh\beta(l-d)\}}{t_0\,\beta\,\{\cosh\beta(l-d) + \beta d\sinh\beta(l-d)\}} = \tau_{max}. \tag{7.126}$$

The minimum shear stress occurs at $x = d$:

$$\tau(d) = \frac{(\alpha_1 - \alpha_2)\,\Delta T\,G_0\,d}{t_0\,\{\cosh\beta(l-d) + \beta d\sinh\beta(l-d)\}} = \tau_{min}. \tag{7.127}$$

As a special case of Eq. (7.126), take $d = 0$, i.e., adhesive layer spans the entire zone $(2\,l)$. Then,

$$\tau_{max} = \tau(l) = \frac{(\alpha_1 - \alpha_2)\,\Delta T\,G_0\,\tanh\beta l}{\beta\,t_0}. \qquad (7.128)$$

If the chip is thin compared with the substrate,

$$\beta^2 \approx \frac{G_0}{t_0}\,\frac{1}{E_1^0\,t_1}, \qquad (7.129)$$

$$\tau_{max} = \tau(l) = \frac{(\alpha_1 - \alpha_2)\,\Delta T\,G_0}{\sqrt{\dfrac{G_0}{t_0\,t_1\,E_1^0}}\,t_0}\,\tanh\beta l$$

$$= \sqrt{\frac{G_0 t_1 E_1^0}{t_0}}\,(\alpha_1 - \alpha_2)\,\Delta T\,\tanh\beta l. \qquad (7.130)$$

Similarly τ_{min}, which takes place at $d = 0$, vanishes:

$$\tau_{min} = 0 \qquad (7.131)$$

Computation of normal stresses, σ_1, σ_2

We now calculate the normal in-plane stresses in chip (material 1) and substrate (material 2), Fig. 7.23. Integrating F_1, Eq. (7.108a) from $x = d$ to $x = l$, we obtain

$$F_1(x) = \int_{x'=d}^{x'} \tau(x')\,\mathrm{d}x' + \underbrace{F_1(x = d)}_{\textstyle F_1^d}$$

$$= \frac{(\alpha_1 - \alpha_2)\Delta T\,G_0}{A'}\,\frac{1}{\beta}\,\{\cosh\beta(x - d) + \beta d \sinh\beta(x - d)\} + F_1^d, \qquad (7.132)$$

where

$$A' = \beta\,t_0\{\cosh\beta(l - d) + \beta d \sinh\beta(l - d)\}. \qquad (7.133)$$

Similarly, the force F_2 is obtained by integrating Eq. (7.108b):

$$F_2(x) = -\int_{x'=d}^{x} \tau(x')\,\mathrm{d}x' + \underbrace{F_2(x=d)}_{\qquad\rightarrow F_2^d}$$

$$= -\frac{(\alpha_1 - \alpha_2)\,\Delta T\,G_0}{A'}\frac{1}{\beta}\{\cosh\beta(x-d) + \beta d \sinh\beta(x-d)\}$$

$$+ F_2^d. \tag{7.134}$$

Applying the boundary condition at $x = l$, $F_1 = F_2 = 0$, we have

$$F_1(l) = \frac{(\alpha_1 - \alpha_2)\,\Delta T\,G_0}{A'\,\beta}\{\cosh\beta(l-d) + \beta d \sinh\beta(l-d)\} + F_1^d = 0,$$

$$\tag{7.135}$$

$$F_2(l) = -\frac{(\alpha_1 - \alpha_2)\,\Delta T\,G_0}{A'\,\beta}\{\cosh\beta(l-d) + \beta d \sinh\beta(l-d)\} + F_2^d = 0.$$

$$\tag{7.136}$$

The in-plane forces F_1 and F_2 are in equilibrium,

$$F_1 + F_2 = 0. \tag{7.137}$$

Therefore,

$$F_1^d = -F_2^d. \tag{7.138}$$

From Eqs. (7.132) and (7.134) we can solve for F_1^d and F_2^d as

$$F_1^d = -\frac{(\alpha_1 - \alpha_2)\,\Delta T\,G_0}{\beta^2\,t_0}\frac{\{\cosh\beta(l-d) + \beta d \sinh\beta(l-d)\}}{\{\cosh\beta(l-d) + \beta d \sinh\beta(l-d)\}}$$

$$= -\frac{(\alpha_1 - \alpha_2)\,\Delta T\,G_0}{\beta^2\,t_0} = -F_2^d. \tag{7.139}$$

Upon substitution of Eq. (7.139) into (7.132) and (7.134), we obtain

$$F_1(x) = \frac{(\alpha_1 - \alpha_2)\,\Delta T\,G_0}{\beta^2\,t_0}\left\{\frac{\cosh\beta(x-d) + \beta d \sinh\beta(x-d)}{\cosh\beta(l-d) + \beta d \sinh\beta(l-d)} - 1\right\},$$

$$\tag{7.140a}$$

$$F_2(x) = -\frac{(\alpha_1 - \alpha_2) \, \Delta T \, G_0}{\beta^2 \, t_0} \left\{ \frac{\cosh \beta(x-d) + \beta d \sinh \beta(x-d)}{\cosh \beta(l-d) + \beta d \sinh \beta(l-d)} - 1 \right\}.$$

$$(7.140b)$$

The in-plane stresses σ_1 and σ_2 are then caluculated from F_1 and F_2 as

$$\sigma_1(x) = \frac{F_1}{t_1}$$

$$= \frac{(\alpha_1 - \alpha_2) \, \Delta T \, G_0}{\beta^2 \, t_0 t_1} \left\{ \frac{\cosh \beta(x-d) + \beta d \sinh \beta(x-d)}{\cosh \beta(l-d) + \beta d \sinh \beta(l-d)} - 1 \right\},$$

$$(7.141a)$$

$$\sigma_2(x) = \frac{F_2}{t_2}$$

$$= -\frac{(\alpha_1 - \alpha_2) \, \Delta T \, G_0}{\beta^2 \, t_0 t_2} \left\{ \frac{\cosh \beta(x-d) + \beta d \sinh \beta(x-d)}{\cosh \beta(l-d) + \beta d \sinh \beta(l-d)} - 1 \right\}.$$

$$(7.141b)$$

Figure 7.24 illustrates the distributions of in-plane stress σ_1 in the chip and shear stress τ in the adhesive layer along the interface bonding line (x-direction). Figure 7.24 indicates that σ_1 becomes a maximum at the center ($x=0$) while τ reaches its maximum value at a free end ($x=l$).

For $d=0$ (adhesive layer covers the entire gap between chip and substrate), at $x=0$ (at the center) where normal stress σ becomes a maximum

$$\sigma_{max} = \sigma_1(0) = \frac{(\alpha_1 - \alpha_2) \, \Delta T \, G_0}{\beta^2 \, t_0 t_1} \left\{ \frac{1}{\cosh \beta l} - 1 \right\}. \qquad (7.142)$$

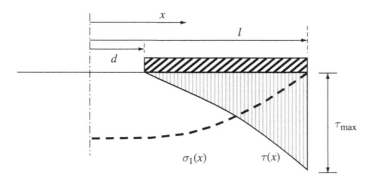

Fig. 7.24   Distribution of in-plane stress σ_1 in chip and shear stress τ in adhesive along x-direction.

We assume here a chip that is thin compared with the thick substrate, i.e., using Eq. (7.129):

$$\beta^2 t_0 t_1 \approx G_0 \frac{t_1}{E_1^0 t_1} = \frac{G_0}{E_1^0} = \frac{(1 - \nu_1) G_0}{E_1}, \qquad (7.143a)$$

$$\frac{G_0}{\beta^2 t_0 t_1} = \frac{E_1}{(1 - \nu_1)}. \qquad (7.143b)$$

Therefore, the maximum in-plane stress $\sigma_{\max}$ in the thin chip with $d=0$ is calculated as

$$\sigma_{\max} = \sigma_1(0) = \frac{(\alpha_1 - \alpha_2) \Delta T E_1}{(1 - \nu_1)} \left\{ \frac{1}{\cosh \beta l} - 1 \right\} \qquad (7.144)$$

For a thin-chip/thick-substrate system, $\beta l \gg 1$, thus $\cosh \beta l \to \infty$, and $\sigma_{\max}$ is approximated as

$$\sigma_{\max} = -\frac{(\alpha_1 - \alpha_2) \Delta T E_1}{(1 - \nu_1)} = \frac{E_1 \Delta T (\alpha_2 - \alpha_1)}{(1 - \nu_1)}. \qquad (7.145)$$

The maximum stress σ_1 in thin chip, given by Eq. (7.145), coincides with that derived by the previous 1D and 2D models, Eqs. (7.87b) and (7.93), if materials 1 and 2 are seen as coating (c) and substrate (s).

Elastic–plastic solutions

As ΔT increases, if the adhesive material is solder metal, it yields from the ends, $x=l$, and the elastic–plastic (e–p) boundary moves toward the center under increasing ΔT.

Suppose the location of the e–p boundary is at $x=x_c$. Then, at an arbitrary point x in the plastic region, referring to the free-body diagram of Fig. 7.25, we have

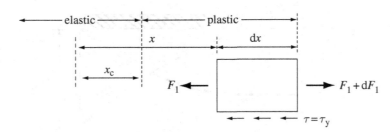

Fig. 7.25   Model for elastic–plastic analysis of thin-chip/solder/substrate problem.

$$dF_1 = \tau_y dx,$$

$$\frac{dF_1}{dx} = \tau_y, \tag{7.146}$$

where τ_y is the yield stress in shear of the solder. By integrating Eq. (7.146) from $x = l$ to x ($\geq x_c$) with the boundary condition $F_1 = 0$ at $x = l$, we solve for F_1 as

$$F_1(x) = \tau_y(x - x_c) - \tau_y(l - x_c) = \tau_y(x - l). \tag{7.147a}$$

From Eq. (7.135), the in-plane force in the substrate is obtained as

$$F_2(x) = -\tau_y(x - x_c) + \tau_y(l - x_c) = -\tau_y(x - l). \tag{7.147b}$$

The in-plane strains in materials 1 (chip) and 2 (substrate) are calculated from Eqs. (7.109) and (7.147).

$$e_1 = \frac{du_1}{dx} = \frac{F_1}{E_1^0 t_1} + \alpha_1 \Delta T$$

$$= \frac{\tau_y}{E_1^0 t_1}(x - l) + \alpha_1 \Delta T, \tag{7.148a}$$

$$e_2 = \frac{du_2}{dx} = -\frac{\tau_y}{E_2^0 t_2}(x - l) + \alpha_2 \Delta T. \tag{7.148b}$$

Rearranging Eqs. (7.148), we obtain

$$\frac{d(u_1 - u_2)}{dx} = \tau_y(x - l)\left(\frac{1}{E_1^0 t_1} + \frac{1}{E_2^0 t_2}\right) + (\alpha_1 - \alpha_2)\Delta T$$

$$\frac{d}{dx}\left(\frac{u_1 - u_2}{t_0}\right) = (x - l)\frac{\tau_y}{t_0}\left(\frac{1}{E_1^0 t_1} + \frac{1}{E_2^0 t_2}\right) + \frac{\Delta\alpha\,\Delta T}{t_0}, \tag{7.149}$$

where

$$\Delta\alpha = (\alpha_1 - \alpha_2). \tag{7.150}$$

Noting that shear strain γ in the adhesive (solder) layer is given by

$$\gamma = \frac{u_1 - u_2}{t_0} \tag{7.151a}$$

and is related to shear stress τ by Hooke's law

$$\tau_y = G_0\,\gamma_y, \tag{7.151b}$$

we obtain the shear strain gradient from Eqs. (7.149) and (7.151) as

$$\frac{d\gamma}{dx} = (x - l) \frac{G_0 \gamma_y}{t_0} \left(\frac{1}{E_1^0 t_1} + \frac{1}{E_2^0 t_2} \right) + \frac{\Delta\alpha \Delta T}{t_0}. \qquad (7.152a)$$

By using Eq. (7.116), Eq. (7.152a) is rewritten as

$$\frac{d\gamma}{dx} = (x - l) \beta^2 \gamma_y + \frac{\Delta\alpha \Delta T}{t_0}. \qquad (7.152b)$$

Integrating Eq. (7.152b) from the e–p boundary, i.e., $x = x_c$ ($\gamma = \gamma_y$), to an arbitrary point x, we obtain the shear strain in the plastic region of the solder,

$$\gamma(x) = \beta^2 \gamma_y \left\{ \frac{1}{2} \left(x^2 - x_c^2 \right) - l(x - x_c) \right\} + \frac{\Delta\alpha \Delta T}{t_0} (x - x_c) + \gamma_y. \qquad (7.153)$$

The maximum shear strain, γ_{max}, occurs at $x = l$:

$$\gamma_{max} = \gamma(x = l) = \beta^2 \gamma_y \left\{ \frac{1}{2} \left(l^2 - x_c^2 \right) - l(l - x_c) \right\} + \frac{\Delta\alpha \Delta T}{t_0} (l - x_c) + \gamma_y. \qquad (7.154)$$

To find the location x_c of the e–p boundary, we use the continuity of normal force at $x = x_c$ from the elastic solution, Eq. (7.140a), and the plastic solution, Eq. (7.147a):

$$\frac{(\alpha_1 - \alpha_2) \Delta T G_0}{\beta^2 t_0} \left\{ \frac{\cosh \beta(x_c - d) + \beta d \sinh \beta(x_c - d)}{\cosh \beta(l - d) + \beta d \sinh \beta(l - d)} - 1 \right\}$$
$$= \tau_y(x_c - l). \qquad (7.155)$$

Solving Eq. (7.155) numerically, we obtain x_c (the location of the elastic–plastic boundary).

7.6 Electromagnetic waves in laminated composites

Electromagnetic wave propagation through a laminated composite provides us with several interesting applications. Here we shall treat two cases: (1) design of a switchable window for EM waves; and (2) surface plasmon resonance in a layered medium.

7.6.1 *Design of switchable window for EM waves*

A switchable window that can pass or block incoming or outgoing electromagnetic (EM) waves plays a key function in designing a smart radome and other related EM control devices. Consider a laminated composite that is composed of a number of alternate layers of two different materials. There

are two approaches to designing switchable windows, one based on the switch-
able material properties (σ, ε, μ), the other based on the switchable spacing of
each layer (thickness of the ith layer being d_i). Here we consider the case of the
switchable window based on the second concept, where the laminated composite is composed of electroactive polymer layers with thickness d_p and ferrite
layers with thickness d_f, Fig. 7.26. It is to be noted here that the thickness of the
electroactive polymer can be changed by switching an applied voltage on and
off while the thickness of the ferrite layer remains unchanged. This problem of
Fig. 7.26 can be solved by using the transmission-line (TL) theory discussed in
Chapter 3. For a given layer with thickness d and material constants μ and ε,
one can relate the incoming voltage $V(0)$ and current $I(0)$ to the outgoing ones,
$V(d)$ and $I(d)$, Fig. 7.27.

According to TL theory, we have

$$\begin{pmatrix} V(0) \\ I(0) \end{pmatrix} = \begin{bmatrix} A_1 & B_1 \\ C_1 & D_1 \end{bmatrix} \begin{pmatrix} V(d) \\ I(d) \end{pmatrix}, \tag{7.156}$$

where

$$\begin{cases} A_1 = D_1 &= \cos \beta_1 d, & (7.157a) \\ B_1 &= jZ_1 \sin \beta_1 d, & (7.157b) \\ C_1 &= \frac{j}{Z_1} \sin \beta_1 d, & (7.157c) \\ A_1 D_1 - B_1 C_1 &= 1. & (7.157d) \end{cases}$$

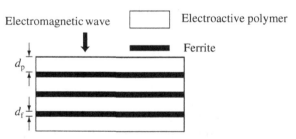

Fig. 7.26   Switchable window made of alternate layers of electroactive
polymer and ferrite.

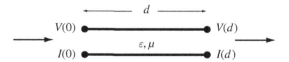

Fig. 7.27   Transmission-line theory applied to a single layer with thickness d
and material constants ε, μ.

The propagation constant β_1 is given by

$$\beta_1 = k_0 n_1 = k_0 (\varepsilon_{r1} \mu_{r1})^{1/2} \tag{7.158a}$$

where n_1 is the refractive index, and the wave parameter k_0 is related to wavelength λ_0, frequency ω, and speed of light c_0 by

$$k_0 = \frac{2\pi}{\lambda_0} = \frac{\omega}{c_0}, \tag{7.158b}$$

$$c_0 = 3 \times 10^8 \quad [\text{m/s}]. \tag{7.158c}$$

The characteristic impedance Z_1 of material 1 (μ_1, ε_1) is given by

$$Z_1 = \sqrt{\frac{\mu_1}{\varepsilon_1}} = \sqrt{\frac{\mu_0}{\varepsilon_0}} \sqrt{\frac{\mu_{r1}}{\varepsilon_{r1}}} = \eta_0 \sqrt{\frac{\mu_{r1}}{\varepsilon_{r1}}} \tag{7.159a}$$

where

$$\eta_0 = 120\pi \quad [\Omega]. \tag{7.159b}$$

For cascading networks made of a laminated composite, Fig. 7.26, the incident voltage and current (V_i, I_i) are related to the transmitted ones (V_t, I_t) by

$$\begin{pmatrix} V_i \\ I_i \end{pmatrix} = \begin{bmatrix} A_1 & B_1 \\ C_1 & D_1 \end{bmatrix} \begin{bmatrix} A_2 & B_2 \\ C_2 & D_2 \end{bmatrix} \cdots \begin{bmatrix} A_n & B_n \\ C_n & D_n \end{bmatrix} \begin{pmatrix} V_t \\ I_t \end{pmatrix} = \begin{bmatrix} A & B \\ C & D \end{bmatrix} \begin{pmatrix} V_t \\ I_t \end{pmatrix}, \tag{7.160}$$

where the transmission coefficient T of the laminated composite is given by

$$T = \frac{2}{A + \dfrac{B}{Z_0} + CZ_0 + D}, \tag{7.161}$$

and $0 \le |T| \le 1$.

Based on the TL model for a laminated composite, Kuga (personal communication) calculated the transmission coefficient T and the results are plotted as a function of the thickness ratio d_p/d_f of electroactive polymer to ferrite in Fig. 7.28. The material parameters used in this calculation are as follows:

electroactive polymer, $\mu_r = 1$, $\varepsilon_r = 2.3$
ferrite, $\mu_r = 1500$, $\varepsilon_r = 1.4$
incoming wavelength, $\lambda = 3$ cm
thickness of ferrite layer, $d_f = 6$ mm
number of electroactive polymer layers, $n_p = 8$
number of ferrite layers, $n_f = 7$.

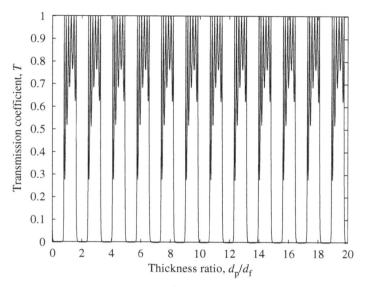

Fig. 7.28   Transmission coefficient T vs. thickness ratio d_p/d_f for laminated composite of Fig.7.26.

It is clear from Fig.7.28 that T can be changed between nearly 1 (full transmission mode) and 0 (full blocking mode) if the thickness ratio d_p/d_f (i.e., d_p) can be changed for a specific range. A change in the thickness of the electroactive polymer can be realized by applying a voltage to the electroactive polymer. Examples of electroactive polymers are solid electrolyte polymers (Popovic *et al.*, 2001; Uchida and Taya, 2001; Le Guilly *et al.*, 2002; Le Guilly *et al.*, 2003).

7.6.2 Surface plasmon resonance in layered medium

An incident light beam can give rise to a surface plasmon wave (SPW) at the interface of a metal/dielectric laminate composed of at least three layers. Use of SPW-based sensors is increasingly popular in the determination of the nanostructure of biomolecules, particularly the dynamic behavior of molecular interactions on the sensor surface such as interactions between antibody and antigen. We shall review the modeling of SPW in a two-layer laminate first, then a three-layer laminate, followed by multi-layer laminate analysis based on transmission theory.

SPW analysis in two-layer laminate

We shall consider here the two-layer laminate shown in Fig. 7.29, where a transverse magnetic (TM) wave incident in medium 1 is both reflected at the

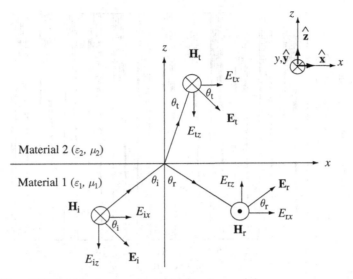

Fig. 7.29   SPW analysis; two-layer laminate model.

interface and transmitted through medium 2. The electric fields $\mathbf{E}_i$ (incident), $\mathbf{E}_r$ (reflected), $\mathbf{E}_t$ (transmitted) are given by

$$\mathbf{E}_i = (\hat{\mathbf{x}}\cos\theta_i - \hat{\mathbf{z}}\sin\theta_i)E_i^0 e^{-jk_1(x\sin\theta_i + z\cos\theta_i)}, \qquad (7.162a)$$

$$\mathbf{E}_r = (\hat{\mathbf{x}}\cos\theta_r + \hat{\mathbf{z}}\sin\theta_r)E_r^0 e^{-jk_1(x\sin\theta_r - z\cos\theta_r)}, \qquad (7.162b)$$

$$\mathbf{E}_t = (\hat{\mathbf{x}}\cos\theta_t - \hat{\mathbf{z}}\sin\theta_t)E_t^0 e^{-jk_2(x\sin\theta_t + z\cos\theta_t)}, \qquad (7.162c)$$

where $\hat{\mathbf{x}}, \hat{\mathbf{y}}$ and $\hat{\mathbf{z}}$ are basis vectors along the x-, y-, and z-axes, respectively, and where k_1 and k_2 are wave parameters of materials 1 and 2, defined by

$$k_1 = \omega\sqrt{\mu_1\varepsilon_1}, \qquad (7.163a)$$

$$k_2 = \omega\sqrt{\mu_2\varepsilon_2}. \qquad (7.163b)$$

First, we shall briefly examine the Maxwell equations relevant to SPW problems, i.e., Eqs. (3.153) and (3.151) with $\mathbf{J}=0$. For TM waves, the non-vanishing equations are

$$-\frac{\partial H_y}{\partial z} = j\omega\varepsilon E_x, \qquad (7.164a)$$

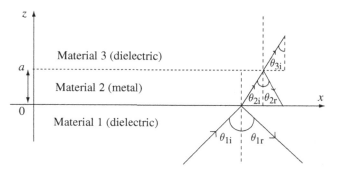

Fig. 7.30   SPW analysis; three-layer laminate model.

$$\frac{\partial H_y}{\partial x} = j\omega\varepsilon E_z, \tag{7.164b}$$

$$\frac{\partial E_x}{\partial z} - \frac{\partial E_z}{\partial x} = -j\omega\mu H_y. \tag{7.164c}$$

Equations (7.164) will be applied to the following three-layer problem.

SPW analysis in three-layer laminate

Here we consider SPW in a three-layer laminate composed of material 1 (dielectric layer), material 2 (metal layer), and material 3 (another dielectric layer), Fig. 7.30.

The electric field E_x in the x-direction is the sum of E_{ix} and E_{rx}. Thus, from Eqs. (7.162), we have

$$\begin{aligned}
E_x &= E_{ix} + E_{rx} \\
&= E_i^0 \cos\theta_i \; e^{-jk_1(x\sin\theta_i + z\cos\theta_i)} + E_r^0 \cos\theta_r \; e^{-jk_1(x\sin\theta_r - z\cos\theta_r)} \\
&= A\cos\theta_i \, e^{-jk_{1x}x - jk_{1z}z} + B\cos\theta_r \, e^{-jk_{1x}x + jk_{1z}z}.
\end{aligned} \tag{7.165}$$

Let the angles of the incident and reflected TM waves be redefined by

$$\theta_1 = \theta_{1i} = \theta_{1r}, \tag{7.166a}$$

$$\theta_2 = \theta_{2i} = \theta_{2r}, \tag{7.166b}$$

$$\theta_3 = \theta_{3i}. \tag{7.166c}$$

Then the E-fields in materials 1, 2, and 3 are rewritten as

$$E_{1x} = A\cos\theta_1 \, e^{-jk_{1x}x - jk_{1z}z} + B\cos\theta_1 \, e^{-jk_{1x}x + jk_{1z}z}, \tag{7.167a}$$

$$E_{2x} = C \cos \theta_2 \, e^{-jk_{2x}x - jk_{2z}z} + D \cos \theta_2 \, e^{-jk_{2x}x + jk_{2z}z}, \tag{7.167b}$$

$$E_{3x} = F \cos \theta_3 \, e^{-jk_{3x}x - jk_{3z}z} + G \cos \theta_3 \, e^{-jk_{3x}x + jk_{3z}z}, \tag{7.167c}$$

where A, B, C, D, E, F, and G are amplitudes of the TM waves. Since there is no reflection in material 3, $G = 0$. We set the amplitude of the incident wave as 1, so $A = 1$.

For the case of surface plasmon resonance (SPR), all the wave vectors **k** should be equal to that of SPR, i.e., k_{SPR},

$$k_{1x} = k_{2x} = k_{3x} = k_{SPR}. \tag{7.168}$$

Therefore, Eqs. (7.167) are further reduced to

$$E_{1x} = \cos \theta_1 \, e^{-jk_{SPR}x - jk_{1z}z} + B \cos \theta_1 \, e^{-jk_{SPR}x + jk_{1z}z}, \tag{7.169a}$$

$$E_{2x} = C \cos \theta_2 \, e^{-jk_{SPR}x - jk_{2z}z} + D \cos \theta_2 \, e^{-jk_{SPR}x + jk_{2z}z}, \tag{7.169b}$$

$$E_{3x} = F \cos \theta_3 \, e^{-jk_{SPR}x - jk_{3z}z}. \tag{7.169c}$$

Applying the boundary conditions at the interface between materials 1 and 2 at $z = 0$, $E_{1x} = E_{2x}$, and that between materials 2 and 3 at $z = a$, $E_{2x} = E_{3x}$, we have

$$\cos \theta_1 \, (1 + B) = \cos \theta_2 \, (C + D), \tag{7.170a}$$

$$\cos \theta_2 \, (Ce^{-jk_{2z}a} + De^{jk_{2z}a}) = F \cos \theta_3 \, e^{-jk_{3z}a}. \tag{7.170b}$$

Integrating Eq. (7.164a) with respect to z, we obtain

$$H_y = \int (-j\omega \varepsilon E_x)\mathrm{d}z. \tag{7.171}$$

For material 1

$$H_{1y} = \int (-j\omega \varepsilon_1 E_{1x})\mathrm{d}z$$

$$= \int -j\omega \varepsilon_1 (\cos \theta_1 \, e^{-jk_{SPR}x - jk_{1z}z} + B \cos \theta_1 \, e^{-jk_{SPR}x + jk_{1z}z})\mathrm{d}z. \tag{7.172}$$

By introducing parameters R_1, R_2, and R_3 defined by

$$R_1 = \frac{k_{1z}}{\omega\varepsilon_1}, \, R_2 = \frac{k_{2z}}{\omega\varepsilon_2}, \, R_3 = \frac{k_{3z}}{\omega\varepsilon_3}, \tag{7.173}$$

Eq. (7.172) can be rewritten as

$$H_{1y} = \frac{1}{R_1}\cos\theta_1 \, e^{-jk_{SPR}x-jk_{1z}z} - \frac{B}{R_1}\cos\theta_1 \, e^{-jk_{SPR}x+jk_{1z}z}. \tag{7.174a}$$

Similarly, for materials 2 and 3, the *H*-fields are given by

$$H_{2y} = \frac{C}{R_2}\cos\theta_2 \, e^{-jk_{SPR}x-jk_{2z}z} - \frac{D}{R_2}\cos\theta_2 \, e^{-jk_{SPR}x+jk_{2z}z}, \tag{7.174b}$$

$$H_{3y} = \frac{F}{R_3}\cos\theta_3 \, e^{-jk_{SPR}x-jk_{3z}z}. \tag{7.174c}$$

Applying the boundary conditions of continuity of H_y-fields across the interfaces at $z = 0$, $H_{1y} = H_{2y}$, at $z = a$, $H_{2y} = H_{3y}$ we obtain

$$\frac{\cos\theta_1}{R_1}(1 - B) = \frac{\cos\theta_2}{R_2}(C - D), \tag{7.175a}$$

$$\frac{\cos\theta_2}{R_2}(Ce^{-jk_{2z}a} - De^{jk_{2z}a}) = \frac{F}{R_3}\cos\theta_3 \, e^{-jk_{3z}a}. \tag{7.175b}$$

Now we shall solve Eqs. (7.170) and (7.175) for the amplitude *B* of the reflected wave in material 1. From Eqs. (7.170b) and (7.175b) we eliminate *F* to obtain

$$C = \frac{R_3 + R_2}{R_3 - R_2}e^{2jk_{2z}a}D. \tag{7.176}$$

Let r_{mn} be the parameter related to both R_m and R_n, defined by

$$\frac{1}{r_{mn}} = \frac{R_n + R_m}{R_n - R_m}. \tag{7.177a}$$

For example, for materials 2 and 3,

$$\frac{1}{r_{23}} = \frac{R_3 + R_2}{R_3 - R_2}. \tag{7.177b}$$

Then Eq. (7.176) is rewritten as

$$C = \frac{1}{r_{23}}e^{2jk_{2z}a}D. \tag{7.178}$$

From Eq. (7.175a)

$$D = \frac{(1 - B)\dfrac{R_2 \cos\theta_1}{R_1 \cos\theta_2}}{\left(\dfrac{1}{r_{23}}e^{2jk_{2z}a} - 1\right)}. \tag{7.179}$$

From Eqs. (7.170a), (7.178), (7.179), we can eliminate C and D:

$$1 + B = \frac{\cos\theta_2}{\cos\theta_1}(C + D)$$

$$= (1 - B)\frac{R_2}{R_1}\frac{\left(\dfrac{1}{r_{23}}e^{2jk_{2z}a} + 1\right)}{\left(\dfrac{1}{r_{23}}e^{2jk_{2z}a} - 1\right)}. \tag{7.180}$$

Solving (7.180) for B, we obtain the reflection coefficient

$$B = \frac{r_{12} + r_{23}e^{-2jk_{2z}a}}{1 + r_{12}r_{23}e^{-2jk_{2z}a}}. \tag{7.181}$$

Equation (7.181) provides the amplitude B of the reflected wave in material 1 for a given set of parameters θ_i, $k_{i,}$, ε_i of the ith material. Please note that

$$r_{12} = \frac{R_2 - R_1}{R_2 + R_1} = \frac{k_{2z}\varepsilon_1 - k_{1z}\varepsilon_2}{k_{2z}\varepsilon_1 + k_{1z}\varepsilon_2}, \tag{7.182a}$$

$$r_{23} = \frac{R_3 - R_2}{R_3 + R_2} = \frac{k_{3z}\varepsilon_2 - k_{2z}\varepsilon_3}{k_{3z}\varepsilon_2 + k_{2z}\varepsilon_3}. \tag{7.182b}$$

With Eqs. (7.182), Eq. (7.181) is reduced to

$$B = \frac{\dfrac{k_{2z}\varepsilon_1 - k_{1z}\varepsilon_2}{k_{2z}\varepsilon_1 + k_{1z}\varepsilon_2} + \dfrac{k_{3z}\varepsilon_2 - k_{2z}\varepsilon_3}{k_{3z}\varepsilon_2 + k_{2z}\varepsilon_3}e^{-2jk_{2z}a}}{1 + \dfrac{k_{2z}\varepsilon_1 - k_{1z}\varepsilon_2}{k_{2z}\varepsilon_1 + k_{1z}\varepsilon_2}\dfrac{k_{3z}\varepsilon_2 - k_{2z}\varepsilon_3}{k_{3z}\varepsilon_2 + k_{2z}\varepsilon_3}e^{-2jk_{2z}a}}. \tag{7.183}$$

By using $k_z = k\cos\theta = k_0\sqrt{\varepsilon_r}\cos\theta = \dfrac{2\pi}{\lambda}\sqrt{\varepsilon_r}\cos\theta$, the wavevectors are given by

$$k_{1z} = \frac{2\pi}{\lambda}\sqrt{\varepsilon_{r1}}\cos\theta_1, \tag{7.184a}$$

$$k_{2z} = \frac{2\pi}{\lambda} \sqrt{\varepsilon_{r2}} \cos \theta_2, \tag{7.184b}$$

$$k_{3z} = \frac{2\pi}{\lambda} \sqrt{\varepsilon_{r3}} \cos \theta_3, \tag{7.184c}$$

where the angle θ_t of the transmitted wave is related to that of the incident wave for two adjacent materials 1 and 2 by

$$\frac{\sin \theta_t}{\sin \theta_i} = \frac{n_1}{n_2} = \frac{\sqrt{\mu_1 \varepsilon_1}}{\sqrt{\mu_2 \varepsilon_2}} = \frac{\sqrt{\varepsilon_1}}{\sqrt{\varepsilon_2}}. \tag{7.185}$$

As light passes from material 1 to material 2, we have

$$\sin \theta_{2t} = \sin \theta_{2i} = \sin \theta_{1i} \frac{n_1}{n_2}, \tag{7.186}$$

$$\cos \theta_{2i} = (1 - \sin^2 \theta_{2i})^{1/2} = \left(1 - \frac{\varepsilon_1}{\varepsilon_2} \sin^2 \theta_{1i}\right)^{1/2}. \tag{7.187}$$

Similarly,

$$\cos \theta_{3i} = (1 - \sin^2 \theta_{3i})^{1/2} = \left(1 - \frac{\varepsilon_1}{\varepsilon_3} \sin^2 \theta_{1i}\right)^{1/2}. \tag{7.188}$$

SPW analysis in multi-layer laminate

For multi-layer laminates with number of layers more than three, we shall use transmission-line (TL) theory where voltage V is set equal to tangential E-field (E_x) and current I to tangential H-field (H_y):

$$V = E_x, \tag{7.189a}$$

$$I = H_y. \tag{7.189b}$$

Let us consider transmission-line theory, Fig.7.31. Voltage and current at an arbitrary point x are given by (Ishimaru,1991)

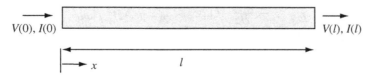

Fig. 7.31   Transmission-line theory.

$$V(x) = \frac{I_L}{2}[(Z_L + Z_0)e^{\gamma(l-x)} + (Z_L - Z_0)e^{-\gamma(l-x)}], \qquad (7.190a)$$

$$I(x) = \frac{I_L}{2Z_0}[(Z_L + Z_0)e^{\gamma(l-x)} - (Z_L - Z_0)e^{-\gamma(l-x)}]. \qquad (7.190b)$$

For $x=0$ (input) and $x=l$ (output), we obtain

$$V(0) = \frac{I_L}{2}[(Z_L + Z_0)e^{\gamma l} + (Z_L - Z_0)e^{-\gamma l}], \qquad (7.191a)$$

$$I(0) = \frac{I_L}{2Z_0}[(Z_L + Z_0)e^{\gamma l} - (Z_L - Z_0)e^{-\gamma l}], \qquad (7.191b)$$

$$V(l) = \frac{I_L}{2}[(Z_L + Z_0) + (Z_L - Z_0)] = I_L Z_L, \qquad (7.192a)$$

$$I(l) = \frac{I_L}{2Z_0}[(Z_L + Z_0) - (Z_L - Z_0)] = I_L, \qquad (7.192b)$$

where Z_0 and Z_L are characteristic impedance and load impedance corresponding to the medium between $x=0$ and $x=l$, I_L is the current flowing through the medium with impedance Z_L, and

$$\gamma = jk_z. \qquad (7.193)$$

With Eq. (7.193), $V(0)$ is written as

$$\begin{aligned}
V(0) &= \frac{I_L}{2}[(Z_L + Z_0)e^{jk_z l} + (Z_L - Z_0)e^{-jk_z l}] \\
&= I_L(Z_L \cos k_z l + jZ_0 \sin k_z l) \\
&= V(l) \cos k_z l + I(l)jZ_0 \sin k_z l. \qquad (7.194)
\end{aligned}$$

The characteristic impedance Z_0 is the ratio of the incident voltage V_0^+ to the incident current I_0^+, which are set equal to E_x^+ and H_y^+, respectively. Then, using Eqs. (7.164a) and (7.174), we obtain Z_0 as

$$Z_0 = \frac{V_0^+}{I_0^+} = \frac{E_x^+}{H_y^+} = \frac{-\dfrac{\partial H_y^+}{\partial z}\Big/j\omega\varepsilon}{H_y^+} = \frac{-(-jk_z)H_y^+/j\omega\varepsilon}{H_y^+} = R_m, \qquad (7.195)$$

where R_m ($m = 1, 2, \ldots, N$), the parameter related to the mth layer, is defined by Eq. (7.173).

Equation (7.194) is reduced to

$$V(0) = V(l) \cos k_z l + I(l) j R_m \sin k_z l. \qquad (7.196)$$

Similarly, for current, we have

$$
\begin{aligned}
I(0) &= \frac{I_L}{2Z_0} [(Z_L + Z_0)e^{jk_z l} - (Z_L - Z_0)e^{-jk_z l}] \\
&= \frac{I_L}{2Z_0} [(Z_L + Z_0)(\cos k_z l + j \sin k_z l) - (Z_L - Z_0)(\cos k_z l - j \sin k_z l)] \\
&= \frac{I_L}{2Z_0} (2Z_0 \cos k_z l + 2Z_L j \sin k_z l) \\
&= I_L \cos k_z l + \frac{V(l)}{Z_0} j \sin k_z l.
\end{aligned}
\qquad (7.197)
$$

With Eq. (7.195), $I(0)$ is reduced to

$$I(0) = V(l) \frac{1}{R_m} j \sin k_z l + I(l) \cos k_z l. \qquad (7.198)$$

Equations (7.196) and (7.198) are rearranged in the following matrix form:

$$
\begin{bmatrix} V(0) \\ I(0) \end{bmatrix} = \begin{bmatrix} A_m & B_m \\ C_m & D_m \end{bmatrix} \begin{bmatrix} V(l) \\ I(l) \end{bmatrix}, \qquad (7.199)
$$

where

$$A_m = D_m = \cos k_{mz} l, \qquad (7.200a)$$

$$B_m = j R_m \sin k_{mz} l, \qquad (7.200b)$$

$$C_m = \frac{j \sin k_{mz} l}{R_m}. \qquad (7.200c)$$

If the multi-layer laminate is composed of N layers, then $m = 1, 2, \ldots, N$, Fig. 7.32.

The relation between incident (i) and transmitted (t) voltage and current is now given by

$$
\begin{bmatrix} V_i \\ I_i \end{bmatrix} = \begin{bmatrix} A & B \\ C & D \end{bmatrix} \begin{bmatrix} V_t \\ I_t \end{bmatrix}, \qquad (7.201)
$$

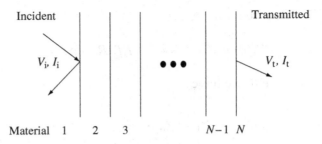

Fig 7.32   SPW model in N-layer laminate.

$$\begin{bmatrix} A & B \\ C & D \end{bmatrix} = \begin{bmatrix} A_2 & B_2 \\ C_2 & D_2 \end{bmatrix} \begin{bmatrix} A_3 & B_3 \\ C_3 & D_3 \end{bmatrix} \dots \begin{bmatrix} A_N & B_N \\ C_N & D_N \end{bmatrix}. \qquad (7.202)$$

It is to be noted in transmission lines that

$$\frac{V_0^+}{I_0^+} = -\frac{V_0^-}{I_0^-} = Z_0, \qquad (7.203)$$

where

$$V_i = V_0^+ (\text{incident}) + V_0^- (\text{reflected}) = E_0 + E_0 \Gamma, \qquad (7.204a)$$

$$I_i = I_0^+ (\text{incident}) + I_0^- (\text{reflected}) = \frac{V_0^+}{Z_0} - \frac{V_0^-}{Z_0} = \frac{E_0}{Z_0} - \frac{E_0 \Gamma}{Z_0} = \frac{E_0}{Z_1}(1 - \Gamma),$$

$$\qquad (7.204b)$$

where $Z_0 = Z_1$ and Γ is the reflection coefficient defined by

$$\Gamma = \frac{V_0^-}{V_0^+}. \qquad (7.205)$$

Also,

$$V_t = \tau E_0, \qquad (7.206a)$$

$$I_t = \frac{\tau}{Z_t} E_0, \qquad (7.206b)$$

where τ is the transmission coefficient defined by

$$\tau = 1 + \Gamma = \frac{V_t}{V_i}. \qquad (7.207)$$

The relation between the incident wave (V_i, I_i) and the transmitted one (V_t, I_t) is given in matrix form:

$$\begin{bmatrix} E_0(1 + \Gamma) \\ \dfrac{E_0}{Z_1}(1 - \Gamma) \end{bmatrix} = \begin{bmatrix} A & B \\ C & D \end{bmatrix} \begin{bmatrix} \tau E_0 \\ \dfrac{\tau}{Z_t} E_0 \end{bmatrix}. \tag{7.208}$$

Solving the matrix equation for τ and Γ, we obtain

$$\tau = \frac{2}{A + \dfrac{B}{Z_t} + Z_1\left(C + \dfrac{D}{Z_t}\right)}, \tag{7.209}$$

$$\Gamma = \tau\left(A + \frac{B}{Z_t}\right) - 1 = \frac{A + \dfrac{B}{Z_t} - Z_1\left(C + \dfrac{D}{Z_t}\right)}{A + \dfrac{B}{Z_t} + Z_1\left(C + \dfrac{D}{Z_t}\right)}. \tag{7.210}$$

It is to be noted in Eqs. (7.209) and (7.210) that Z_1 and Z_t are the impedances associated with layer 1 and layer N, Fig. 7.32, and they are equal to R_1 and R_N, respectively.

Therefore, the reflection coefficient Γ is now rewritten as

$$\Gamma = \frac{A + \dfrac{B}{R_N} - R_1\left(C + \dfrac{D}{R_N}\right)}{A + \dfrac{B}{R_N} + R_1\left(C + \dfrac{D}{R_N}\right)}, \tag{7.211}$$

where

$$\begin{bmatrix} A & B \\ C & D \end{bmatrix} = \prod_{m=2}^{N-1} \begin{bmatrix} A_m & B_m \\ C_m & D_m \end{bmatrix}$$

$$= \prod_{m=2}^{N-1} \begin{bmatrix} \cos(k_{mz} t_m) & j R_m \sin(k_{mz} t_m) \\ j\dfrac{\sin(k_{mz} t_m)}{R_m} & \cos(k_{mz} t_m) \end{bmatrix}, \tag{7.212}$$

and where t_m and n_m are the thickness and refractive index of the mth layer, respectively. The parameter R_m is

$$R_m = \frac{k_{mz}}{\omega \varepsilon_m} = \frac{k_{mz} \lambda}{2\pi c_0 n_m^2}, \tag{7.213a}$$

and

$$k_{mz} = k_m \cos \theta_m = k_0 n_m \cos \theta_m = k_0 n_m \left[1 - \left(\frac{n_1}{n_m}\right)^2 \sin^2 \theta_1\right]^{1/2}. \qquad (7.213b)$$

Surface plasmon resonance (SPR) is widely used as a key mechanism in designing a number of biosensors. There are two major types of SPR equipment. One is "angle modulation," where we input a single wave with a specific wavelength, such as a He–Ne laser, and find the resonance angle, and the other is "wavelength modulation," where we input an optical wave with many wavelengths such as white light and observe a certain wavelength which is absorbed. Here we shall present the numerical results based on the SPR equations developed earlier in this section.

Before the numerical calculations based on SPR model, we have to set the input data of multi-layer laminates. The refractive index of metals such as gold (Au) and chromium (Cr) is a complex number which depends on the wavelength of the incident light.

$$\boldsymbol{n}(\lambda) = n(\lambda) - jk(\lambda), \qquad (7.214)$$

where $\boldsymbol{n}(\lambda)$ is a complex-valued refractive index, $n(\lambda)$ and $k(\lambda)$ are real and imaginary parts, respectively.

Let us assume that $n(\lambda)$ and $k(\lambda)$ are approximated by

$$n(\lambda) = a_1 + a_2\lambda + \frac{a_3}{\lambda^2} + \frac{a_4}{\lambda^4} + \frac{a_5}{\lambda^6}, \qquad (7.215a)$$

$$k(\lambda) = b_1 + b_2\lambda + \frac{b_3}{\lambda^2} + \frac{b_4}{\lambda^4} + \frac{b_5}{\lambda^6}. \qquad (7.215b)$$

The a-parameters $(a_1, a_2, \ldots, a_5)$ and b parameters $(b_1, b_2, \ldots, b_5)$ of gold and chromium are given by

$$a(\mathrm{Au}) = (0.0705, 0.2562, -0.0530, -0.0362, 0.0198),$$

$$b(\mathrm{Au}) = (2.7742, 4.5739, 0.1185, -0.8814, 0.1752),$$

$$a(\mathrm{Cr}) = (1.7048, 2.4846, 0.6029, -0.2369, 0.0166),$$
$$\qquad (7.216)$$

$$b(\mathrm{Cr}) = (1.5322, 2.1596, 0.6694, -0.0028, -0.0126).$$

Figure 7.33 shows an approximate curve of $n(\lambda)$ of gold (Au) as a function of wavelength λ.

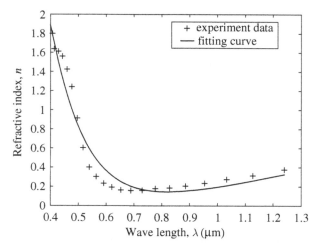

Fig. 7.33   Experimental data and approximate curve of $n(\lambda)$ of gold (Au).

Fig. 7.34 illustrates an SPR model of a five-layer laminate model. We shall calculate the reflection coefficient (reflectivity) Γ of the SPR in this multi-layer system. We are interested in the effect of the polymer layer (of thickness 10 nm and refractive index $n = 1.4$). It is to be noted in Fig. 7.34 that the Cr layer is needed as an adhesive bonding layer between the prism (BK7, $n = 1.5$) and Au. Use of water and the polymer layer is to simulate a realistic SPR characterization of unknown molecules in a flow cell (mostly water).

Figure 7.35 shows the results of the numerical simulation of the reflectivity–incident-angle (θ_i) relation for the multi-layer model of Fig. 7.34 with (solid

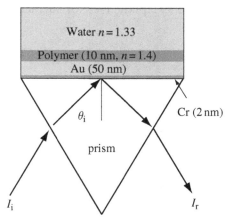

Fig. 7.34   SPR model of five-layer laminate composed of prism, Cr (2 nm), Au (50 nm), polymer (10 nm), and water.

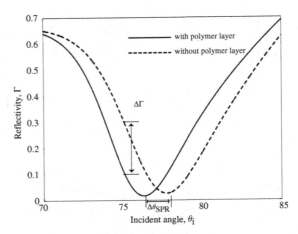

Fig. 7.35   Reflectivity Γ vs. incident angle θ_i for the multi-layer laminate of Fig. 7.34 with polymer layer (solid curve) and without polymer layer (dashed line).

curve) and without (dashed curve) polymer layer. Figure 7.35 demonstrates the sensitivity of an SPR sensor for sensing the existence of the polymer layer. To quantify the sensitivity, one can use the change $\Delta\theta_i$ in the SPR angle or the change $\Delta\Gamma$ in the reflectivity at a specific incident angle ($\theta_i = 75°$ was used in the figure). If the polymer layer is to simulate unknown molecules in more realistic sets of SPR characterizations, the thickness of the unknown molecules can be as small as a few nm or so. Even for such an extremely thin layer, the SPR sensor system based on either $\Delta\Gamma$ or $\Delta\theta_i$ can detect the existence of the unknown molecules in very small amounts. The SPR sensor, however, cannot distinguish the in-plane resolution (i.e., x–y plane) as the unknown layer is assumed to be uniform in the x–y plane (sensor surface).

8

Engineering problems

The preceding chapters are mostly aimed at modeling and design of electronic composites. Here we shall look at some of the engineering problems related to electronic composites: (1) processing, (2) standard measurement methods of properties, and (3) electromigration.

8.1 Processing

The processing routes of electronic composites for use in electronic packaging and in micro-electromechanical systems (MEMS) have common processing steps, which are

(1) photolithography,
(2) metallization,
(3) cladding,
(4) injection molding,
(5) compression molding,
(6) deposition of organic films.

We shall outline these processing methods below. There are many more processing routes that are available, but in this chapter we focus only on the major ones. Detailed accounts of the processing of electronic packaging and MEMS are given elsewhere, for example in a book by Madou (2002).

Photolithography

The most widely used processing technique in the electronic packaging and MEMS industry is photolithography, which serves as the primary step before the final IC and MEMS packages are assembled. The standard method of photolithography is illustrated in Fig. 8.1 where two cases of photoresist

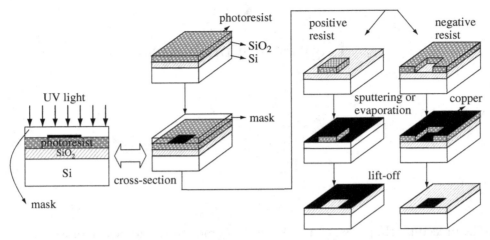

Fig. 8.1   Standard photolithography used for IC and MEMS packages, followed by Cu metallization.

removal are shown, positive and negative, resulting in the corresponding metal (Cu) circuit line or pad.

The first step in Fig. 8.1 is oxidation of the topmost Si wafer, to convert it to SiO_2 whose function is to protect the Si wafer from the etchant during the removal (lift-off) of the photoresist material and serves as a sacrificial layer in the MEMS process. Recent demands for higher spatial resolution and higher aspect ratios have led to alternative lithography technology, such as x-ray lithography.

Metallization

Deposition of a thin layer of metal (such as Cu, Au, or Al) in a specific pattern on a substrate, Si, or printed circuit board (PCB), is another key step in forming the desired circuit pattern. Popular metallization methods that are used for electronic and MEMS packaging include

(i) physical vapor deposition (PVD),
(ii) chemical vapor deposition (CVD),
(iii) screen printing,
(iv) plasma polymerization,
(v) self-assembled monolayer (SAM).

The cost-effectiveness of processing increases from (i) toward (iii), which are primarily for Si-based substrates, while (iv) and (v) are aimed at coating on both Si-based and polymer substrates. PVD includes thermal evaporation and

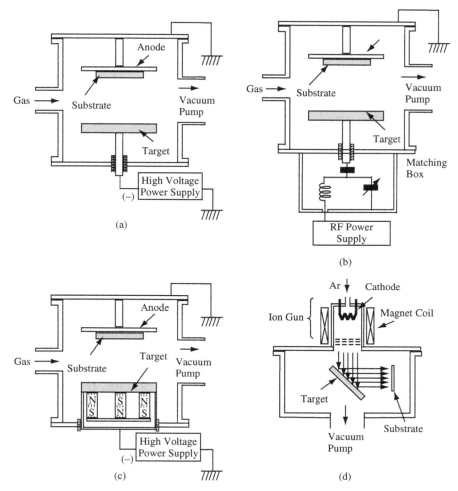

Fig. 8.2   Selected PVDs: (a) DC-diode glow discharge sputtering, (b) RF diode glow charge sputtering, (c) DC planar magnetron sputtering, and (d) ion-beam sputtering. (After Enoki, 1995.)

sputtering, where the former is used only with metal species that can be evaporated at high temperatures while the latter method does not have a limitation as to which material can be deposited, but requires a sputtering source material. Some PVD setups are shown in Fig. 8.2.

PVD is also used for etching some parts of the electronic packages and MEMS surfaces. There are two modes of etching by PVD, glow discharge and ion-beam methods, where the main principle of material removal is due to the kinetic energy carried by bombarding charged ions (Ar^+), with and without chemical reactions. This etching normally takes place on the substrate

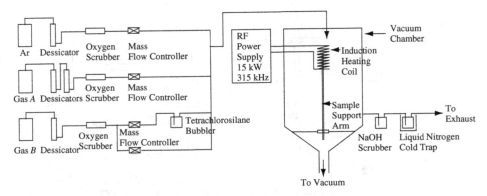

Fig. 8.3   Plasma-enhanced CVD (Richards, 1996).

material, with a thin masking coating on top, which is placed in the target source location i.e., opposite mode of operation to deposition of the target source.

There are several different variations in CVD: plasma-enhanced CVD, low-pressure CVD, and metal–organic CVD. Compared with PVD, most of CVD techniques require less power, lower deposition temperatures, provide less damage on the surface of a substrate and faster deposition rates on larger substrate areas, but the types of material that can be deposited on a substrate are limited. A typical CVD system requires a set of gases that flow into a chamber which is initially evacuated to a high vacuum, followed by input of Ar gas to create a plasma, which enhances the deposition of the reactant gases on the substrate that can be heated to a desired temperature. Figure 8.3 illustrates an example of a plasma-enhanced CVD system.

Screen printing is a cost-effective deposition method for 2D electric circuit patterns where inks made of conductive metals with an organic binder and solvent are inserted through a stainless-steel sheet with patterned openings which will become circuit lines for electronic and MEMS packages. The minimum width of the opening (slit) in the stainless-steel plate, however, is relatively large ($\sim$100 μm) although recent progress has been made down to 20 μm with some reliability.

Cladding

Cladding is the bonding of two thin plates at low temperature after etching both surfaces of the plates, providing low-cost and high-volume processing of thin laminated composite sheets. Figure 8.4(a) shows the principle of bonding and (b) is an example of a high-volume processing setup.

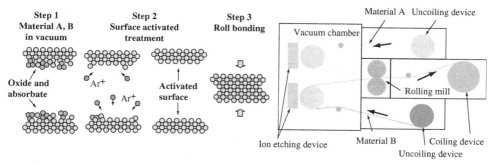

Fig. 8.4 Cladding of two thin metal sheets: (a) concept of clear fine cladding, (b) continuous process setup. (Courtesy of Mr. K. Saijo of Toyo Kohan.)

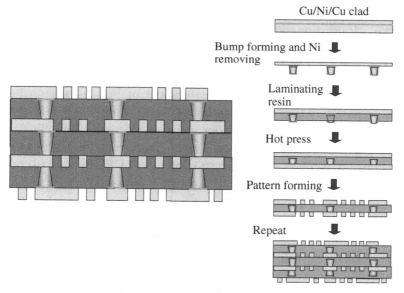

Fig. 8.5 Processing of ML-PWB using Cu/Ni/Cu clad laminate (courtesy of Mr. K. Saijo of Toyo Kohan).

It is to be noted that more than two sheets can be processed easily by including more sheet feeding and cladding systems. It is also noted here that the cladding can include the case of polymer film bonding to metal sheet. In addition to low-cost, high-volume processing, another advantage of the clean etching–cladding method of Fig. 8.4 is to provide a distinct interface between two material sheets with no diffusion layer and no compounds formed at the interface because bonding is at room temperature after etching the surfaces by Ar^+ ion bombardment. The thin laminate composite plates produced by cladding play a key role in processing multi-layered printed wiring board (ML-PWB) as shown in Fig. 8.5.

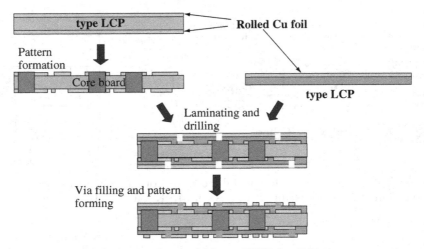

Fig. 8.6   Processing route for multi-layer PWB for RF components (courtesy of Mr. K. Saijo of Toyo Kohan).

The Cu/Ni/Cu clad laminate in Fig. 8.5 is bonded between the Cu/Ni plate and the Cu plate by the cladding process equipment of Fig. 8.4 where the Cu/Ni plate was prepared by electrolytic plating of Ni (typically ∼0.5–few µm) on the Cu plate (typically ∼10–30 µm). Under increasing demand for cellular phones and other portable antenna devices, use of lightweight material with a low dielectric constant (such as a liquid crystal polymer (LCP)) bonded to conductive metals has been accelerated. The cladding of LCP film and Cu foil provides a key RF PWB component. Figure 8.6 illustrates the processing route for multi-layer PWB for RF components.

The clean etching–cladding method is also used to process multi-layer sheets with embedded resistor lines, where the resistor lines are typically Ni–Cr and the conductor lines are Cu. Figure 8.7 illustrates the processing route for a resistor line with conductor circuit embedded in the insulator (LCP) where the key starting material is Cu/Cr–Ni/LCP clad laminate.

Injection molding

Injection molding has been used extensively for low-cost processing of plastic parts, some of them electronic composites, such as insulator parts made of a mica-filled polymer matrix. An injection-molding machine is composed of two parts, (1) a mixer, to prepare semi-liquid polymers (or polymer compounds), and send them to (2) a compression unit to form a shaped final part, Fig. 8.8.

Due to the demand for electronic packaging and MEMS components with micron-sized high tolerance, the conventional injection molding machines

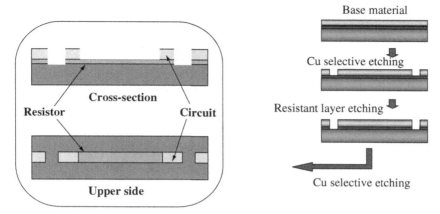

Fig. 8.7   Processing of the clean etching–cladding method for a resistor–conductor network in an LCD insulator (courtesy of Mr. K. Saijo of Toyo Kohan).

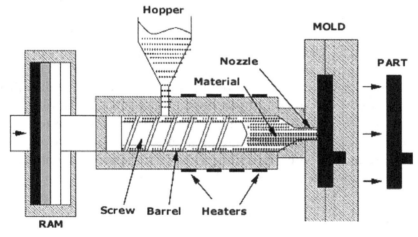

Fig. 8.8   Injection molding machine. Diagram downloaded from the Intelligent Systems Laboratory Website of Michigan State University (http://islnotes.cps.msu.edu/trp/inj/inj_what.html).

have been modified to allow flat shaped parts with some low-aspect-ratio bumps, for example, compact discs (CDs) and housing for micro-fluidics. The micron-sized cavities in the mold can be filled with the plastic mass with its temperature equal to or slightly higher than its glass transition temperature (T_g). If the flowing plastic mass is replaced by plastic compounds made of fillers and polymer, the viscosity increases, and getting such plastic compounds to penetrate into narrow channels of higher aspect ratio in the mold presents a challenge. This is the case with short fibers with higher aspect ratio. A carbon

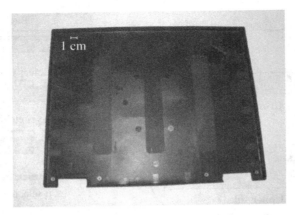

Fig. 8.9   Injection molded laptop housing made of carbon short-fiber/
polymer matrix composite where circular areas are access gates for injected
flow (courtesy of Mr. Matsuhisa, Toray).

short-fiber/polymer matrix composite has been successfully injection molded
into a housing for laptop PCs, as shown in Fig. 8.9.

It is to be noted that high mechanical strength is required for the laptop PC
housing, which can be achieved by uniform distribution of stiff carbon short
fibers. This uniform distribution of short fibers can be realized by using
multiple gates for the mold. The optimum design of the mold with such gates
requires 3D flow analysis of injection molding as well as the detailed analysis
of the short-fiber orientations and the filler spatial concentration in a 3D
mold, followed by predictions of the thermomechanical property data for all
spatial locations. Analytical modeling of the thermomechanical properties of
short-fiber composites has been made by using the Eshelby model (Takao
et al., 1982; Taya and Arsenault, 1989; Kataoka and Taya, 2000).

Compression molding

Compression molding has been used as extensively as injection molding in
processing electronic packaging and MEMS components. Unlike injection
molding, compression molding uses polymer or polymer compound sheets
as a starting material, which is heated close to or at T_g, followed by compression of the sheet between upper and lower dies (molds). A typical compression-
molding machine is illustrated in Fig. 8.10.

The compression molding sequence of Fig. 8.10 is a batch process; its
productivity is limited. To increase productivity, a continuous compression-
molding system is better suited where a plastic sheet or a plastic compound
sheet is sent continuously into the hot-pressing unit. For the processing of

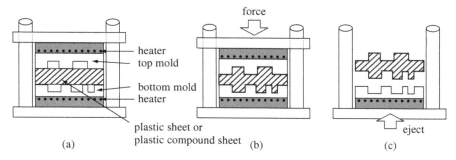

Fig. 8.10   Compression-molding machine. (a) Load/heat, (b) compress, and
(c) eject.

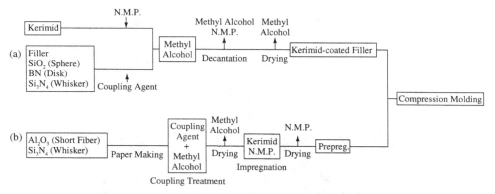

Fig. 8.11   Continuous processing involving preform process and compression molding. Two types of preform process are shown: (a) premix, and
(b) paper making, which is further detailed in Fig. 8.12. (Takei *et al.*, 1991a,
with permission from Elsevier Ltd.)

printed circuit board (PCB), Takei *et al.* (1991a,b) proposed a continuous
processing route involving preform making and compression molding, Figure
8.11 shows the routes for two types of preform: (a) premix, and (b) paper
making. The paper-making process is further illustrated in Fig. 8.12.

Takei *et al.* (1991b) processed a hybrid composite plate for a PCB using the
two-stage processing of Fig. 8.11(b). The microstructure of the preform is
shown in Fig. 8.13. Micro-fibridized cellulose was used to produce a tightly
bound skeletal Si_3N_4 whiskers/Al_2O_3 short-fiber preform.

Deposition of organic films

Deposition of organic films on substrates can be made by some of the preceding coating methods for use in electronic packages and MEMS. More recently,
the deposition of organic films on the surfaces of BioMEMS sensors and
actuators has become an increasingly important technology. Here we shall

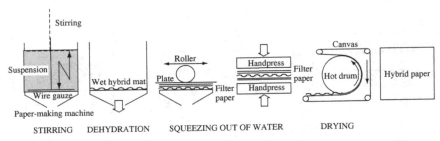

Fig. 8.12 Paper-making process to form preform made of two different fillers, Al$_2$O$_3$ short fibers and Si$_3$N$_4$ whiskers (Takei *et al.*, 1991b, with permission from Elsevier Ltd).

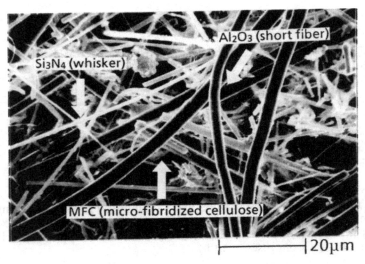

Fig. 8.13 Preform of a hybrid composite where two types of fillers (Si$_3$N$_4$ whiskers and Al$_2$O$_3$ short fibers) are tightly bonded by micro-fibridized cellulose (Takei *et al.*, 1991b, with permission from Elsevier Ltd).

state briefly three methods of deposition of organic films: (1) plasma polymerization, (2) Langmuir–Blodgett (LB) method, and (3) self-assembled monolayer (SAM).

Plasma polymerization has been used extensively to modify the surface of a polymer substrate and deposit organic film on the as-modified polymer. A typical plasma polymerization apparatus is illustrated in Fig. 8.14.

In the PVD process, a gaseous or volatile compound is introduced into a reaction chamber, fragmented and/or ionized in a glow discharge plasma, and reassembled on a surface. The processes that simultaneously occur during the plasma reaction are ionization, neutralization, recombination, polymerization, etching, implantation and, deposition. Under appropriate

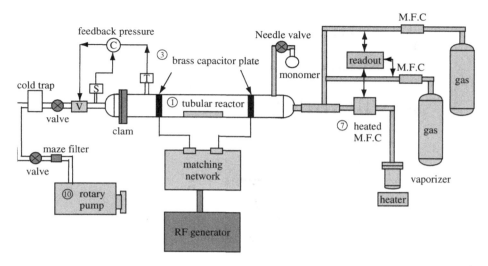

Fig. 8.14   Diagram of plasma polymerization apparatus. M.F.C. is mass flow control; S,T: vacuum gauges (Pirani and Baratron sensors).

conditions, a deposit will form on a substrate placed in the reactor. If the molecule introduced into the gas phase is an organic compound, polymerization is possible, but the structure of the polymer may be far from that of the precursor (Johnston and Ratner, 1996). In the last 10 years, methods have been developed to obtain a plasma deposition that closely resembles the structure of the precursor, and some of the parameters (RF power, precursor flow rate, pressure, reaction time, pulsing of the power, etc.) that control the structure of the deposit have been elucidated. Butoi *et al.* (2000) pointed out the influence of substrate position relative to the RF coil. In their research, CF_2-rich films deposited close to the RF coil were highly amorphous and cross-linked, whereas films deposited 28 cm downstream from the coil showed a high degree of order and less cross-linking. Plasma polymerization is a flexible strategy to modify the surface of materials. The resulting coatings are strongly adherent and many surface chemistries are possible. The applications of plasma polymer coatings range from corrosion protection to biomaterials that can interact with proteins. Zhang *et al.* (2003) plasma polymerized allylamine on pretreated glass slides to immobilize DNA. They found that low-power plasma deposits were able to swell in solution, making the amine groups within the bulk of the polymer accessible to DNA.

Langmuir (1938) introduced a horizontal lifting method, useful for deposition of very rigid films. This method, however, was invented about 1000 years ago and is called "Sumi Nagashi": a Japanese artist spread Chinese

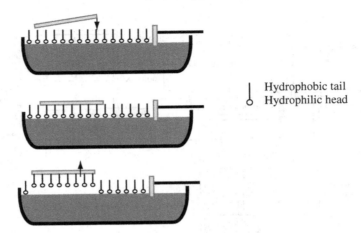

Hydrophobic tail
Hydrophilic head

Fig. 8.15   Langmuir–Blodgett film deposition.

ink (a suspension of carbon particles in a protein solution) on a water surface. The protein monolayer thus formed was picked up by paper. The films are layers of molecules transferred from the water–air interface onto a solid substrate. A bath of water is covered with a film and a substrate is dipped horizontally into the bath. The film bonds to the surface of the solid substrate during the dipping. The film usually consists of molecules having a hydrophobic "tail" and a hydrophilic "head." The molecules align with the hydrophobic tail away from the water to bind readily to the substrate. These layers can be in a monolayer or multi-layer form. Multi-layers are produced by repeating the dipping process. Common materials used for the films are liquid-crystal compounds, porphyrin, phthalocyanine, and polymers. There are many potential applications, for example applying a polyamide monolayer on graphite (Schedel-Niedrig *et al.*, 1991).

Figure 8.15 illustrates the case of the deposition of a single layer of an organic film with hydrophobic tail and hydrophilic head which is deposited on one side of a substrate. If the need is to deposit on both sides of the substrate, the "Y-type" LB method illustrated in Fig. 8.16(a) can be used. LB deposition methods based on a non-centrosymmetric system, Fig 8.16(b), (c), are also shown as X- and Z-types.

Self-assembly is a process by which molecules spontaneously assemble on a surface. The molecules chemisorb on the surface of a solid, leaving a functional group on the exposed end. The surface-active head is chosen to be attracted to the specific surface. The outward facing functional group is then tailored to produce the desired surface properties on the assembled layer. One requirement of the functional group is that it cannot be highly attracted to the substrate, to ensure that the functional group points away from the surface. A common

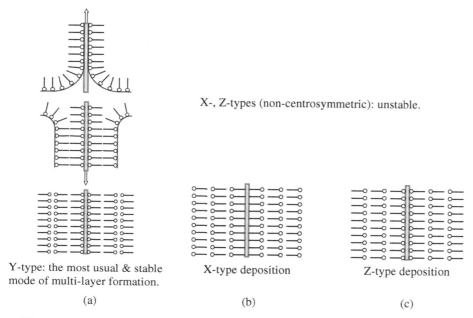

X-, Z-types (non-centrosymmetric): unstable.

Y-type: the most usual & stable mode of multi-layer formation.

X-type deposition

Z-type deposition

(a)

(b)

(c)

Fig. 8.16   Various types of LB film deposition methods.

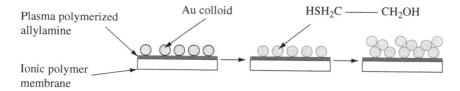

Plasma polymerized allylamine

Au colloid

HSH_2C ——— CH_2OH

Ionic polymer membrane

Fig. 8.17   Stepwise assembly of Au on ionic polymer membrane. (adapted from Musick *et al.*, 1997, © 1997, with permission from Amer. Chem. Soc.)

instance of this process uses an alkanethiol head to attach to a gold surface (Whitesides *et al.*, 1991; Kawasaki *et al.*, 2000).

Self-assembled monolayer (SAM) deposition is "borrowed" from Nature, where it is used by biological cells; it is a room-temperature, low-cost process. Figure 8.17 illustrates SAM applied to Au colloidal deposition on an ionic polymer substrate. In this example, both plasma polymerization and self-assembly are used. First, a thin coating of allylamine is deposited on the surface of an ionic polymer membrane by plasma polymerization. Then Au colloids assemble on the allylamine coating because of the high affinity of gold for amine groups. If the allylamine coating is patterned, a patterned Au deposition is obtained. Figure 8.18 shows an example of Au electrode pattern made by SAM (Le Guilly, 2004).

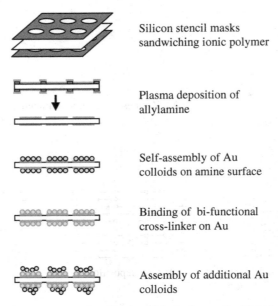

Silicon stencil masks
sandwiching ionic polymer

Plasma deposition of
allylamine

Self-assembly of Au
colloids on amine surface

Binding of bi-functional
cross-linker on Au

Assembly of additional Au
colloids

Fig. 8.18   Micro-patterned self-assembled electrodes fabrication scheme
(Le Guilly, 2004).

8.2 Standard measurement methods of properties

Properties of electronic composites need to be accurately measured. The key
properties are mechanical (stiffness tensor C_{ij}), thermal (thermal conductivity
coefficient of thermal expansion α, and specific heat c), and electromagnetic
(electrical conductivity σ, dielectric constant ε, and magnetic permeability μ).
For materials with coupled behavior such as piezoelectric materials, we have
also to measure the piezoelectric constants (d_{ij} or e_{ij}). Here we shall cover the
standard measurement methods for these properties.

8.2.1 Mechanical properties

Electronic composites are often anisotropic due to the presence of fillers, see
Fig. 3.4. Therefore, their elastic constants (stiffness tensor C_{ijkl}, Eq. (3.17a), or
compliance tensor S_{ijkl}, Eq. (3.17b)) need rigorous measurement methods. If a
composite is orthotropic, thus having nine elastic constants, Eq. (3.20) in terms
of compliance tensor, one can use a set of strain gauges to monitor various
strain components (e_{ij}) under increasing or decreasing applied stress (σ_{ij}). Then
the nine elastic constants are calculated from the measured strains using Eq.
(3.20). An electrical resistance strain gauge system is the most popular method
of measuring strains, first proposed by Thompson (1856). The electrical

resistance R of a metal wire of cross-sectional area A, length L, and electrical resistivity ρ is given by

$$R = \frac{\rho L}{A}. \tag{8.1}$$

Taking the total derivative of Eq. (8.1) and dividing it by R results in

$$\frac{dR}{R} = \frac{d\rho}{\rho} + \frac{dL}{L} - \frac{dA}{A}, \tag{8.2}$$

where the change dA in area divided by A is given by

$$\frac{dA}{A} = -2\nu\frac{dL}{L} + \nu^2\left(\frac{dL}{L}\right)^2, \tag{8.3}$$

and ν is Poisson's ratio of the metal. The strain sensitivity (S_A) is defined by

$$S_A = \frac{dR/R}{dL/L} = \frac{d\rho/\rho}{dL/L} + 1 + 2\nu, \tag{8.4}$$

where the higher-order term is neglected. The values of S_A for common strain gauge materials are given in Table 8.1 (Dally *et al.*, 1987).

S_A is very sensitive to temperature, and the mechanical property measurements with strain gauges are often done in a changing temperature environment. Thus we have to account for the temperature sensitivity of resistivity ρ. Then the resistance change dR/R is expressed by two terms

$$\frac{dR}{R} = S_e\Delta e + S_T\Delta T, \tag{8.5}$$

where S_e is the sensitivity factor to strain change Δe (gauge factor), and S_T is the sensitivity factor to temperature change ΔT. Both factors are well documented for commercially available strain gauges.

Table 8.1 *Strain gauge sensitivity S_A, and composition of common materials for strain gauges* (Dally *et al.*, 1987).

Material	Composition	S_A
Advance (or Constantan)	45Ni–55Cu	2.1
Karma	74Ni–20Cr–3Al–3Fe	2.0
Isoelastic	55.5Fe–36Ni–8Cr–0.5Mo	3.6
Nichrome V	80Ni–20Cr	2.1
Platinum–tungsten	92Pt–8W	4.0
Armour D	70Fe–20Cr–10Al	2.0

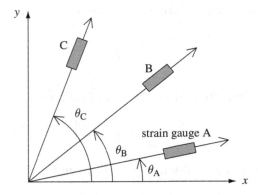

Fig. 8.19 Three strain gauges placed with different orientation angles for measurement of 2D strain components e_{xx}, e_{yy} and e_{xy} in the x-, y-coordinates.

In order to obtain the state of multi-axial strains (mostly up to two-dimensional in the x–y plane), one has to use multiple strain gauges, Fig. 8.19, where three strain gauges, A, B, and C are placed with different orientation angles with respect to the x-axis, θ_A, θ_B, and θ_C, respectively. Then the strain components along these orientations are expressed in terms of the strain components in the x-, y-coordinates:

$$e_A = e_{xx} \cos^2 \theta_A + e_{xy} \sin^2 \theta_A + 2e_{xy} \sin \theta_A \cos \theta_A, \tag{8.6a}$$

$$e_B = e_{xx} \cos^2 \theta_B + e_{yy} \sin^2 \theta_B + 2e_{xy} \sin \theta_B \cos \theta_B, \tag{8.6b}$$

$$e_C = e_{xx} \cos^2 \theta_C + e_{yy} \sin^2 \theta_C + 2e_{xy} \sin \theta_C \cos \theta_C. \tag{8.6c}$$

The unknowns e_{xx}, e_{yy}, and e_{xy} can be solved for from Eqs. (8.6) for measured values of e_A, e_B, and e_C.

If the orientation angles are specified as

$$\theta_A = 0°, \theta_B = 45°, \text{ and } \theta_C = 90°, \tag{8.7}$$

then Eqs. (8.6) are simplified as

$$e_A = e_{xx}, \tag{8.8a}$$

$$e_B = \frac{(e_{xx} + e_{yy} + 2e_{xy})}{2}, \tag{8.8b}$$

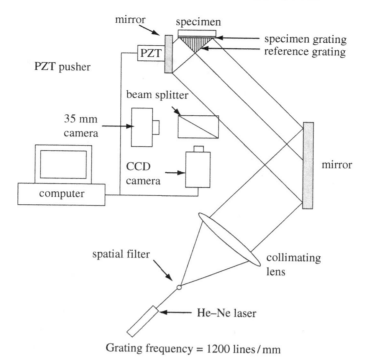

Fig. 8.20   Laser moiré interferometric system (Tran *et al.*, 1999. Reprinted from *Experimental Mechanics*, with permission of Soc. Exp. Mech. Inc.).

$$e_{\text{C}} = e_{yy}. \qquad (8.8c)$$

From Eqs. (8.8), we can easily obtain the three strain components, e_{xx}, e_{yy}, and e_{xy}. A strain gauge arrangement with the orientation angles specified by Eq. (8.7) is called a "rosette" and is commercially available.

When the specimen size is decreased, typical of electronic packages and MEMS components, use of the strain gauge method is difficult, and optical methods are better suited for the measurement of the strain field. The use of laser moiré interferometry is becoming increasingly popular in direct measurements of the strain field on a 2D specimen surface, because of its high spatial resolution (down to the submicron range). Figure 8.20 illustrates a typical setup of the optical system for moiré interferometry. Moiré interferometry, however, requires rigorous optical equipment and careful preparation of the specimen surface, including application of a grating.

The specimen grating is prepared from a master grating which is processed by photolithography onto a metal film which is then used as a mold in a compression-molding setup. The specimen, coated with uncured polymer adhesive, and the metal film are pressed together to produce a grated specimen, see Fig. 8.21.

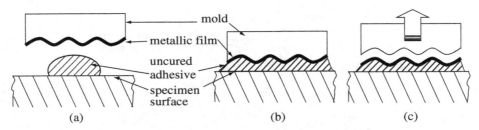

Fig. 8.21 Sequence of preparing a specimen with grating. (a) Master grating with metal film on top of polymer adhesive mounted on a specimen, (b) pressing, and (c) specimen with grating. (Post, 1987. Reprinted from *Handbook of Experimental Mechanics*, with permission of Society. Exp. Mech. Inc.)

The theoretical background of moiré interferometry is given by Post and his co-workers (Post, 1987; Post *et al.*, 1994). A high-frequency diffraction grating is replicated on the surface of the specimen. Upon loading of the specimen into the interferometer, the specimen grating, which deforms, interacts with the reference grating, which does not deform, forming a moiré fringe pattern. The moiré fringe pattern provides the displacements along the x-axis (U-field) and y-axis (V-field), which are calculated from

$$U = \frac{N_x}{f},$$ (8.9a)

$$V = \frac{N_y}{f},$$ (8.9b)

where f is the frequency of the virtual reference grating, and N_x and N_y are the fringe orders in the U- and V-fields of the fringe pattern. Electronic packaging components which are often made of electronic composites are of small size. Applications of moiré interferometry to these electronic packaging components requires use of a microscope able to resolve the fringe pattern. Han (1992) developed an experimental–numerical method to increase spatial resolution by using optical/digital fringe multiplication (O/DFM). Then the U- and V-fields are modified to

$$U = \frac{N_x^*}{f\beta},$$ (8.10a)

$$V = \frac{N_y^*}{f\beta},$$ (8.10b)

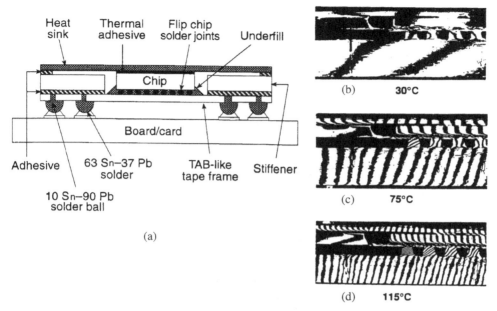

Fig. 8.22   (a) Tape ball grid array. (b)–(d) *U*-field moiré fringe pattern at (b) 30 °C, (c) 75 °C, and (d) 115 °C. (Han, 1998. Reprinted from *Experimental Mechanics* with permission of Soc. Exp. Mech. Inc.)

where N_x^* and N_y^* are the contour numbers, which are proportional to the fringe orders in the moiré fringe patterns, β is a fringe multiplication factor, and f is normally set equal to 4800 lines/mm. The spatial resolution based on O/DFM can be of submicron order. Han and co-workers applied the above O/DFM to observe several electronic packages under thermal loading – tape ball grid array (TBGA) package (Han, 1998) and wire bond–plastic ball grid array (WB-PBGA) package (Cho *et al.*, 2001). Figure 8.22(a) shows the profile of a TBGA which is subjected to various temperatures. The *U*-field fringe patterns of the TBGA subjected to temperatures of 30 °C, 75 °C, and 115 °C are shown in Fig. 8.22(b), (c), and (d), respectively. It is clear from Fig. 8.22(b)–(d) that the horizontal relative displacement in PCB and solder balls increases with temperature, resulting in higher shear strain in the solder balls. Figure 8.22(d) demonstrates that the shear strain increases toward the left edge of the solder ball array, which is coincident with the distribution of shear stress illustrated in Fig. 7.24.

Cho *et al.* (2001) observed the *U*- and *V*-fields of wire bond–plastic ball grid array (WB-PBGA). Figure 8.23(a) shows the entire WB-PBGA package which is dissected and bonded by grating. Figure 8.23(b) and (c) denote the *U*- and *V*-fields, respectively, when subjected to a temperature of 80 °C.

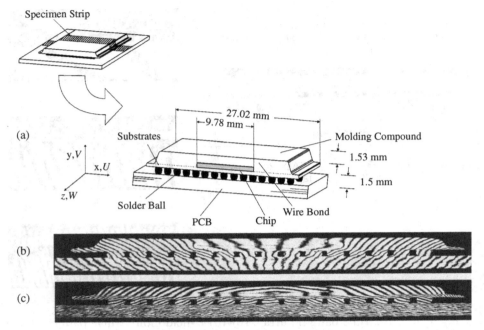

Fig. 8.23   Wire bond–plastic ball grid array (WB-PBGA). (a) Entire view
before dissecting, (b) *U*-field and (c) *V*-field fringe pattern of WB-PBGA
subjected to 80 °C. (Cho *et al.*, 2001. Reprinted from *Experimental
Techniques*, with permission of Soc. Exp. Mech. Inc.)

8.2.2 Thermal properties

The key thermal properties of electronic composites are the coefficient of
thermal expansion (CTE) (1/K), thermal conductivity K (W/m·K) and specific
heat capacity c (J/kg·K).

  Electronic composites are often composed of several different constituents –
metals, polymers, and ceramics – and they are subjected to a temperature
change ΔT during processing and in service. The CTEs of different constituents combined with ΔT lead to relatively large misfit strains, expressed in $\Delta \alpha$
ΔT where $\Delta \alpha$ denotes the difference of CTE (α) among different constituents.
Even if the misfit strain is small, under repeated temperature change (thermal
cycling) due to on/off of chips and other heat-emitting electronic components,
thermal stress and strain in electronic composites result in degradation of the
composites and their components. As the power of chips and other electronic
components increases rapidly, removal of the heat emitted by these is a key
thermal management issue. In view of the above, determination of the basic
thermal properties (α, K, and c) is a prerequisite for design engineers of
electronic packages and MEMS.

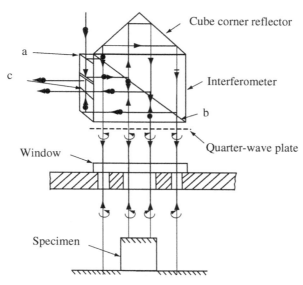

Fig. 8.24   The double-pass interferometer where a, b and c are beam splitters (Bennett, 1984, with permission from Kluwer/Plenum Publishers).

The CTE of an electronic composite specimen (typically a rod shape of length l) is normally measured by a dilatometer, which consists of a heating source, a quartz rod to convey the thermal expansion displacement (Δl), and a displacement measurement unit which can be of two types: a Linear Voltage Displacement Transformer (LVDT), and an optical method. Figure 8.24 shows an illustrative setup for a dilatometer with an optical method using a double-pass interferometer (Bennett, 1984). Measured data in terms of Δl vs. temperature T are obtained. The slope of the Δl–T curve provides the CTE, which is normally a function of T.

The CTEs of electronic composites are often anisotropic, having the property of a second-order tensor α_{ij}. Takei *et al.* (1991a) measured α_{ij} of electronic composites for use in a printed circuit board (PCB) using an LVDT dilatometer. Values of α_{ij} for a Kerimid® matrix loaded with four fillers are shown in Fig. 8.25(a)–(d), where i and j take values of x, y or z, with x, y being in plane and z being perpendicular to the plane. The values of α_{xy} and α_z of SiO_2 particle/Kerimid® matrix composites almost coincide with each other, indicating that this composite is nearly isotropic due to spherical SiO_2 particles, while for composites with other types of filler α_z exceeds α_{xy}, indicating that these are anisotropic due to the non-spherical shape of the fillers.

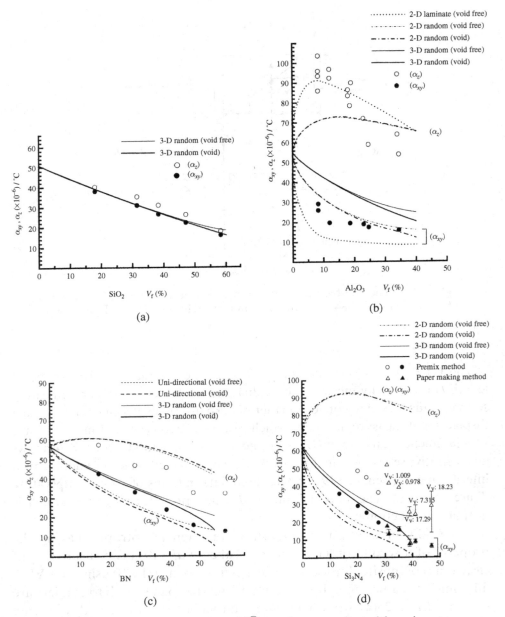

Fig. 8.25   *Coefficients* α_{ij} *of* Kerimid® *matrix composites with various types of filler as a function of filler volume fraction:* (a) SiO_2 particle, (b) Al_2O_3 short-fiber, (c) BN flake, *and* (d) Si_3N_4 whisker. (Takei *et al.*, 1991a, with permission from Elsevier Ltd.)

Thermal conductivity

Thermal conductivity K can be measured under the assumption of steady-state one-dimensional heat conduction (see American Society for Testing of Materials, ASTM C177-97 or ASTM D5470-01). ASTM C177-97 uses a symmetric heat flux arrangement requiring two identical specimens with a heat source between them, and is best suited for measurements of thermally less conductive materials or insulators, while ASTM D5470-01 is based on one-directional heat flow with only one specimen required, this being better suited for measurements of thermally more conductive materials. Hatta *et al.* (1992) used the former method to measure the diffusivity λ of selected electronic composites for use in PCBs where the environmental temperature is controlled. Once λ is measured, K is calculated from $K = \lambda/\rho c$ where ρ is mass density and c is specific heat capacity. Their apparatus is illustrated in Fig. 8.26.

Once steady-state heat conduction is established, one can measure the heat flow Q (W) by

$$Q = K_r S \frac{(T_1 - T_2)}{t_r}, \tag{8.11}$$

where K_r, t_r, and S are the thermal conductivity, thickness, and surface area of the reference material, respectively, and T_1 and T_2 are the temperatures of the top and bottom of the reference material (such as copper), respectively. Then the thermal resistance of a specimen R (K/W) is calculated from

$$R_t = \frac{T_2 - T_3}{Q}, \tag{8.12}$$

where T_3 is the temperature just underneath the specimen.

Since R_t is related to thermal conductivity K by Eq. (3.52), K (W/m·K) is calculated from

$$K = \frac{t}{R_t S}. \tag{8.13}$$

If the thermal properties are not expected to change much with environmental temperature, we can use a simpler apparatus with only a central heat source, i.e, without the heater to control the environmental temperature, but with a symmetric arrangement of two samples that sandwiches the central heater, Fig. 8.27.

The thermal conductivity of Al_2O_3 short-fiber/Kerimid® matrix composites measured by the apparatus of Fig. 8.27 is shown in Fig. 4.8. For reasonably conductive materials, we can use the apparatus based on ASTM D5470-01

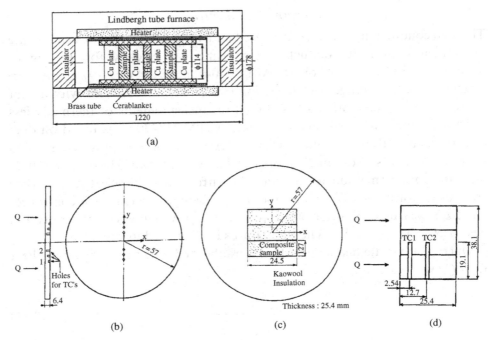

(a)

(b)   (c)   (d)

Fig. 8.26   Thermal property measurements for in-plane (λ_{xy}, K_{xy}) and out-of-plane (λ_z, K_z) properties. (a) Overall setup, (b) disk-shaped specimen with thermocouple locations for out-of-plane thermal properties (λ_z, K_z), (c) three identical specimens for measurements of in-plane thermal properties (λ_{xy}, K_{xy}), and (d) thermocouple locations of (c) Dimensions in mm. (Hatta *et al.*, 1992, with permission from Sage Publications Ltd.)

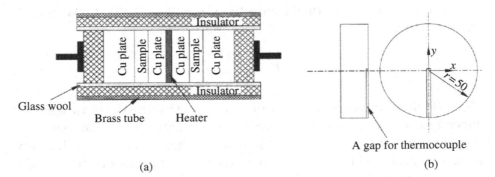

(a)   (b)

Fig. 8.27   Apparatus for measurement of thermal conductivity K of electronic composites: (a) furnace, (b) specimen geometry with thermocouple groove (Dunn *et al.*, 1993, with permission from Sage Publications Ltd.).

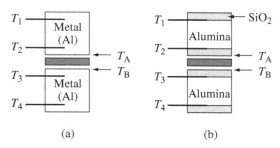

(a)   (b)

Fig. 8.28   One-dimensional steady-state heat conduction: (a) ASTM D5470-01, (b) a new design based on thermistor. (Park and Taya, 2003.)

with one-directional heat flow, Fig. 8.28(a), where four temperatures are measured by thermocouples embedded in a conductive metal, such as aluminum, from which the temperature T_A at the top and T_B at the bottom of the specimen are obtained by the extrapolation, and the thermal resistance R_t is calculated as

$$R_t = \frac{(T_A - T_B)}{Q},$$ (8.14)

where Q is the heat flow which is estimated by

$$Q = K_r S \frac{(T_1 - T_2)}{t_r}$$ (8.15)

where T_1 and T_2 are the temperatures at two locations of the reference metal (Al), and S is the surface area of the reference metal, which is equal to that of the specimen. Thermal impedance θ is sometimes used instead of thermal resistance R_t. Then, θ is defined by

$$\theta = R_t S.$$ (8.16)

In the electronic packaging industry, the unit of θ is often $cm^2 \cdot K/W$.

In all the above apparatuses, thermocouples are inserted in the reference materials through grooves. If more accurate measurements are warranted, temperature sensor patterns can be embedded in a reference material (alumina), Fig. 8.28(b), where the reference material (Al_2O_3) that sandwiches the specimen is composed of angled Au lines, as shown in Fig. 8.29.

By using the Al_2O_3 reference plates embedded with the new temperature sensor (Au–TiW lines) of Fig. 8.29, Park and Taya (2003) measured the thermal impedance θ ($cm^2 \cdot K/W$) of a commercial thermal interface material (TIM). The measured results of θ vs. temperature are shown in Fig. 8.30. It is noted that this TIM is composed of a phase-changeable polymer (PCP) matrix

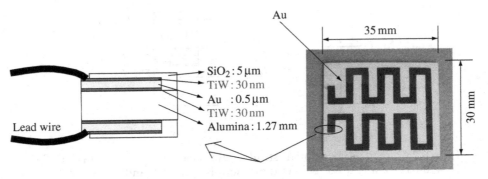

Fig. 8.29   A new temperature sensor embedded in Al_2O_3 reference material
(Park and Taya, 2003).

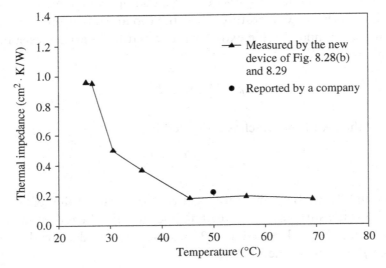

Fig. 8.30   Thermal impedance θ vs. temperature of a commercial TIM with
phase-changeable polymer matrix (Park and Taya, 2003).

and embedded conductive fillers, such as silver flakes. The PCP softens as the
temperature increases and the thickness of the TIM can be reduced by applying a vertical pressure to the package containing the TIM (such as TIM +
chip). As the thickness of the TIM is reduced, the thermal resistance R_t (thus
impedance θ) is reduced linearly as

$$R_t = \rho\frac{t}{S} \tag{8.17a}$$

or

$$\theta = \rho t, \tag{8.17b}$$

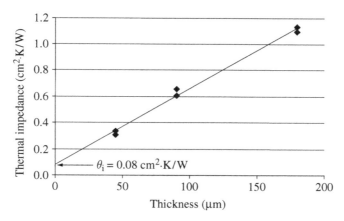

Fig. 8.31   Thickness dependence of thermal impedance θ of a commercial TIM with phase-changeable polymer matrix (Park and Taya, 2003).

where ρ is intrinsic volumetric thermal resistivity, and t and S are the thickness and surface area of the TIM. To study the thickness dependence of R_t (or θ), θ vs. t relationship of a commercial TIM is plotted in Fig. 8.31 where a spacer of known thickness was used in the experiment. The straight line passing through the three sets of data intersects the vertical axis at $\theta_i = 0.08\ \mathrm{cm^2 \cdot K/W}$, which is considered as the thermal impedance of the interface.

Specific heat capacity

Specific heat capacity $c\,(\mathrm{J/kg \cdot K})$ is defined by

$$\Delta Q = c\Delta T \tag{8.18}$$

where $\Delta Q\,(\mathrm{J})$ is the change in heat per unit mass (kg) under a uniform temperature change in a specimen $\Delta T(\mathrm{K})$. The property c has two values depending on the condition of measurements, constant volume of a specimen (c_v) and constant pressure (c_p), which are related to each other by

$$c_p - c_v = \frac{\alpha_{ii}^2 V_0 T}{\beta}, \tag{8.19}$$

where the volumetric CTE (1/K), $\alpha_{ii} = \alpha_{11} + \alpha_{22} + \alpha_{33}$, V_0 is volume per unit mass $(\mathrm{m^3/kg})$, and β is compressibility (1/Pa). Two methods of measuring specific heat are described in detail in ASTM D2766-95 and ASTM E1269-01. Both ASTM methods are aimed at measurement of the specific heat capacity of liquids and solids, but E1269-01 is based on use of a differential scanning calorimetry (DSC). Since DSC equipment is normally easy to access in a modern laboratory, E1269-01 appears to be the easier method, hence it is described

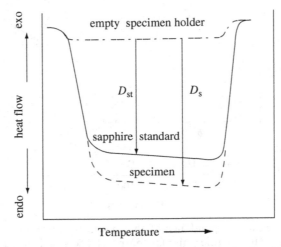

Fig. 8.32   Typical heat flow vs. temperature curve obtained by differential scanning calorimetry (DSC) where solid, dashed, and dash–dot curves denote the results of sapphire standard placed in specimen holder, specimen in specimen holder, and empty specimen holder, respectively (ASTM E1269-01, 2001).

here briefly. Accurate measurement of specific heat capacity requires calibration based on synthetic sapphire (α-alumina) whose specific heat capacity values are given as a function of temperature in the table of ASTM E1269-01. If calibration is made prior to DSC measurement, the specific heat capacity of a specimen under a constant pressure c_p is given by

$$c_p = c_p^* \frac{(D_s \cdot W_{st})}{(D_{st} \cdot W_s)} \tag{8.20}$$

where c_p^*, W_{st} and D_{st} are the constant-pressure specific heat capacity, weight, and DSC displacement (heat flow) of standard sapphire, and W_s and D_s are the weight and DSC displacement of a specimen, respectively. Fig. 8.32 shows a typical DSC displacement chart as a function of testing temperature where the definitions of D_s and D_{st} are given.

The thermal properties (α, K, and c_p) of popular electronic packaging materials are listed in Appendix B2.

8.2.3 Piezoelectric constants

Let us look at two methods of measurement of piezoelectric constants: (1) static, and (2) dynamic. The static technique is easier to understand, but the accuracy of the measured piezoelectric constants is within a few percent error,

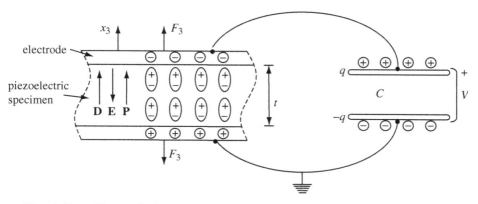

Fig. 8.33   Signs of charges and fields for static test of sample with d_{33} greater than 0 under tension (ANSI/IEEE, 1987).

while the dynamic (resonance) method is now the standard measurement method for piezoelectric constants. There are two methods of static measurement, one based on the direct effect, the other on the converse effect. We shall examine here the static method based on the direct effect, Fig. 8.33, in which a given stress σ_3 (force F_3) is applied across the thickness t of a piezoelectric specimen connected to a potentiostat which measures the voltage induced in the capacitor with capacitance C. To this end we will rewrite the constitutive equation of a piezoelectric material given by Eq. (3.218) in terms of the electric flux density D vs. stress σ relation with the isothermal condition, $\theta = 0$,

$$D_i = d_{ikl}\sigma_{kl} + \varepsilon_{in}E_n, \qquad (8.21)$$

$$d_{nij} = e_{nkl}S_{klij}, \qquad (8.22)$$

where d_{ikl} is a third-order tensor of the piezoelectric d-constant. Following the 1987 IEEE convention, i.e., short notation of piezoelectric constant and stress and strain components (6×1 column vector form), the above equation is further expressed by

$$D_i = d_{im}\sigma_m + \varepsilon_{in}E_n. \qquad (8.23)$$

If the thickness of the piezoelectric specimen is taken in the x_3-axis in which tensile stress σ_3 is applied, Eq. (8.23) is reduced to

$$D_3 = d_{33}\sigma_3 + \varepsilon_{33}E_3, \qquad (8.24)$$

where ε_{33} is the dielectric constant in the thickness direction of the piezoelectric specimen and E_3 is the electric field given by

Engineering problems

$$E_3 = -\frac{V}{t}. \tag{8.25}$$

Please note that E_3 is pointing toward the negative x_3-axis, according to the coordinate system used in Fig. 8.33. In the capacitor with capacitance C, the induced voltage V is given by

$$V = \frac{q}{C}, \tag{8.26}$$

where

$$q = D_3 S, \tag{8.27}$$

and S (length×width) is the surface area of the specimen. Substituting Eqs. (8.25)–(8.27) into Eq. (8.24), we can obtain

$$V = \frac{d_{33}\sigma_3 S}{\left(C + \frac{\varepsilon_{33} S}{t}\right)}. \tag{8.28}$$

For a given input of applied tensile force F_3 ($= \sigma_3 S$), by measuring the induced voltage V we can obtain the piezoelectric d-constant d_{33} in the thickness direction, provided that the dielectric constant ε_{33} of the piezoelectric specimen and the capacitance C of the voltage-measurement unit are known together with geometrical data.

The static method based on the converse piezoelectric effect can be obtained from the constitutive equation Eq. (3.219), with the second term modified and $\theta = 0$, becomes

$$e_{ij} = S_{ijkl}\sigma_{kl} + g_{kij}\varepsilon_{kl}E_l. \tag{8.29}$$

(Substitute Eq. (3.96) in index form into Eq. (3.219) to obtain Eq. (8.29).)
Noting that the product $g_{kij}\varepsilon_{kl}$ is equal to the d-constant,

$$d_{lij} = g_{kij}\varepsilon_{kl}, \tag{8.30}$$

Eq. (8.29) is rewritten as

$$e_{ij} = S_{ijkl}\sigma_{kl} + d_{lij}E_l. \tag{8.31}$$

In the absence of an applied stress, and using the short notation of the 1987 IEEE convention, Eq. (8.31) is reduced to

$$e_i = d_{li}E_l, \tag{8.32}$$

where e_i is a 6×1 column vector and E_l is a 3×1 column vector, see Eqs. (3.19) and (3.223).

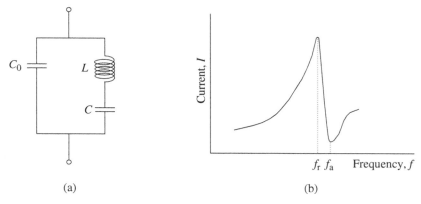

(a)   (b)

Fig. 8.34   Dynamic resonance measurement of piezoelectric d-constant and electrical–mechanical coupling coefficient k: (a) equivalent circuit, (b) current vs. frequency.

The measurement of the d-constant based on Eq. (8.32) provides a more accurate value of d than the direct static method, as control of a stress-free state in the piezoelectric specimen and measurement of the induced strain for a given applied electric field are relatively easy. However, both static methods are now more usually replaced by the dynamic resonance method, which provides much more accurate data.

Here we will discuss briefly a simple dynamic resonance method to measure the piezoelectric d-constant. Consider a piezoelectric specimen with thickness t subjected to an alternating current with variable frequency f under a constant voltage V. The equivalent circuit of the measurement setup is shown in Fig. 8.34(a) where C and L are the capacitance and inductance, respectively, of the piezoelectric specimen and other measurement components. A typical current I vs. frequency f relation is shown in Fig. 8.34(b) where the first peak (f_r) and second peak (f_a) are called the resonance frequency and antiresonance frequency, respectively, and they are defined by

$$f_r = \frac{1}{2\pi\sqrt{LC}},\tag{8.33a}$$

$$f_a = \frac{1}{2\pi\sqrt{LC}}\sqrt{1 + \frac{C}{C_0}}.\tag{8.33b}$$

From Eqs. (8.33) we obtain

$$1 - \left(\frac{f_r}{f_a}\right)^2 = \frac{C}{C + C_0} \equiv k^2,\tag{8.34}$$

where k denotes the electrical–mechanical coupling coefficient and k^2 is given by

$$k^2 = \frac{U_m}{U_e + U_m},\tag{8.35}$$

where U_m is mechanical energy and U_e is electrical energy.

The electrical energy U_e that is stored in a piezoelectric specimen with dielectric constant ε, thickness t, and surface area S is given by

$$U_e = \frac{1}{2}C_0 V^2 = \frac{1}{2}\frac{\varepsilon S}{t} V^2.\tag{8.36}$$

Under an electric field of V/t, strain is induced through the piezoelectric d-constant, see Eq. (8.32). Thus, the strain energy U_m in a piezoelectric specimen is given by

$$U_m = \frac{1}{2}E\left(d\frac{V}{t}\right)^2 St,\tag{8.37}$$

where E is Young's modulus of the piezoelectric specimen. Substitution of Eqs. (8.36) and (8.37) into Eq. (8.35) provides

$$k^2 = \frac{d^2 E}{\varepsilon + d^2 E}.\tag{8.38}$$

With Eq. (8.34), Eq. (8.38) can be written as

$$\frac{k^2}{1 - k^2} = \left(\frac{f_a}{f_r}\right)^2 - 1 = d\sqrt{\frac{E}{\varepsilon}}.\tag{8.39}$$

Equation (8.39) provides the d-constant from the measured values of f_r and f_a, and known material constants E and ε.

8.2.4 Electromagnetic properties

Accurate measurements of magnetic permeability μ, permittivity (dielectric constant) ε, and electric conductivity σ are very important for electronic packaging, RF engineers, and MEMS designers. For example, the dielectric constant of a substrate material of a PCB must be known in order to design microwave and high-frequency circuits. Another example is in modern fighter planes, which are designed to reduce their radar cross section (RCS) by using composite materials. Because the amount of radar reflection (cross section) is strongly dependent on the materials used, the permeability and permittivity of composite materials such as a carbon composite must be measured at all the radar operating frequencies (from VHF to millimeter wavelength). Here we shall review the methods of determining the permeability, permittivity, and resistivity (or conductivity).

Dielectric constant ε and permeability μ

First, we will study the transmission techniques. We assume that the test sample is placed inside a coaxial transmission line or waveguide. From the measured S_{11} and S_{21}, the material characteristics are obtained using an inversion algorithm (Nicolson and Ross, 1970). It will be shown that the Nicolson–Ross method diverges when the sample length becomes a multiple of $\lambda/2$.

For DC or very low-frequency cases, we can measure ε ($= \varepsilon_0 \varepsilon_r$) of a dielectric specimen with surface area S and thickness d from Eq. (3.147) provided that $d \ll \sqrt{S}$, as in these cases the fringe field can be neglected.

For high-frequency cases, the transmission method is the most popular method to obtain the relative dielectric constant ε_r and relative permeability μ_r, which are complex-valued:

$$\varepsilon_r = \varepsilon'_r - j\varepsilon''_r, \tag{8.40a}$$

$$\mu_r = \mu'_r - j\mu''_r, \tag{8.40b}$$

where ε'_r, ε''_r are real and imaginary parts of the relative dielectric constant, and μ'_r and μ''_r are real and imaginary parts of the relative permeability.

Let us use the Nicolson–Ross method and assume that the sample holder is a coaxial TL, as illustrated in Fig. 8.35.

For three different regions, 1, 2, and 3 (from left to right in the figure), voltage and current are given by

$$V_1 = V_{in} e^{-j\gamma_0 l} + V_1^- e^{j\gamma_0 l}, \tag{8.41a}$$

$$I_1 = \frac{1}{Z_0} \left(V_{in} e^{-j\gamma_0 l} - V_1^- e^{j\gamma_0 l} \right), \tag{8.41b}$$

$$V_2 = V_2^+ e^{-j\gamma_1 l} + V_2^- e^{j\gamma_1 l}, \tag{8.42a}$$

$$I_2 = \frac{1}{Z_s} \left(V_2^+ e^{-j\gamma_1 l} - V_2^- e^{j\gamma_1 l} \right), \tag{8.42b}$$

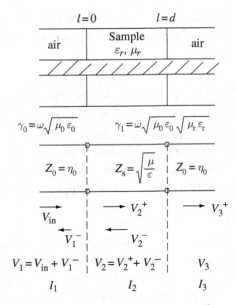

Fig. 8.35   Transmission method for measurement of ε_r and μ_r.

$$V_3 = V_3^+ e^{-j\gamma_0(l-d)}, \tag{8.43a}$$

$$I_3 = \frac{1}{Z_0}\left(V_3^+ e^{-j\gamma_0(l-d)}\right), \tag{8.43b}$$

where the boundary conditions at $l=0$ are:

$$V_1 = V_2 \text{ and } I_1 = I_2, \tag{8.44a}$$

and at $l=d$ are:

$$V_2 = V_3 \text{ and } I_2 = I_3. \tag{8.44b}$$

The reflection coefficient Γ is given by

$$\Gamma = \frac{Z_s - Z_0}{Z_s + Z_0} = \frac{\sqrt{\dfrac{\mu_r}{\varepsilon_r}} - 1}{\sqrt{\dfrac{\mu_r}{\varepsilon_r}} + 1}, \tag{8.45}$$

and the transmission coefficient T by

$$T = e^{-j\gamma_1 d} = e^{-j\omega\sqrt{\mu_0\varepsilon_0}\sqrt{\mu_r\varepsilon_r}d}. \tag{8.46}$$

The *S*-parameters are given by

$$S_{11} = \frac{V_1^-}{V_{in}} = \frac{(1 - T^2)\Gamma}{1 - T^2\Gamma^2},$$ (8.47a)

$$S_{21} = \frac{V_3^+}{V_{in}} = \frac{(1 - \Gamma^2)T}{1 - T^2\Gamma^2}.$$ (8.47b)

It is to be noted that S_{11} and S_{21} are measured values, they provide four equations, while Γ and T are related to ε_r and μ_r, with four unknowns

$$\Gamma = K \pm \sqrt{K^2 - 1},$$ (8.48)

$$K = \frac{(S_{11}^2 - S_{21}^2) + 1}{2S_{11}},$$ (8.49)

$$T = \frac{(S_{11} + S_{21}) - \Gamma}{1 - (S_{11} + S_{21})\Gamma}.$$ (8.50)

From Eqs. (8.45) and (8.46), μ_r and ε_r are solved as

$$\mu_r = \frac{1 + \Gamma}{\Lambda(1 - \Gamma)\sqrt{\left(\frac{1}{\lambda_0^2} - \frac{1}{\lambda_c^2}\right)}},$$ (8.51)

$$\varepsilon_r = \frac{\left(\frac{1}{\Lambda^2} + \frac{1}{\lambda_c^2}\right)\lambda_0^2}{\mu_r},$$ (8.52)

where

$$\frac{1}{\Lambda^2} = -\left[\frac{1}{2\pi d}\ln\left(\frac{1}{T}\right)\right]^2,$$ (8.53)

and λ_0, λ_c are the free-space wavelength and the waveguide cut-off wavelength, respectively. For a transverse electromagnetic (TEM) wave $\lambda_c \rightarrow \infty$ (coaxial TL)

This transmission method becomes troublesome if S_{11} becomes small. One example is a low-reflection material such as Styrofoam ($\varepsilon_r \approx 1.05$). Another example is when the effective length of the sample becomes $m\lambda/2$ where m is an integer. In this case, $Z_{in} = Z_0$ and S_{11} becomes 0.

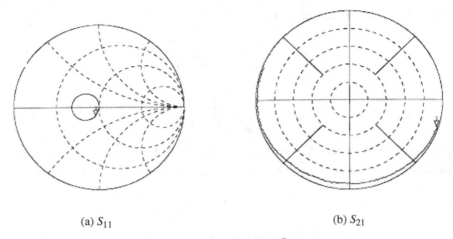

(a) S_{11} (b) S_{21}

Fig. 8.36 Measured S-parameters of a Teflon® sample: (a) S_{11} and (b) S_{21}. The x- and y-axes are real and imaginary values, respectively, of S_{ij}.

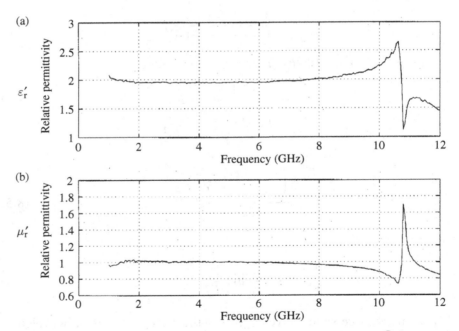

Fig. 8.37 Calculated real parts of (a) ε_r and (b) μ_r for a Teflon® sample.

Measured S-parameters of a Teflon® material are shown in Fig. 8.36. Calculated values of ε_r and μ_r using the Nicolson–Ross method (real part only) are shown in Fig. 8.37 where instability around 11 GHz due to the $\lambda/2$ condition is clearly visible.

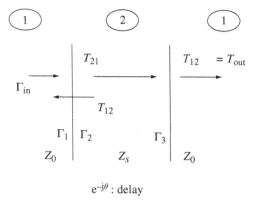

$e^{-j\theta}$: delay

Fig. 8.38   Reflection and transmission coefficients of three-layer model.

In order to overcome the above problems associated with the Nicolson–Ross method, Barker-Jarvis *et al.* (1990) proposed an improved technique to measure ε_r and μ_r which will not be detailed here. Appendix B3 lists the relative dielectric constants ε_r of popular materials where $\tan \delta$ is defined by $\tan \delta = \varepsilon_r''/\varepsilon_r'$ and $\varepsilon_r = \varepsilon_r' - j\varepsilon_r''$.

Derivation of reflection and transmission coefficients of a layered structure based on multiple reflection.

Referring to Fig. 8.38, the reflection coefficients are

$$\Gamma_1 = \frac{Z_s - Z_0}{Z_s + Z_0},$$ (8.54a)

$$\Gamma_2 = \frac{Z_0 - Z_s}{Z_0 + Z_s} = -\Gamma_1,$$ (8.54b)

$$\Gamma_3 = \frac{Z_0 - Z_s}{Z_0 + Z_s} = \Gamma_2 = -\Gamma_1$$ (8.54c)

and the transmission coefficients are

$$T_{21} = 1 + \Gamma_1,$$ (8.55a)

$$T_{12} = 1 + \Gamma_2 = 1 - \Gamma_1.$$ (8.55b)

The total reflection can be considered as a summation of all multiple reflections. Then, a reflection coefficient Γ_{in} is defined by

$$\Gamma_{in} = \frac{Z_{in} - Z_0}{Z_{in} + Z_0}$$

$$= \Gamma_1 + T_{12}T_{21}\Gamma_3 e^{-2j\theta} + T_{12}T_{21}\Gamma_3^2\Gamma_2 e^{-4j\theta} + \cdots$$

$$= \Gamma_1 + T_{12}T_{21}\Gamma_3 e^{-2j\theta} \sum_{n=0}^{\infty} \Gamma_2^n \Gamma_3^n e^{-2jn\theta}, \qquad (8.56)$$

where the second- and higher-order terms represent the multiple reflections.

Since $\sum_{n=0}^{\infty} x^n = \frac{1}{1-x} |x|$, Γ_{in} is reduced to

$$S_{11}(\text{measured}) = \Gamma_{in} = \frac{\Gamma_1\left(1 - e^{-2j\theta}\right)}{1 - \Gamma_1^2 e^{-2j\theta}}. \qquad (8.57)$$

Similarly, the transmission coefficient T_{out} is given by

$$S_{21}(\text{measured}) = T_{out} = \frac{\left(1 - \Gamma_1^2\right)e^{-j\theta}}{1 - \Gamma_1^2 e^{-2j\theta}}. \qquad (8.58)$$

Electrical resistivity

Two types of measurement method are often used, a four-point probe for low resistance (below $100\,\Omega$) and a two-point probe for high resistance. Normally, the measurement of low resistance is more difficult than that of high resistance.

The four-point probe is the most popular method to measure a low-resistance material as this can avoid the problem of contact resistance inherent in a two-point probe. Figure 8.39 shows a four-point probe test fixture with spring-loaded probe tips fixed into an insulating acrylic plate, where the outer (current) probes are 2 inches (5.08 cm) apart from each other and the inner (voltage) probes are 0.5 inch (1.27 cm) apart from the current probes, giving a voltage probe spacing of 1 inch (2.54 cm) (Kim, 1998). The relatively long voltage probe spacing enhances the specimen resistance when using a relatively low DC source and allows an acceptable accuracy to be achieved. After a specimen is placed onto the four-point probe test fixture, a constant current is applied through the outer probes using a Hewlett Packard (HP) 6186C DC source and the voltage is measured between the inner probes with a Keithley 2000 digital multimeter.

The electrical resistance R of a specimen based on the four-point probe is obtained in terms of a V–I relation, from which electrical resistance R is calculated. Then the volume resistivity ρ is obtained from

$$\rho = R\frac{A}{L} \qquad (8.59)$$

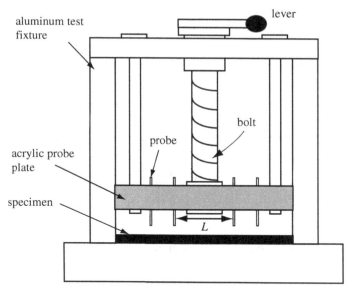

Fig. 8.39   Four-point probe system for low-resistance measurement (Kim, 1998).

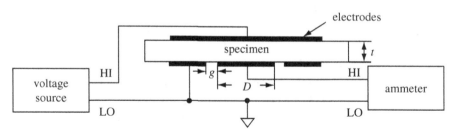

Fig. 8.40   Two-point probe system for high-resistance measurement (Kim, 1998). "HI" and "LO" are high and low applied voltage, respectively. (Not to scale.)

where A is the cross-sectional area (width × thickness) of a specimen and L is the distance between the inner probes, see Fig. 8.39. This four-point probe normally provides reasonably accurate value of resistance (error within 0.02%).

The method for high-resistance measurement, which is applicable for an insulator or a semiconductor, is well described in ASTM D257. Figure 8.40 illustrates a typical system where an HP 16008B resistivity cell is used with an electrode 1 inch (25.4 mm) in diameter and a guard electrode ring 1.5 inch (38.1 mm) in diameter. The guard electrode ring is located together with the bottom electrode, giving a guard gap g of typically 0.25 inch (6.35 mm). The guard gap should be twice as large as the specimen thickness t (Note that Fig. 8.40 is not drawn to scale in this respect). A disk-shaped specimen with diameter D and thickness t is mounted between the top and bottom electrode plates of the HP 16008B resistivity cell. To minimize the interfacial resistance between the specimen and electrodes, conductive silver paint is applied onto the specimen surfaces. A high DC voltage, ranging from 100 to 900 volts, is applied to the specimen using a Bertan 205B-01R high-voltage supply, and the corresponding current is measured using a Keithley 485 autoranging picoammeter or a Keithley 2000 multimeter after an electrification time of 60 seconds. Then the volume resistivity of the specimen is calculated by

$$\rho = \frac{S}{t}\frac{V}{I},$$ (8.60)

where S is the surface area of the specimen, given by

$$S = \frac{\pi(D+g)^2}{4}.$$ (8.61)

Magnetization measurement methods

The magnetization vector $\mathbf{M}$ of a ferromagnetic material is a key magnetic property, which is normally expressed in terms of magnetic field $\mathbf{H}$. Thus, determination of the $\mathbf{M}$–$\mathbf{H}$ curve of a ferromagnetic material is a prerequisite for experimental work. Here we shall review three basic methods of measurement of the $\mathbf{M}$–$\mathbf{H}$ curve of ferromagnetic materials which include "soft" and "hard" magnetic materials. The three methods are based on (1) magnetic force, (2) magnetic field, and (3) electric current or voltage (Chikamizu, 1964). The measurement of magnetization vector $\mathbf{M}$ based on force F stems from the following equation:

$$F_x = \mu_0 \mathbf{M} \cdot \frac{\partial \mathbf{H}}{\partial x} V,$$ (8.62)

where μ_0 is the magnetic permeability in a vacuum defined by Eqs. (3.104)–(3.106), $\mathbf{M}$ and V are the magnetization vector and the volume of a ferromagnetic specimen, respectively, and F_x is the force along the x-direction. If the orientation of $\mathbf{H}$ is specified, for example along the x-direction with $H_y = H_z = 0$, Eq. (8.62) is reduced to

$$F_x = \mu_0 M_x \frac{\partial H}{\partial x} V.$$ (8.63)

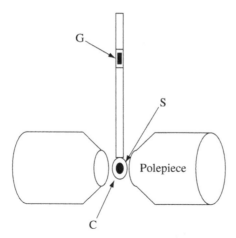

Fig. 8.41   Vertical magnetic pendulum to measure the magnetic force, from which the magnetization of a ferromagnetic specimen is calculated (Bozorth and Williams, 1956).

Therefore the magnetization M_x is calculated from Eq. (8.63) by measuring the magnetic force F_x and knowing $\partial H/\partial x$. Bozorth and Williams (1956) designed an apparatus based on a vertical magnetic pendulum, Fig. 8.41, where a ferromagnetic specimen (S) surrounded by a coil (C) is attached at the bottom of a pendulum subjected to an inhomogeneous H-field producing a specified gradient $\partial H/\partial x$. The deflection of the pendulum is measured by a strain gauge (G) located in the top portion of the pendulum. This deflection, a measure of the magnetic force acting on the specimen, allows the magnetization to be calculated.

The magnetization of a ferromagnetic specimen can also be obtained using a vibrating specimen magnetometer (VSM) by measuring the AC signal emitted from a specimen which is subjected to a forced vibration by a voice coil. The resulting alternating current in the two set of coils is fed into an amplifier for detection and measurement, Fig. 8.42. The VSM is the most popular equipment to measure the magnetization of a ferromagnetic material; a typical commercial setup is shown in Fig. 8.43.

A ballistic galvanometer with an inductance/resistance (LR) circuit can be used to measure the total electric charge Q (which is related to the magnetization M) of a specimen subjected to a known magnetic field H. Figure 8.44 shows the magnetization measurement setup where a specimen (S) is attached to a rod (E) which is pulled out from two solenoids (A and B; each having n turns) connected to a ballistic galvanometer (G). The rod with coils (A and B) and specimen (S) is placed inside a solenoid (C) that provides a known H-field.

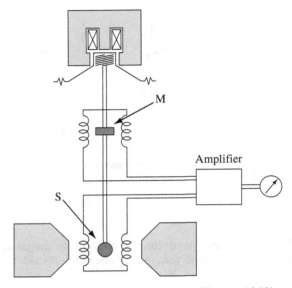

Fig. 8.42   Vibrating specimen magnetometer (Foner, 1959).

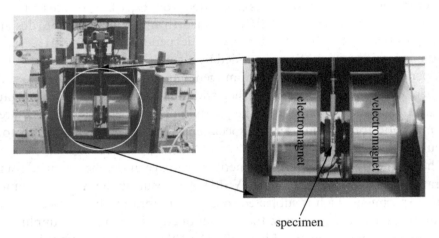

Fig. 8.43   Vibrating specimen magnetometer (VSM) (courtesy of Dr. Sutou, Tohoku University).

The total electric charge Q stored in the galvanometer (G) is proportional to the displacement u of the galvanometer needle:

$$Q = ku. \tag{8.64}$$

A specimen with magnetization M and volume v has a total magnetization of Mv. When a ferromagnetic specimen with magnetization M is moved through coils A and B, a magnetic flux Φ is generated:

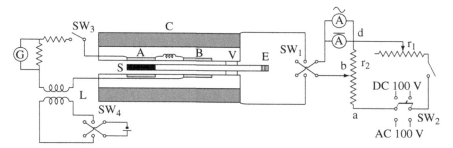

Fig. 8.44   Magnetization measurement method based on ballistic galvanometer (Tsutsui, 1960).

$$\Phi = 2Mvn. \tag{8.65}$$

The electric current I that flows spontaneously in the ballistic galvanometer
is given by

$$I = \frac{dQ}{dt} = \frac{V}{R} \tag{8.66}$$

while voltage V is related to the rate of change with time of the magnetic flux Φ
(Faraday's law, Eq. (3.80) with $N = 1$),

$$V = \frac{d\Phi}{dt}. \tag{8.67}$$

From Eqs. (8.66) and (8.67)

$$Q = \frac{1}{R}\Phi. \tag{8.68}$$

Substituting Eqs. (8.64) and (8.65) into (8.68), we obtain

$$M = \left(\frac{kR}{2n}\right)u. \tag{8.69}$$

Equation (8.69) indicates that the magnetization of a ferromagnetic specimen is proportional to the displacement u of the ballistic galvanometer needle
with a proportionality constant given by $kR/2n$.

8.3 Electromigration

Electromigration is a phenomenon where metallic atoms flow toward an
anode, resulting in depletion of metal and formation of voids and hillocks,
posing serious degradation in conductive metallic lines. This electromigration

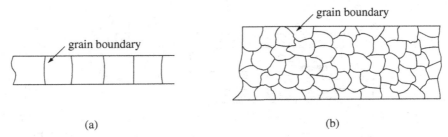

Fig. 8.45   Microstructure of a metal line: (a) narrow line with bamboo-like structure, (b) wider line with polycrystalline structure with many grains across line width.

phenomenon was first studied quantitatively by Huntington and Grone (1961) who proposed a formula to predict the atomic flux which is dependent on many parameters, but most dominantly on current density j where the mechanism at the micro-level is the diffusion of atoms along the grain boundaries of a polycrystalline metal. As the width of metallic lines is being reduced rapidly due to miniaturization of IC and MEMS packages, designers are much concerned with the degradation of such metallic lines due to electromigration. For narrow metallic lines, current density j is expected to increase, resulting in an increase in atomic flux according to the Huntington–Grone formula. But the microstructure of narrow metallic lines is bamboo-like, Fig. 8.45(a), while that of wider lines is composed of a typical polycrystalline structure with many grains that bridge across the line width, Fig. 8.45(b).

Although the narrower lines were expected to be degraded more severely than the wider lines, Vaidya *et al.* (1980) found experimentally that narrower lines of Al–0.5% Cu exhibited a longer life because the bamboo-like microstructure with grain boundaries perpendicular to the line direction has a lower total length of boundaries than the wider polycrystalline metal line. The above two studies prompted further studies to clarify the exact mechanisms associated with electromigration. In parallel with such studies, Black (1969) proposed a simple formula to predict the lifetime τ of a metal line

$$\tau = A j_{in}^{-n} \exp\left(\frac{Q}{k_B T}\right), \qquad (8.70)$$

where j_{in} is input current density, k_B is the Boltzmann constant, n is a parameter (normally 1–3) (Ghate, 1983), T is absolute temperature, Q is the activation energy (0.4–0.8 eV) and A is a constant dependent on the metal-line

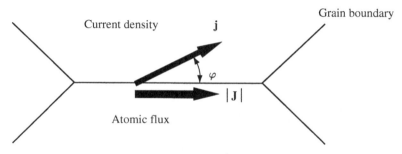

Fig. 8.46   Atomic flux along grain boundaries. (Sasagawa *et al.*, 1999a, with permission from Amer. Inst. Phys.)

characteristics. In Black's formula, constants A and n are obtained empiri-cally by accelerated testing which is performed at a higher-than-use tempera-ture. Thus the validity of using A and n for analysis of a metal line with different line geometry and use temperatures is not sure-footed. To increase the accuracy of the lifetime prediction model, Sasagawa *et al.*(1998) proposed a more rigorous model focusing on atomic flux divergence (AFD) by account-ing for all factors that contribute to void formation in a metal line, i.e., current density, temperature, their gradients, and metal-line characteristics. In their model, an atomic flux vector $\mathbf{J}$ is given by (Huntington and Grone, 1961)

$$\mathbf{J} = \frac{ND_0}{k_{\mathrm{B}}T}\exp\left(-\frac{Q_{\mathrm{gb}}}{k_{\mathrm{B}}T}\right)Z^*e\rho|\mathbf{j}|\cos\varphi \qquad (8.71)$$

where N is the atomic density, D_0 is a prefactor, Q_{gb} is the activation energy for grain boundary diffusion, Z^* is the effective valence, e is electric charge, $\mathbf{j}$ is the current density vector, φ is the angle between the current density vector and the grain boundary line, Fig. 8.46, ρ is the electrical resistivity, dependent on temperature as $\rho = \rho_0\{1 + \alpha(T - T_{\mathrm{s}})\}$, ρ_0 is the resistivity at T_{s}, which is the substrate temperature, T is the temperature of the metal line, and α is the temperature coefficient. Sasagawa *et al.* (1999a) applied the above equation to predict atomic flux divergence for a given unit cell made of a polycrystalline metal line, Fig. 8.47, where hexagonal grains are periodically arranged and the effects of current density $\mathbf{j}$ and the temperature field T are both considered. The atomic fluxes along the grain boundaries within the unit cell are summed, multiplied by the cross-sectional area of the grain boundary (width δ) and then divided by the volume of the unit cell, providing atomic flux divergence per unit time (AFD$_{\mathrm{gb}\theta}$), which is given by

Engineering problems

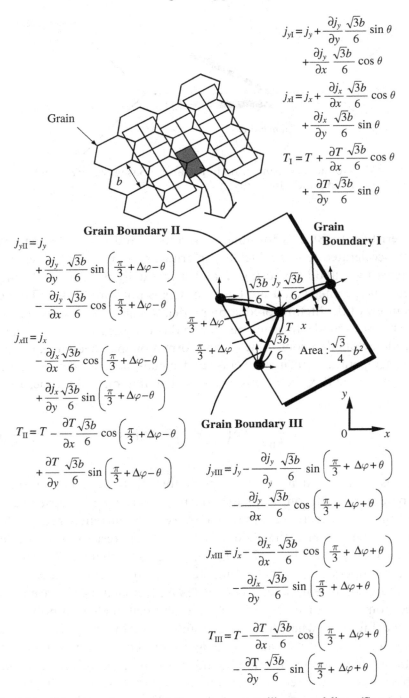

Fig. 8.47 Line structure model for a polycrystalline metal line. (Sasagawa *et al.*, 1999a, with permission from Amer. Inst. Phys.)

$$
\begin{aligned}
\mathrm{AFD}_{\mathrm{gb}\theta} = C_{\mathrm{gb}}\rho \frac{4}{\sqrt{3}b^2}\frac{1}{T}\exp\left(-\frac{Q_{\mathrm{gb}}}{k_{\mathrm{B}}T}\right) & \left[\sqrt{3}\Delta\varphi\ (j_x\cos\theta + j_y\sin\theta) \right. \\
& -\frac{b}{2}\Delta\phi\left\{\left(\frac{\partial j_x}{\partial x}-\frac{\partial j_y}{\partial y}\right)\cos 2\theta + \left(\frac{\partial j_x}{\partial y}+\frac{\partial j_y}{\partial x}\right)\sin 2\theta\right\} \\
& \left. +\frac{\sqrt{3}b}{4T}\times\left(\frac{Q_{\mathrm{gb}}}{k_{\mathrm{B}}T}-1\right)\left(\frac{\partial T}{\partial x}j_x + \frac{\partial T}{\partial y}j_y\right)\right],
\end{aligned}
\tag{8.72}
$$

where $\mathrm{AFD}_{\mathrm{gb}\theta}$ is dependent on the orientation angle θ of grain boundary I with respect to the x-axis, and C_{gb} is given by

$$
C_{\mathrm{gb}} = \frac{ND_0 Z^* e\delta}{k_{\mathrm{B}}}.
\tag{8.73}
$$

It is to be noted in Eq. (8.72) that positive values of $\mathrm{AFD}_{\mathrm{gb}\theta}$ mean depletion of metal atoms, i.e., void formation, while negative values of $\mathrm{AFD}_{\mathrm{gb}\theta}$ mean increase in metal atoms. Since θ takes all possible values from 0 to 2π, we obtain the average $\mathrm{AFD}_{\mathrm{gen}}$ in a unit cell of a polycrystallic metal line per unit time as

$$
\mathrm{AFD}_{\mathrm{gen}} = \frac{1}{4\pi}\int_0^{2\pi}\left(\mathrm{AFD}_{\mathrm{gb}\theta} + |\mathrm{AFD}_{\mathrm{gb}\theta}|\right)\,d\theta.
\tag{8.74}
$$

$\mathrm{AFD}_{\mathrm{gen}}$ defined by Eq. (8.74) represents the number of metal atoms that are depleted per unit volume and per unit time, thus it is equivalent to the volume of voids that are formed in a unit volume of a metal line per unit time. The distributions of current density $\mathbf{j}$ and temperature T are calculated by using the finite element method, where the following governing equations must be satisfied along with the boundary conditions:

$$
\nabla^2\phi_{\mathrm{e}} = 0,
\tag{8.75}
$$

$$
\mathbf{j} = -\frac{1}{\rho_0}\nabla\phi_{\mathrm{e}},
\tag{8.76}
$$

$$
\lambda\nabla^2 T + \rho_0\mathbf{j}\cdot\mathbf{j} + (\rho_0\alpha\mathbf{j}\cdot\mathbf{j} - H)(T - T_{\mathrm{s}}) = 0,
\tag{8.77}
$$

where H is a constant for the heat flow from a metal line to a substrate, λ is thermal conductivity of the metal line, ϕ_{e} is the electric potential, and ∇^2 is a Laplacian.

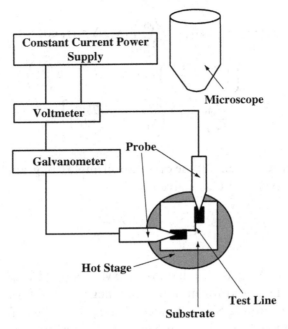

Fig. 8.48   Experimental setup. (Sasagawa *et al.*, 1999b, with permission from Amer. Inst. Phys.)

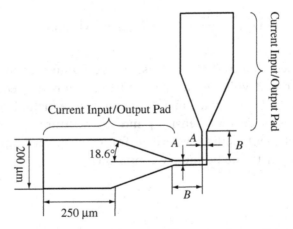

Fig. 8.49   Detail of angled polycrystalline aluminum lines used in the experiment. (Sasagawa *et al.*, 1999a, with permission from Amer. Inst. Phys.)

Experiments by Sasagawa *et al.* (1999a,b) on polycrystalline and bamboo-like aluminum lines are reproduced in Figs. 8.48 to 8.57. The microscope in their experimental setup, Fig. 8.48, is a scanning electron microscope (SEM). The test line used for polycrystalline samples is shown in more detail in Fig. 8.49. This setup was to determine the parameters Q_{gb}, $\Delta\varphi$ and C_{gb} used in Eq. (8.72).

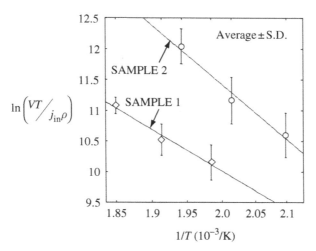

Fig. 8.50 $\ln(VT/j_{in}\rho)$ vs. $1/T$ relation of two different aluminum lines, see Fig. 8.49. (Sasagawa *et al.*, 1999a, with permission from Amer. Inst. Phys.)

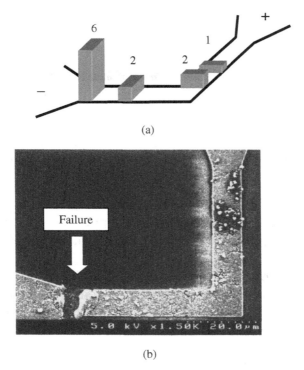

Fig. 8.51 (a) Predicted frequency distribution of failure site, and (b) SEM photomicrograph of failure for Sample 1 (A [see Fig. 8.49] $= 8.9$ μm). (Sasagawa *et al.*, 1999a, with permission from Amer. Inst. Phys.)

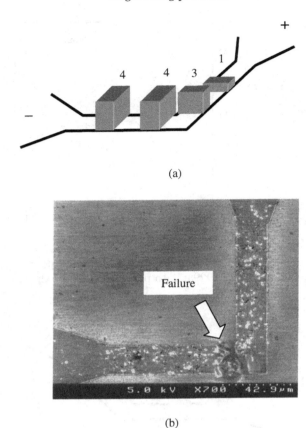

(a)

(b)

Fig. 8.52   (a) Predicted frequency distribution of failure site, and (b) SEM photomicrograph of failure for Sample 2 (A [see Fig. 8.49] $= 19\,\mu$m). (Sasagawa *et al.*, 1999a, with permission from Amer. Inst. Phys.)

The activation energy Q_{gb} is obtained experimentally from the graph of $\ln(VT/j_{in}\rho)$ vs. $1/T$, Fig. 8.50; the slope of the curve provides $-Q_{gb}/k_B$. From these results they were able to predict the mean time to failure of the polycrystalline samples.

Sasagawa *et al.* (1999a) performed accelerated testing on these polycrystalline aluminum lines to determine failure location and mean time to failure. Figure 8.51(a) shows the predicted frequency distribution of failure sites, and Fig. 8.51(b) a typical failure for Sample 1. Results for Sample 2 are shown in Fig. 8.52. Figures 8.51 and 8.52 demonstrate the validity of the model of Sasagawa *et al.*

Electromigration in a metal line with a bamboo-like structure (see Fig. 8.45(a)), is predominantly through grains (i.e., lattice diffusion), not

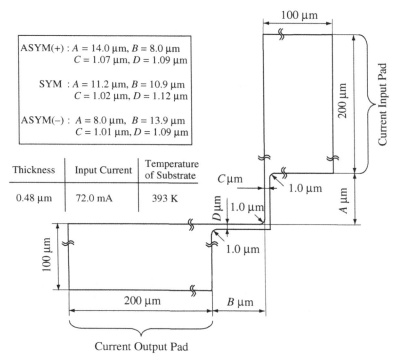

Fig. 8.53   Dimensions and experimental conditions of angled bamboo-like metal lines. (Sasagawa *et al.*, 1999b, with permission from Amer. Soc. Mech. Eng.)

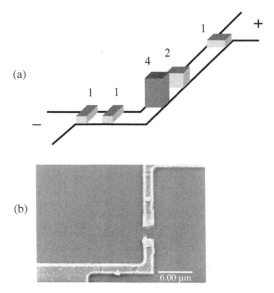

Fig. 8.54   Experimental results in ASYM(+). (a) Frequency distribution of failure site. Mean time to failure (experiment): 9160s [7965s], lifetime (prediction): 7100s. (b) Example of SEM observation of failure location. (Sasagawa *et al.*, 1999b, with permission from Amer. Soc. Mech. Eng.)

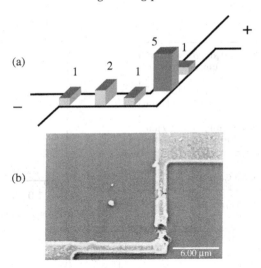

Fig. 8.55   Experimental results in SYM. (a) Frequency distribution of failure site. Mean time to failure (experiment): 7836s [7344s], lifetime (prediction): 7000s. (b) Example of SEM observation of failure location. (Sasagawa
et al., 1999b, with permission from Amer. Soc. Mech. Eng.)

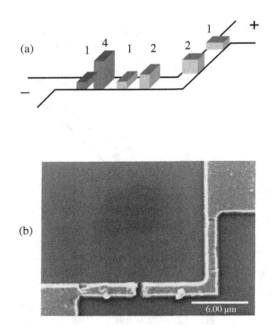

Fig. 8.56   Experimental results in ASYM(−). (a) Frequency distribution of
failure site. Mean time to failure (experiment): 6769s [6072s], lifetime (prediction): 5800s. (b) Example of SEM observation of failure location.
(Sasagawa *et al.*, 1999b, with permission from Amer. Soc. Mech. Eng.)

 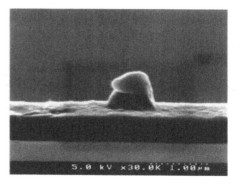

(a) (b)

Fig. 8.57 SEM micrographs of hillocks in angled aluminum bamboolike lines: (a) swelling mode, (b) lump mode. (Sasagawa *et al.*, 2000, with permission from Elsevier Ltd.)

along the grain boundaries which are oriented perpendicular to the metal-line direction. For the bamboo-like case, the model of Sasagawa *et al.*, Eq. (8.72), is simplified to

$$\text{AFD}_{\text{lat}} = C_{\text{lat}}\rho\frac{1}{T}\left(\frac{Q_{\text{lat}}}{k_B T} - 1\right)\exp\left(-\frac{Q_{\text{lat}}}{k_B T}\right)\left(\frac{\partial T}{\partial x}j_x + \frac{\partial T}{\partial y}j_y\right), \tag{8.78}$$

$$\text{AFD}_{\text{gen}} = \frac{1}{2}(\text{AFD}_{\text{lat}} + |\text{AFD}_{\text{lat}}|), \tag{8.79}$$

where AFD_{lat} is the atomic flux divergence by the lattice diffusion mechanism, C_{lat} is defined by

$$C_{\text{lat}} = \frac{ND_0 Z^* e}{k_B}, \tag{8.80}$$

and AFD_{gen} is the average atomic flux divergence in a unit cell per unit time.

Sasagawa *et al.* (1999b) applied the above model to angled aluminum lines with a bamboo-like structure (i.e., narrower line width). The dimensions and experimental conditions for angled bamboo-like lines they used are given in Fig. 8.53 where three different types of angled aluminum bamboo-like lines are processed for the experimental verification. Figures 8.54, 8.55 and 8.56 show (a) predictions of mean time to failure at different locations, and (b) SEM photos showing failure location in the same line, for the asymmetric (+), symmetric, and asymmetric (−) lines, respectively. These figures demonstrate the validity of their model for bamboo-like lines.

Electromigration can take place also in terms of hillock formation (accumulation of atoms in a confined area). Examples of hillocks are shown in Fig. 8.57.

Sasagawa *et al.* have applied the above model to other practically important cases, such as in passivated lines (2002a,b).

Appendix A
Eshelby tensors

A1 Eshelby tensor S_{ijkl} for elasticity

The Eshelby tensors S_{ijkl} for an isotropic matrix are a function of the geometry of an ellipsoid with axes a_1, a_2, and a_3, and the Poisson ratio ν of the matrix. Here we present the Eshelby tensors for ellipsoids with a simple geometry. The reader can find complete information on the Eshelby tensors for general cases of an ellipsoid in Mura (1987).

The domain of an ellipsoidal inclusion is bounded by

$$\frac{x_1^2}{a_1^2} + \frac{x_2^2}{a_2^2} + \frac{x_3^2}{a_3^2} = 1,$$

where a_1, a_2, and a_3 are the principal radii of the ellipsoid and coincide with the x_1-, x_2-, and x_3-axes, respectively. S_{ijkl} satisfies the symmetry $S_{ijkl} = S_{jikl} = S_{ijlk}$.

(1) Sphere: $a_1 = a_2 = a_3$ (Fig. A.1)

$$S_{1111} = S_{2222} = S_{3333} = \frac{7 - 5\nu}{15(1 - \nu)},$$

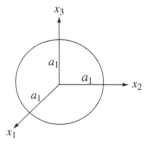

Fig. A.1   Sphere, radius $a_1 = a_2 = a_3$.

$$S_{1122} = S_{2233} = S_{3311} = S_{1133} = S_{2211} = S_{3322} = \frac{5\nu - 1}{15(1 - \nu)},$$

$$S_{1212} = S_{2323} = S_{3131} = \frac{4 - 5\nu}{15(1 - \nu)}.$$

(2) Elliptical cylinder: $(a_1 \neq a_2) \ll a_3 \to \infty$ (Fig. A.2)

$$S_{1111} = \frac{1}{2(1 - \nu)} \left\{ \frac{a_2^2 + 2a_1 a_2}{(a_1 + a_2)^2} + (1 - 2\nu) \frac{a_2}{(a_1 + a_2)} \right\},$$

$$S_{2222} = \frac{1}{2(1 - \nu)} \left\{ \frac{a_1^2 + 2a_1 a_2}{(a_1 + a_2)^2} + (1 - 2\nu) \frac{a_1}{(a_1 + a_2)} \right\},$$

$$S_{3333} = 0,$$

$$S_{1122} = \frac{1}{2(1 - \nu)} \left\{ \frac{a_2^2}{(a_1 + a_2)^2} - (1 - 2\nu) \frac{a_2}{(a_1 + a_2)} \right\},$$

$$S_{2211} = \frac{1}{2(1 - \nu)} \left\{ \frac{a_1^2}{(a_1 + a_2)^2} - (1 - 2\nu) \frac{a_1}{(a_1 + a_2)} \right\},$$

$$S_{2323} = \frac{a_1}{2(a_1 + a_2)},$$

$$S_{2233} = \frac{1}{2(1 - \nu)} \cdot \frac{2\nu a_1}{(a_1 + a_2)},$$

$$S_{3311} = 0,$$

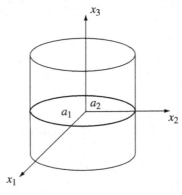

Fig. A.2   Elliptical cylinder, $(a_1 \neq a_2) \ll a_3 \to \infty$.

$$S_{1133} = \frac{1}{2(1-\nu)} \cdot \frac{2\nu a_2}{(a_1 + a_2)},$$

$$S_{3322} = 0,$$

$$S_{1212} = \frac{1}{2(1-\nu)} \left\{ \frac{a_1^2 + a_2^2}{2(a_1 + a_2)^2} + \frac{1}{2}(1 - 2\nu) \right\},$$

$$S_{3131} = \frac{a_2}{2(a_1 + a_2)}.$$

Please note that continuous fibers are usually considered to be cylinders with a circular cross-section, $a_1 = a_2 \ll a_3 = \infty$, for which the above S_{ijkl} are further simplified, i.e., a function only of Poisson's ratio ν.

(3) *Discus shape:* $a_1 = a_2 \gg a_3$ (*Fig. A.3*)

$$S_{1111} = S_{2222} = \frac{\pi(13 - 8\nu)}{32(1-\nu)} \frac{a_3}{a_1},$$

$$S_{3333} = 1 - \frac{\pi(1 - 2\nu)}{4(1-\nu)} \frac{a_3}{a_1},$$

$$S_{1122} = S_{2211} = \frac{\pi(8\nu - 1)}{32(1-\nu)} \frac{a_3}{a_1},$$

$$S_{1133} = S_{2233} = \frac{\pi(2\nu - 1)}{8(1-\nu)} \frac{a_3}{a_1},$$

$$S_{3311} = S_{3322} = \frac{\nu}{(1-\nu)} \left\{ 1 - \frac{\pi(1 + 4\nu)}{8\nu} \frac{a_3}{a_1} \right\},$$

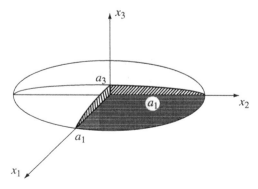

Fig. A.3   Discus with shape $a_1 = a_2 \gg a_3$.

$$S_{1212} = \frac{\pi(7 - 8\nu)}{32(1 - \nu)} \frac{a_3}{a_1},$$

$$S_{1313} = S_{2323} = \frac{1}{2}\left\{1 + \frac{\pi(\nu - 2)}{4(1 - \nu)}\frac{a_3}{a_1}\right\}.$$

(4) Oblate spheroid: $a_1 = a_2 > a_3$ *(Fig. A.4)*

$$S_{1111} = S_{2222} = -\frac{3}{8(1 - \nu)} \cdot \frac{\alpha^2}{(1 - \alpha^2)} + \frac{1}{4(1 - \nu)}\left\{1 - 2\nu + \frac{9}{4(1 - \alpha^2)}\right\}g,$$

$$S_{3333} = \frac{1}{2(1 - \nu)}\left\{4 - 2\nu - \frac{2}{(1 - \alpha^2)}\right\} + \frac{1}{2(1 - \nu)}\left\{-4 + 2\nu + \frac{3}{1 - \alpha^2}\right\}g,$$

$$S_{1122} = S_{2211} = \frac{1}{8(1 - \nu)}\left\{1 - \frac{1}{(1 - \alpha^2)}\right\}$$
$$+ \frac{1}{16(1 - \nu)}\left\{-4(1 - 2\nu) + \frac{3}{(1 - \alpha^2)}\right\}g,$$

$$S_{1133} = S_{2233} = \frac{1}{2(1 - \nu)} \cdot \frac{\alpha^2}{(1 - \alpha_2)} - \frac{1}{4(1 - \nu)}\left\{1 - 2\nu + \frac{3\alpha^2}{(1 - \alpha^2)}\right\}g,$$

$$S_{3311} = S_{3322} = \frac{1}{2(1 - \nu)}\left\{-(1 - 2\nu) + \frac{1}{1 - \alpha^2}\right\}$$
$$+ \frac{1}{4(1 - \nu)}\left\{2(1 - 2\nu) - \frac{3}{(1 - \alpha^2)}\right\}g,$$

$$S_{1212} = -\frac{1}{8(1 - \nu)} \cdot \frac{\alpha^2}{(1 - \alpha^2)} + \frac{1}{16(1 - \nu)}\left\{\frac{3}{(1 - \alpha^2)} + 4(1 - 2\nu)\right\}g,$$

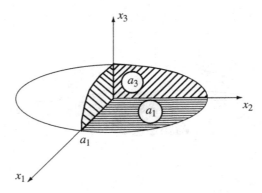

Fig. A.4   Oblate spheroid, $a_1 = a_2 > a_3$.

$$S_{1313} = S_{2323} = \frac{1}{4(1-\nu)}\left\{1 - 2\nu + \frac{1+\alpha^2}{1-\alpha^2}\right\}$$

$$- \frac{1}{8(1-\nu)}\left\{1 - 2\nu + 3\left(\frac{1+\alpha^2}{1-\alpha^2}\right)\right\}g,$$

where

$$\alpha = \frac{a_3}{a_1} < 1,$$

$$g = \frac{\alpha}{(1-\alpha^2)^{3/2}}\left\{\cos^{-1}\alpha - \alpha(1-\alpha^2)^{1/2}\right\}.$$

(5) Prolate spheroid: $a_1 = a_2 < a_3$ (Fig. A.5)

$$S_{1111} = S_{2222} = \frac{3}{8(1-\nu)}\frac{\alpha^2}{(\alpha^2-1)} + \frac{1}{4(1-\nu)}\left\{1 - 2\nu - \frac{9}{4(\alpha^2-1)}\right\}g,$$

$$S_{3333} = \frac{1}{2(1-\nu)}\left[1 - 2\nu + \frac{3\alpha^2-1}{\alpha^2-1} - \left\{1 - 2\nu + \frac{3\alpha^2}{(\alpha^2-1)}\right\}g\right],$$

$$S_{1122} = S_{2211} = \frac{1}{4(1-\nu)}\left[\frac{\alpha^2}{2(\alpha^2-1)} - \left\{1 - 2\nu + \frac{3}{4(\alpha^2-1)}\right\}g\right],$$

$$S_{1133} = S_{2233} = -\frac{1}{2(1-\nu)}\frac{\alpha^2}{(\alpha^2-1)} + \frac{1}{4(1-\nu)}\left\{\frac{3\alpha^2}{(\alpha^2-1)} - (1-2\nu)\right\}g,$$

$$S_{3311} = S_{3322} = -\frac{1}{2(1-\nu)}\left\{1 - 2\nu + \frac{1}{\alpha^2-1}\right\}$$

$$+ \frac{1}{2(1-\nu)}\left\{1 - 2\nu + \frac{3}{2(\alpha^2-1)}\right\}g,$$

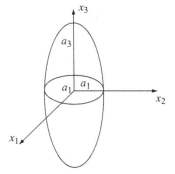

Fig. A.5   Prolate spheroid, $a_1 = a_2 < a_3$.

$$S_{1212} = \frac{1}{8(1-\nu)} \frac{\alpha^2}{(\alpha^2-1)} + \frac{1}{4(1-\nu)} \left\{ 1 - 2\nu - \frac{3}{4(\alpha^2-1)} \right\} g,$$

$$S_{2323} = S_{1313} = \frac{1}{4(1-\nu)} \left\{ 1 - 2\nu - \frac{(\alpha^2+1)}{(\alpha^2-1)} \right\}$$

$$- \frac{1}{8(1-\nu)} \left\{ 1 - 2\nu - \frac{3(\alpha^2+1)}{(\alpha^2-1)} \right\} g,$$

where

$$\alpha = \frac{a_3}{a_1} > 1,$$

$$g = \frac{\alpha}{(\alpha^2-1)^{3/2}} \left\{ \alpha(\alpha^2-1)^{1/2} - \cosh^{-1}\alpha \right\}.$$

A2   Eshelby tensors S_{ij} for uncoupled physical behavior

A semi-detailed description of S_{ij} is given in the Appendix of Hatta and Taya (1986). Please note that $S_{ij} = 0$ for $i \neq j$.

(1) Sphere: $a_1 = a_2 = a_3$ (Fig. A.1)

$$S_{11} = S_{22} = S_{33} = \frac{1}{3}.$$

(2) Elliptical cylinder: $a_1, a_2 \ll a_3 \to \infty$ (Fig. A.2)

$$S_{11} = \frac{a_2}{a_1 + a_2}, \quad S_{22} = \frac{a_1}{a_1 + a_2}, \quad S_{33} = 0.$$

(3) Penny shape: $a_1 = a_2 \gg a_3$ (Fig. A.3)

$$S_{11} = S_{22} = \frac{\pi a_3}{4a_1}, \quad S_{33} = 1 - \frac{\pi a_3}{2a_1}.$$

(4) Oblate spheroid: $a_1 = a_2 > a_3$ (Fig. A.4)

$$S_{11} = S_{22} = \frac{a_1^2 a_3}{2(a_1^2 - a_3^2)^{3/2}} \left\{ \cos^{-1}\frac{a_3}{a_1} - \frac{a_3}{a_1}\left(1 - \frac{a_3^2}{a_1^2}\right)^{1/2} \right\},$$

$$S_{33} = 1 - 2S_{22}.$$

(5) Prolate spheroid: $a_1 = a_2 < a_3$ (Fig. A.5)

$$S_{11} = S_{22} = \frac{a_1^2 a_3}{2(a_3^2 - a_1^2)^{3/2}} \left\{ \frac{a_3}{a_1}\left(\frac{a_3^2}{a_1^2} - 1\right)^{1/2} - \cosh^{-1}\frac{a_3}{a_1} \right\},$$

$$S_{33} = 1 - 2S_{22}.$$

It is to be noted that S_{ij} defined above is also applicable to S tensors for the uncoupled electromagnetic Eshelby problem, Chapter 4, which include a demagnetization factor discussed in Subsections 1.2.3 and 4.4.4.

A3 Eshelby tensors for ellipsoidal inclusions in piezoelectric matrix

For a two-dimensional elliptical cylinder inclusion ($a_3 \to \infty$, $\alpha = a_2/a_1$, see Fig. A.2) in a transversely isotropic matrix (axis of symmetry being the x_3-axis), the non-zero components of the Eshelby tensors are explicitly obtained as (Dunn, 1994; Mikata, 2000):

$$S_{1111} = \frac{\alpha}{2(\alpha+1)^2}\left[\frac{2C_{11}+C_{12}}{C_{11}} + (2\alpha+1)\right],$$

$$S_{1212} = S_{2121} = S_{1221} = S_{2112} = \frac{\alpha}{2(\alpha+1)^2}\left[\frac{\alpha^2+\alpha+1}{\alpha} - \frac{C_{12}}{C_{11}}\right],$$

$$S_{1313} = S_{3131} = S_{1331} = S_{3113} = \frac{\alpha}{2(\alpha+1)},$$

$$S_{1122} = \frac{\alpha}{2(\alpha+1)^2}\left[\frac{(2\alpha+1)C_{12}-C_{11}}{C_{11}}\right],$$

$$S_{1133} = \frac{C_{13}}{C_{11}}\frac{\alpha}{\alpha+1},$$

$$S_{1143} = \frac{e_{31}}{C_{11}}\frac{\alpha}{\alpha+1},$$

$$S_{2211} = \frac{\alpha}{2(\alpha+1)^2}\left[\frac{(\alpha+2)C_{12}-C_{11}}{C_{11}}\right],$$

$$S_{2222} = \frac{\alpha}{2(\alpha+1)^2}\left[\frac{2C_{11}+C_{12}}{C_{11}} + \frac{\alpha+2}{\alpha}\right],$$

$$S_{2323} = S_{3232} = S_{2332} = S_{3223} = \frac{1}{2(\alpha+1)},$$

$$S_{2233} = \frac{C_{13}}{C_{11}}\frac{1}{\alpha+1},$$

$$S_{2243} = \frac{e_{31}}{C_{11}}\frac{1}{\alpha+1},$$

$$S_{4141} = \frac{\alpha}{\alpha+1},$$

$$S_{4242} = \frac{1}{\alpha+1},$$

where C_{ij}, e_{ij} are elastic and piezoelectric constants based on the ANSI/IEEE 1987 convention, see Subsection 3.5.1.

For a circular cylindrical inclusion ($\alpha = 1$), the non-zero components of the Eshelby tensors reduce to

$$S_{1111} = S_{2222} = \frac{5C_{11} + C_{12}}{8C_{11}},$$

$$S_{1212} = S_{2121} = S_{1221} = S_{2112} = \frac{3C_{11} - C_{12}}{8C_{11}},$$

$$S_{1313} = S_{3131} = S_{1331} = S_{3113} = S_{2323} = S_{3232} = S_{2332} = S_{3223} = \frac{1}{4},$$

$$S_{1122} = S_{2211} = \frac{3C_{12} - C_{11}}{8C_{11}},$$

$$S_{1133} = S_{2233} = \frac{C_{13}}{2C_{11}}, \quad S_{1143} = S_{2243} = \frac{e_{31}}{2C_{11}},$$

$$S_{4141} = S_{4242} = \frac{1}{2}.$$

For a ribbon-like inclusion ($a_2 \gg a_1$), the non-zero components of the Eshelby tensors reduce to

$$S_{1111} = \frac{(3C_{11} + C_{12})}{2C_{11}}\alpha,$$

$$S_{1212} = S_{2121} = S_{1221} = S_{2112} = \frac{1}{2} - \frac{(C_{11} + C_{12})}{2C_{11}}\alpha,$$

$$S_{1313} = S_{3131} = S_{1331} = S_{3113} = \frac{\alpha}{2},$$

$$S_{1122} = -\frac{(C_{11} - C_{12})}{2C_{11}}\alpha,$$

$$S_{1133} = \frac{C_{13}}{C_{11}}\alpha,$$

$$S_{1143} = \frac{e_{31}}{C_{11}}\alpha,$$

$$S_{2211} = \frac{(2C_{12} - C_{11})}{2C_{11}}\alpha,$$

$$S_{2222} = 1 - \frac{(C_{11} - C_{12})}{2C_{11}}\alpha,$$

$$S_{2323} = S_{3232} = S_{2332} = S_{3223} = \frac{1 - \alpha}{2},$$

$$S_{2233} = \frac{C_{13}}{C_{11}}(1 - \alpha),$$

$$S_{2243} = \frac{e_{31}}{C_{11}}(1 - \alpha),$$

$$S_{4141} = \alpha,$$

$$S_{4242} = 1 - \alpha.$$

The Eshelby tensors are represented in a 9×9 matrix format. It is to be noted, however, that, in general, S is not diagonally symmetric and that care must be taken in the matrix representation of S to correctly account for the factor of two with regard to the shear strains. The Eshelby tensor for a spheroidal inclusion in a piezoelectric matrix with transverse isotropy is given explicitly by Mikata (2001).

A4   Eshelby tensors in matrix form

Here explicit forms of several Eshelby tensors are given in matrix form.

(1) Eshelby tensor for an elastic inclusion, referred to in Appendix A1

$$\mathbf{e} = \mathbf{S} \cdot \mathbf{e}^*,$$

where strain $\mathbf{e}$ and eigenstrain $\mathbf{e}^*$ are defined by

$$\mathbf{e} = \begin{pmatrix} e_{11} \\ e_{22} \\ e_{33} \\ 2e_{23} \\ 2e_{31} \\ 2e_{12} \end{pmatrix}, \qquad \mathbf{e}^* = \begin{pmatrix} e_{11}^* \\ e_{22}^* \\ e_{33}^* \\ 2e_{23}^* \\ 2e_{31}^* \\ 2e_{12}^* \end{pmatrix}$$

and the S matrix by

$$\mathbf{S} = \begin{pmatrix} S_{1111} & S_{1122} & S_{1133} & 0 & 0 & 0 \\ S_{2211} & S_{2222} & S_{2233} & 0 & 0 & 0 \\ S_{3311} & S_{3322} & S_{3333} & 0 & 0 & 0 \\ 0 & 0 & 0 & 2S_{2323} & 0 & 0 \\ 0 & 0 & 0 & 0 & 2S_{3131} & 0 \\ 0 & 0 & 0 & 0 & 0 & 2S_{1212} \end{pmatrix}.$$

(2) Eshelby tensor for an uncoupled inclusion, referred to in Appendix A2

$$\mathbf{Z} = \mathbf{S} \cdot \mathbf{Z}^*,$$

where field vector $\mathbf{Z}$ and eigenfield vector $\mathbf{Z}^*$ are defined by

$$\mathbf{Z} = \left\{ \begin{matrix} Z_1 \\ Z_2 \\ Z_3 \end{matrix} \right\}, \qquad \mathbf{Z}^* = \left\{ \begin{matrix} Z_1^* \\ Z_2^* \\ Z_3^* \end{matrix} \right\},$$

and S is given by

$$\mathbf{S} = \begin{bmatrix} S_{11} & S_{12} & S_{13} \\ S_{21} & S_{22} & S_{23} \\ S_{31} & S_{32} & S_{33} \end{bmatrix}.$$

(3) Eshelby tensor for a ellipsoidal inclusion referred to in Appendix A3

For a spheroidal inclusion in a piezoelectric matrix with geometrical and crystallographic axes of symmetry along x_3, we have

$$\mathbf{Z} = \mathbf{S} \cdot \mathbf{Z}^*,$$

where $\mathbf{Z}$ and $\mathbf{Z}^*$ are defined by

$$\mathbf{Z} = \begin{pmatrix} e_{11} \\ e_{22} \\ e_{33} \\ 2e_{23} \\ 2e_{31} \\ 2e_{12} \\ -E_1 \\ -E_2 \\ -E_3 \end{pmatrix}, \qquad \mathbf{Z}^* = \begin{pmatrix} e_{11}^* \\ e_{22}^* \\ e_{33}^* \\ 2e_{23}^* \\ 2e_{31}^* \\ 2e_{12}^* \\ -E_1^* \\ -E_2^* \\ -E_3^* \end{pmatrix},$$

and the S matrix by

$$\mathbf{S} = \begin{pmatrix} S_{1111} & S_{1122} & S_{1133} & 0 & 0 & 0 & S_{1141} & S_{1142} & S_{1143} \\ S_{2211} & S_{2222} & S_{2233} & 0 & 0 & 0 & S_{2241} & S_{2242} & S_{2243} \\ S_{3311} & S_{3322} & S_{3333} & 0 & 0 & 0 & S_{3341} & S_{3342} & S_{3343} \\ 0 & 0 & 0 & 2S_{2323} & 0 & 0 & 2S_{2341} & 2S_{2342} & 2S_{2343} \\ 0 & 0 & 0 & 0 & 2S_{1313} & 0 & 2S_{1341} & 2S_{1342} & 2S_{1343} \\ 0 & 0 & 0 & 0 & 0 & 2S_{1212} & 2S_{1241} & 2S_{1242} & 2S_{1243} \\ S_{4111} & S_{4122} & S_{4133} & S_{4123} & S_{4113} & S_{4112} & S_{4141} & 0 & 0 \\ S_{4211} & S_{4222} & S_{4233} & S_{4223} & S_{4213} & S_{4212} & 0 & S_{4242} & 0 \\ S_{4311} & S_{4322} & S_{4333} & S_{4323} & S_{4313} & S_{4312} & 0 & 0 & S_{4343} \end{pmatrix}$$

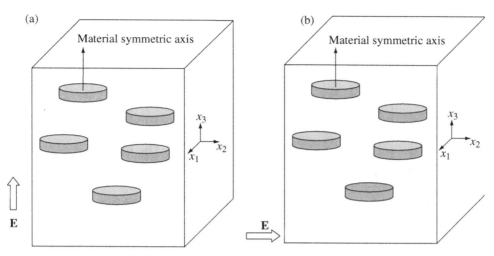

Fig. A.6   Penny-shaped piezoelectric fillers embedded in a matrix subjected to applied field **E**.

Some of the S_{ijkl} components vanish for the following simpler cases of inclusion geometry.

(i) For a flat oblate spheroid (penny-shaped) inclusion embedded in a transversely isotropic piezoelectric matrix subjected to two types of applied electric field **E** as illustrated in Fig. A.6(a) and (b), the **S** tensor is given by

$$S = \begin{pmatrix} S_{1111} & S_{1122} & S_{1133} & 0 & 0 & 0 & 0 & 0 & S_{1143} \\ S_{2211} & S_{2222} & S_{2233} & 0 & 0 & 0 & 0 & 0 & S_{2243} \\ S_{3311} & S_{3322} & S_{3333} & 0 & 0 & 0 & 0 & 0 & S_{3343} \\ 0 & 0 & 0 & 2S_{2323} & 0 & 0 & 0 & 2S_{2342} & 0 \\ 0 & 0 & 0 & 0 & 2S_{3131} & 0 & 2S_{3141} & 0 & 0 \\ 0 & 0 & 0 & 0 & 0 & 2S_{1212} & 0 & 0 & 0 \\ 0 & 0 & 0 & 0 & S_{4131} & 0 & S_{4141} & 0 & 0 \\ 0 & 0 & 0 & S_{4223} & 0 & 0 & 0 & S_{4242} & 0 \\ S_{4311} & S_{4322} & S_{4333} & 0 & 0 & 0 & 0 & 0 & S_{4343} \end{pmatrix}$$

It is to be noted in Fig. A.6 (and Figs. A.7 and A.8, later) that two examples of the applied electric field are shown, but this does not influence the **S** tensors.

(ii) For an elliptical cylindrical inclusion with geometrical and crystallographic symmetry axes along x_3, as illustrated in Fig. A.7(a) and (b)

$$\mathbf{Z} = \mathbf{S} \cdot \mathbf{Z}^*$$

where **Z** and $\mathbf{Z}^*$ are defined as in (i) and the **S** matrix by

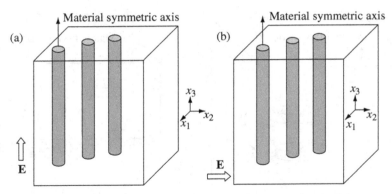

Fig. A.7   Elliptical cylindrical piezoelectric fillers embedded in a matrix subjected to applied electric field **E**.

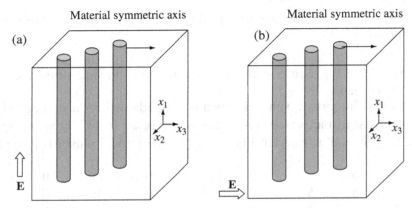

Fig. A.8   Elliptical cylindrical piezoelectric fibers (with material symmetric axis along x_1) embedded in a matrix subjected to applied field **E**.

$$
S = \begin{pmatrix}
S_{1111} & S_{1122} & S_{1133} & 0 & 0 & 0 & 0 & 0 & S_{1143} \\
S_{2211} & S_{2222} & S_{2233} & 0 & 0 & 0 & 0 & 0 & S_{2243} \\
0 & 0 & 0 & 0 & 0 & 0 & 0 & 0 & 0 \\
0 & 0 & 0 & 2S_{2323} & 0 & 0 & 0 & 0 & 0 \\
0 & 0 & 0 & 0 & 2S_{3131} & 0 & 0 & 0 & 0 \\
0 & 0 & 0 & 0 & 0 & 2S_{1212} & 0 & 0 & 0 \\
0 & 0 & 0 & 0 & 0 & 0 & S_{4141} & 0 & 0 \\
0 & 0 & 0 & 0 & 0 & 0 & 0 & S_{4242} & 0 \\
0 & 0 & 0 & 0 & 0 & 0 & 0 & 0 & 0
\end{pmatrix}
$$

(iii) For an elliptical cylindrical inclusion with geometrical symmetry along the x_1-axis and crystallographic symmetry along the x_3-axis. This case is illustrated in Fig. A.8(a) and (b) where the piezoelectric fiber has its material (crystallographic) symmetric axis along the x_1-direction. The Eshelby tensor for this case was explicitly obtained in closed form by Mikata (2000).

$$Z = S \cdot Z^*$$

where Z and Z^* are defined as in (i) and the S matrix by

$$
S = \begin{pmatrix}
0 & 0 & 0 & 0 & 0 & 0 & 0 & 0 & 0 \\
S_{2211} & S_{2222} & S_{2233} & 0 & 0 & 0 & 0 & 0 & S_{2243} \\
S_{3311} & S_{3322} & S_{3333} & 0 & 0 & 0 & 0 & 0 & S_{3343} \\
0 & 0 & 0 & 2S_{2323} & 0 & 0 & 0 & 2S_{2342} & 0 \\
0 & 0 & 0 & 0 & 2S_{3131} & 0 & 2S_{3141} & 0 & 0 \\
0 & 0 & 0 & 0 & 0 & 2S_{1212} & 0 & 0 & 0 \\
0 & 0 & 0 & 0 & 0 & 0 & 0 & 0 & 0 \\
0 & 0 & 0 & S_{4223} & 0 & 0 & 0 & S_{4242} & 0 \\
S_{4311} & S_{4322} & S_{4333} & 0 & 0 & 0 & 0 & 0 & S_{4343}
\end{pmatrix}.
$$

Appendix B
Physical constants and properties of materials

B1 Physical constants

Permittivity of free space, $\varepsilon_0 = 8.854 \times 10^{-12}$ F/m
Permeability of free space, $\mu_0 = 4\pi \times 10^{-7}$ H/m
Impedance of free space, $\eta_0 = 376.7\ \Omega$
Velocity of light in free space, $c_0 = 2.998 \times 10^8$ m/s
Charge of electron, $q = 1.602 \times 10^{-19}$ C
Mass of electron, $m = 9.109 \times 10^{-31}$ kg
Boltzmann's constant, $k_B = 1.380 \times 10^{-23}$ J/K
Planck's constant, $h/2\pi = 1.0545 \times 10^{-34}$ J·s
Stefan–Boltzmann constant, $\sigma = 5.670 \times 10^{-8}$ W/m²·K⁴

B2 Thermal properties of popular packaging materials.

Material	Specific heat capacity, c_p (J/kg·K)	Thermal expansion coefficient, α (1/K×10⁻⁶)	Thermal conductivity, K (W/m·K)
Metals			
Aluminum	900	23.6	247
Copper	386	16.5	398
Gold	130	13.8	315
Iron	448	11.8	80.4
Nickel	443	13.3	89.9
Silver	235	19	428
Tungsten	142	4.5	178
SAE 1025 steel	486	12.5	51.9
316 stainless steel	502	16	16.3

B2 (cont.)

Material	Specific heat capacity, c_p (J/kg·K)	Thermal expansion coefficient, α (1/K×10^{-6})	Thermal conductivity, K (W/m·K)
W–Cu (90–10)	209	6	209.3
Molybdenum	251	3–5.5	138
Brass (70Cu–30Zn)	375	20	120
Silicon	702–718	2.3–4.7	124–148
Alloy 42 (FeNi)	–	6	16
Ceramics			
Alumina (Al_2O_3)	775	8.8	30.1
Beryllia (BeO)	1050	9	220
Magnesia (MgO)	940	13.5	37.7
Spinel ($MgAl_2O_4$)	790	7.6	15.0
Fused silica (SiO_2)	740	0.5	2.0
Soda–lime glass	840	9.0	1.7
SiC	670	4.5–4.9	283
AlN	745	4.3–4.7	82–320
Polymers			
Polyethylene	2100	60–220	0.38
Polypropylene	1880	80–100	0.12
Polystyrene	1360	50–85	0.13
Polytetrafluoroethylene (Teflon)	1050	135–150	0.25
Phenol-formaldehyde (Bakelite)	1650	68	0.15
Nylon 6,6	1670	80–90	0.24
Polyisoprene	–	220	0.14
Polyimide	–	50	0.4
Epoxy	–	15	0.3

Sources: Callister, 1994; Hannemann *et al.*, 1994, with permission of John Wiley and Sons, Inc.,©1994. Lau *et al.*, 1998, reproduced with permission of The McGraw-Hill Companies.

B3 Dielectric constants ε_r of popular materials.

Material	Frequency (GHz)	ε_r'	$\tan \delta$ (at 25 °C)
Alumina (99.5%)	10	9.5–10	0.000 3
Barium tetratitanate	6	37($\pm$5%)	0.000 5
Beeswax	10	2.35	0.005
Beryllia	10	6.4	0.000 3
Ceramic (A-35)	3	5.6	0.004 1
Fused quartz	10	3.78	0.000 1
Gallium arsenide	10	13	0.006
Glass (pyrex)	3	4.82	0.005 4
Glazed ceramic	10	7.2	0.008
Lucite	10	2.56	0.005
Nylon (610)	3	2.84	0.012
Paraffin	10	2.24	0.000 2
Plexiglass	3	2.60	0.005 7
Polyethylene	10	2.25	0.000 4
Polystyrene	10	2.54	0.000 33
Porcelain (dry process)	100	5.04	0.007 8
Rexolite (1422)	3	2.54	0.000 48
Silicon	10	11.9	0.004
Styrofoam (103.7)	3	1.03	0.000 1
Teflon	10	2.08	0.000 4
Titania (D-100)	6	96($\pm$5%)	0.001
Vaseline	10	2.16	0.001
Water (distilled)	3	76.7	0.157

Relative dielectric constant $\varepsilon_r\ (=\varepsilon/\varepsilon_0)=\varepsilon_r'-j\varepsilon_r''$; $\tan \delta = \varepsilon_r''/\varepsilon_r'$; $\varepsilon_0 = 8.854 \times 10^{-12}$ F/m.
Sources: Pozar, 1998, with permission of John Wiley and Sons, Inc.,©1998.

B4 Magnetic properties of soft and hard magnets.

(i) Soft magnets

Material	Composition, % by wt. (remainder Fe)	Initial permeability (gauss/oersted)	Saturation induction (gauss)	Curie temperature (°C)	Density (g/cm³)
Alfenol 12	12 Al	4 500	14 800	650	6.74
Iron (commercial)	0.2 impurity	200	21 580	770	7.87
Oriented Si–Fe	3.25 Si	5 200	20 200	740	7.67
Permalloy 45	45 Ni	3 500	14 500	400	8.27
Permalloy 78	78 Ni	15 000	8 700	460	8.76
Supermalloy	5 Mo, 79 Ni	100 000	7 900	400	8.78
Supermendur	2 V, 49 Co	1 000	24 000	980	8.20

(ii) Hard magnets

Name	Composition, % by wt. (remainder Fe)	Coercive force, H_c (oersted)	Residual flux density, B_r (gauss)	Energy product, (BH_{max}) (10^6 G·Oe)	Density (g/cm³)
5% Tungsten steel	5W, 0.7C	70	10 500	0.33	8.1
36% Cobalt steel	36Co, 3.75W, 5.75Cr, 0.8C	240	9 750	0.93	8.2
Remalloy 2	12Co, 20Mo	340	8 550	1.2	8.4
Vicalloy 2	52Co, 13V	415	9 000	2.3	8.2
Cunife 1	60Cu, 20Ni	500	5 400	1.3	8.6
Alnico 5	8Al, 14Ni, 24Co, 3Cu	620	12 500	5.3	7.3
Samarium–cobalt	SmCo	8 400	8 700	18.5	8.4
Ceramic 5	$Mo.6Fe_2O_3$	2 400	3 800	3.4	4.9
Neodymium 35	NdFeB	10 500	12 300	35	7.4

Sources: Billings *et al.*, 1972; Fink and Beaty, 2000, with permission of The McGraw-Hill Companies.

References

Abeles, B., Pinch, H. L. and Gittleman, J. I., 1975, *Phys. Rev. Lett.,* **35**, 247–250.

Akimoto, T., Ikebukuro, K. and Karube, I., 2003, *Biosensors and Bioelectronics*, **18**, 1447–1453.

Akimoto, T., Sasaki, S., Ikebukuro, K. and Karube, I., 2000, *Analytica Chemica*, **417**, 125–131.

Akiyama, T., Collard, D. and Fujita, H., 1997, *J. MEMS*, **6** (1), 10–17.

Almajid, A. A. and Taya, M., 2001, *J. Intell. Mater. Sys. Struct.*, **12**, 341–351.

Almajid, A. A., Taya, M. and Hudnut, S., 2001, *Intl. J. Solids Struct.*, **38**, 3377–3391.

ANSI/IEEE, 1987, *Standard on Piezoelectricity*, Vol.176, 9–54.

Aslam, M. and Schulz, D., 1995, *Transducers '95 and Eurosensors IX*, Stockholm, Sweden, June 25–29, pp. 222–287.

Aspeden, R. M., 1986, *Proc. Roy. Soc. London*, **A406**, 287–298.

Balberg, I., 1987, *Phys. Rev. Lett.*, **59** (12), 1305–1308.

Balberg, I. and Binenbaum, N., 1983, *Phy. Rev.*, **B28**, 3799–3812.

Balberg, I., Wagner, N., Goldstein, Y., and Weisz, S. Z., 1990, *Mat. Res. Soc. Symp. Proc.*, **195**, 233–238.

Banno, H., 1983, *Ferroelectrics*, **50**, 3–12.

Barker-Jarvis, J., Vanzura, E. J. and Kissick, W. A. 1990, *IEEE Trans. Microwave Theory Tech.*, **38**, 1096–1103.

Bennett, S., 1984, *Thermal Expansions,* ed. Thomas A. Hahn, New York: Plenum Press, pp. 235–243.

Benveniste, Y., 1987, *Mech. Mater.*, **6**, 105–111.

Bhattacharyya, S. K., Basu, S. and De, S. K., 1978, *Composites*, **9**, 177.

Bhattacharyya, S. K. and Chaklader, A. C. D., 1982, *Polym. Plast. Technol. Eng.*, **19**, 21.

Bigg, D. M., 1979, *Polym. Eng. Sci.*, **19**, 1188–1192.

Billings, B. H., Bleil, D. F., Cook, R. K., *et al.*, 1972, *American Institute of Physics Handbook*, 3rd edn., New York: McGraw-Hill, pp. 5–160.

Björkman, H., Rangsten, P. and Hjort, K., 1999a, *Sensors and Actuators*, **A78** (1), 41–47.

Björkman, H., Rangsten, P., Hollman, P. and Hjort, K., 1999b, *Sensors and Actuators*, **A78** (1), 24–29.

Black, J. R., 1969, Proc. IEEE, **57**, 1587–.

Bourouina, T., Lebrasseur, E., Reyne, G., Debray, A. and Fujita, H., 2002, *J. MEMS*, **11** (4), 355–361.

Bozorth, R. M. and Williams, H. J., 1956, *Phys. Rev.*, **103**, 572.

Broadbent, S. R. and Hammersley, J. M., 1957, *Proc. Camb. Phil. Soc.*, **53**, 629–641.

Bruggeman, D. A. G., 1935, *Ann. Phys.* (Leipz.), **24**, 636–664.

Butoi, C. I., Mackie, N. M., Gamble, L. J., *et al.*, 2000, *Chem. Mater.*, **12**, 2014–2024.

Cahill, C. P., Johnston, K. S. and Yee, S. S., 1997, *Sensors and Actuators*, **B45**, 161–166.

Callister, W. D., Jr., 1994, *Materials Science and Engineering: Introduction*, 3rd edn, New York: John Wiley & Sons, Inc.

Carver, K. R. and Mink, J. W., 1981, *IEEE Trans. Antennas and Propagation*, **AP-29**, 1, 2–24.

Chan, H. L. W. and Unsworth, J. 1989, *IEEE Trans. Ultrasonics, Ferroelectrics and Frequency Control*, **36**, 434–441.

Chang, D. D., Crawford, P. A., Fulton, J. A., *et al.*, 1993, *IEEE Trans. Components, Hybrids and Manufacturing Tech.*, **16** (8), 828–835.

Chawla, K. K., 1987, *Composite Materials Science and Engineering*, Springer-Verlag

Chen, Y., Snyder, J. E., Schwichtenberg, C. R., Dennis, K. W., Falzgraf, D. K., McCallum, R. W. and Jiles, D. C., 1999, *Appl. Phys. Lett.*, **74** (8), 1159–1161.

Chikamizu, S., 1964, *Physics of Magnetism*, translated by S. H. Charays, Malabar, Florida: Krieger Publishing Company.

Cho, S. M., Cho, S. Y. and Han, B., 2001, *Experimental Techniques*, **26** (3), 25–29.

Christensen, R. M., 1979, *Mechanics of Composite Materials*, New York: John Wiley.

Chung, D. D., 1995, *Materials for Electronic Packaging*, Newton, MA: Butterworth-Heinemann.

Clark, A. E., 1993, *J. Intell. Mater. Sys. Struct.*, **4** (1), 70–75.

Clausius, R., 1879, *Die Mechanixche Behandlung der Elektrizität*, Brunswick: Vieweg, p. 62.

Clyne, T. W. and Withers, P. J., 1993, *An Introduction to Metal Matrix Composites*, Cambridge: Cambridge University Press.

Coehoorn, R., de Mooji, D., Duchatoau, J. and Buschow, K., 1988, *J. de Phys.*, **C8**, 669.

Cohen, M. H., Jortner, J. and Webman, I., 1978, In *A.I.P. Conf. Proc. No. 40: Electrical Transport and Optical Properties of Inhomogeneous Media*, eds. J. C. Garland and D. B. Tanner, New York: American Institute of Physics, pp. 63–81.

Cox, H. L., 1952, *Brit. J. Appl. Phys.*, **3**, 72.

Dally, J. W., Riley, W. F. and Sirkis, J. S., 1987, "Strain gauges", Chap. 2 of *Handbook on Experimental Mechanics*, 2nd edn, ed. A. S. Kobayashi, New York: Society of Experimental Mechanics, Inc.

Daniel, I. M. and Ishai, O., 1994, *Engineering Mechanics of Composite Materials*, Oxford: Oxford University Press.

Dean, P., 1963, *Proc. Camb. Phil. Soc.*, **59**, 397–410.

Deutscher, G., 1981, In *Lecture Notes in Phys. No. 149: Disordered Systems and Localization*, ed. C: Castellani *et. al.*, Berlin: Springer-Verlag, pp. 26–40.
    1984, In *Percolation and Superconductivity*, eds. A. M. Goldman and S. A., Wolf, New York: Plenum Press, pp. 95–113.

Deutscher, G., Rappaport, M. and Ovadyahn, Z., 1978, *Solid State Commun.*, **28**, 593–595.

Dewa, A. S. and Ko, W. H., 1994, "Biosensors", Chap. 9 of *Semiconductor Sensors*, ed. S. M. Sze, New York: John Wiley & Sons, Inc., pp. 415–472.

Dunn, M., 1993, *J. Appl. Phys.*, **73** (10), 5131–5140.
    1994, *Intl. J. Eng. Sci.*, **32** (1), 119–131.

1995, *J. Intell. Mater. System Structs.*, **6**, 255–265.

Dunn, M. and Taya, M., 1993a, *Proc. Roy. Soc. London*, **A443**, 265–287.

1993b, *Intl. J. Solids Struct.*, **30**, 161–175.

Dunn, M., Taya, M., Hatta, H., Takei, T. and Nakajima, Y., 1993c, *J. Comp. Mater.*, **27**, 1493–1519.

Dunn, M. and Wienecke, H. A., 1997, *Intl. J. Solids Struct.*, **34**, 3571.

Elliot, R. J., Heap, B. R., Morgan, D. J. and Rushbrook, G. S., 1960, *Phys. Rev. Lett.*, **5**, 366–367.

Enoki, T., 1994, *Morphology and optoelectronic properties of transparent electrodes of InO_3, Cd_2SnO_4 and Zn_2SnO_4*, Ph.D. dissertation, Tohoku University, Sendai, Japan.

Ertl, S., Adamschik, M., Schmid, P. Gluche, P., Flöter, A. and Kohn, E., 2000, *Diamond and Related Materials*, **9**, 970–974.

Esashi, M., Sugiyama, S., Ikeda, K., Wang, Y. and Miyashitai, H., 1998, *Proc. IEEE*, **86** (8), 1627–1639.

Eshelby, J. D., 1957, *Proc. Roy. Soc. London*, **241**, 376–396.

1959, *Proc. Roy. Soc. London*, **252**, 561–569.

Essam, J. W., 1972, *Phase Transitions and Critical Phenomena*, eds. C. Domb and M. S. Gree, New York: Academic Press, pp. 197–270.

1980, *Rep. Prog. Phys.*, **43**, 833–912.

Fink, D. G. and Beaty, H. W., 2000, *Standard Handbook for Electrical Engineers*, 14th edn, London: McGraw-Hill, pp. 4–118.

Flory, P. J., 1953, *Principles of Polymer Chemistry*, New York: Cornell University Press.

Foner, S., 1959, *Rev. Sci. Instr.*, **30**, 548.

Frisch, H. L., Sonnenblick, E., Vyssotsky, V. A. and Hammersley, J. M., 1961, *Phys. Rev.*, **124**, 1021–1022.

Freund, L. B. and Suresh, S., 2003, *Thin Film Materials*, Cambridge: Cambridge University Press.

Frost, H. J. and Ashby, M. F., 1982, *Deformation–Mechanism Maps*, Oxford: Pergamon Press.

Fujimori, H., Masumoto, T., Obi, Y. and Kikuchi, M., 1974, *Japan. J. Appl. Phys.*, **13**, 1889.

Fujita, A., Fukamichi, K., Gejima, F., Kainuma, R. and Ishida, K., 2000, *J. Appl. Phys. Lett.* **77**, 3054–3056.

Furukawa, T., Fujiro, K. and Fukada, E., 1990, *Japan. J. Appl. Phys.*, **15**, 2119–2129.

Gamano, C., 1970, *IEEE Trans. Audio. Electro.*, **19**, 6.

Gerteisen, S. R., 1982, 37th *Ann. Tech. Conf. Soc. Plas. Ind.*, **11-E**, 1–7.

Ghate, P. B., 1983, *Solid State Tech.*, **26**, 113.

Gibson, R. F., 1994, *Principles of Composite Material Mechanics*, London: McGraw-Hill.

Gurland, J., 1966, *Trans. Met. Soc. AIME*, **236**, 1966.

Hammersley, J. M., 1961, *J. Math. Phys.*, **2**, 728–733.

Hammersley, J. M. and Handscomb, D. S., 1964, *Monte Carlo Method*, London: Methuen.

Han, B., 1992, *Experimental Mechanics*, **32** (1), 38–41.

1998, *Experimental Mechanics*, **38** (4), 278–288.

Hannemann, R. J., Kraus, A. D. and Pecht, M., 1994, *Physical Architecture of VLSI Systems*, New York: John Wiley & Sons, Inc.

Hartling, G. H., 1994, *Amer. Ceram. Soc. Bull.*, **73**, 93–96.

Hashin, Z. and Shtrikman, S., 1962, *J. Appl. Phys.*, **33** (10), 3125–3131.

1963, *J. Mech. Phys. Solids*, **11**, 127–140.

Hatta, H. and Taya, M., 1985, *J. Appl. Phys.*, **58**, 2478–2486.

1986, *Intl. J. Eng. Sci.*, **24**, 1159–1172.

Hatta, H., Taya, M., Kulacki, F. A. and Harder, J. F., 1992, *J. Comp. Mater.*, **26** (5), 612–625.

Hill, R., 1950, *The Mathematical Theory of Plasticity*, Oxford: Oxford Science Publications.

Hinomura, T. Nasu, S., Kanekiyo, H. and Hirosawa, S., 1997, *J. Japan. Inst. Metals*, **61** (2), 184–190.

Homola, J., Yee, S. S. and Ganglitz, G., 1999, *Sensors and Actuators*, **B54** (1–2), 3–15.

Howe, R. T., Boster, B. E. and Pisano, A. P., 1996, *Sensors and Actuators*, **A56**, 167–177.

Howe, R. T. and Muller, R. S., 1983, *J. Electrochem. Soc.*, **130** (6), 1420–1423.

Huang, J. H. and Yu, J. S., 1994, *Composite Eng.*, **4** (11), 1169–1182.

Hull, D., 1981, *An Introduction to Composite Materials*, Cambridge: Cambridge University Press.

Hull, D. and Clyne, T. W., 1996, *An Introduction to Composite Materials*, 2nd edn, Cambridge: Cambridge University Press.

Hunn, J. D., Withrow, S. P., White, C. W., Clausing, R. E. and Heatherly, L., 1994, *Appl. Phys. Lett.*, **65** (24), 3072–3074.

Huntington, H. B. and Grone, A. R., 1961, *J. Phys. Chem. Solids*, **20** (1/2), 76–87.

IEEE Spectrum, January 2001, *Technology 2001: Analysis and Forecast.*

Ikeda, T., 1990, *Fundamentals of Piezoelectricity*, Oxford: Oxford University Press.

Ishimaru, A., 1991, *Electromagnetic Wave Propagation, Radiation, and Scattering*, New Jersey: Prentice-Hall.

James, R. D. and Wuttig, M., 1998, *Phil. Mag. A*, **77**, 1273–1299.

Johnston, E. E. and Ratner, B. D., 1996, *J Electron Spect. Rel. Phenom.*, **81**, 303–317.

Kakeshita, T., Takeuchi, T., Fukuda, T., *et al.*, 2000, *Appl. Phys. Lett.*, **77**, 1502–1504.

Kapitulnik, A. and Deutscher, G., 1983, *J. Phys. A.*, **16**, L255–L257.

Kataoka, Y. and Taya, M., 2000, *Trans. JSME*, **65**, (631), 523–529.

JSME Intl. J. **43**(1), 46–53.

Kato, H., Liang, Y. and Taya, M., 2000a, *Scripta Mater.*, **46**, 471–475.

Kato, H., Wada, T., Liang, Y., Tagawa, T., Taya, M. and Mori, T., 2002b, *Mat. Sci. Eng. A*, **332**, 134–139.

Kawasaki, M., Sato, T., Tanaka, T., Takao, K., 2000, *Langmuir*, **16** (4), 1719–2000.

Kellog, O. D., 1953, *Foundation of Potential Theory*, New York: Dover.

Kelly, A., 1973, *Strong Solids*, 2nd edn, Oxford: Clarendon Press.

Kelly, A. and Street, K. N., 1972, *Proc. Roy. Soc. London*, **A328**, 283–293.

Kerner, E. H., 1956, *Proc. Roy. Soc. London*, **B69**, 802.

Kim, S. R., Kang, S. Y., Park, J. K., Nam, J. T., Son, D. and Lim, S. H., 1998a, *J. Appl. Phys.*, **83** (11), 7285–7287.

Kim, W. J., 1998, *Design of Electrically and Thermally Conductive Polymer Composites for Electronic Packaging*, Ph.D. dissertation, University of Washington.

Kim, W. J., Taya, M., Yamada, K. and Kamiya, N., 1998b, *J. Appl. Phys.*, **83** (5), 2593–2598.

Kinoshita, N. and Mura, T., 1971, *phys. stat. solidi*, **A5**, 759–768.

Kirkpatrick, S., 1971, *Phys. Rev. Lett.*, **27**, 1722–1725.

1973, *Rev. Mod. Phys.*, **45**, 574–588

Klicker, K. A., Biggers, J. V. and Newnham, R. E., 1981, *J. Amer. Ceram. Soc.*, **64** (1), 5.

Kneller, E. and Hawig, R., 1991, *IEEE Trans. Magn.*, **27**, 3588.

Kohn, E., Gluche, P. and Adamschik, M., 1999, *Diamond and Related Materials*, **8**, 934–940.

Kretschmann, E. and Raether, H., 1968, *Z. Naturforsch*, **23A**, 2135–2136.

Kuga, Y., Lee, S. W., Almajid, A., Taya, M., Li, J. F. and Watanabe, R., 2005, *IEEE Trans. Dielectric and Electrical Insulation*, in press.

Kusaka, H. and Taya, M., 2004, *J. Composite Mater.*, **38**, 1011–1035.

Kusy, R. P., 1977, *J. Appl. Phys.*, **48**, 5301.

Kusy, R. P. and Turner, D. T., 1971, *Nature Phys. Sci.*, **229**, 58.

Lagoudas, D. C. and Zhonghe, B., 1994, *Smart Mater. Struct.*, **3**, 309–317.

Landauer, R., 1952, *J. Appl Phys.*, **23**, 779–784.

 1978, In *Proc. A.I.P. Conf. No. 40: Electrical Transport and Optical Properties of Inhomogeneous Media*, ed. J. C. Garland and D. B. Tanner, New York: American Institute of Physics, pp. 2–43.

Langmuir, S., 1938, *J. Amer. Chem. Soc.*, **57**, 1007.

Last, B. J. and Thouless, D. J., 1971, *Phys. Rev. Lett.*, **27**, 1719–1721.

Lau, J., Wong, C. P., Prince, J. L. and Nakayama, W., 1998, *Electric Packaging: Design, Materials, Process and Reliability*, Washington DC: McGraw-Hill.

Lee, S. I., 2002, *Development of Optically Controlled Microwave Devices and Artificial Materials*, Ph.D. dissertation, University of Washington.

Lee, S. W., Kuga, Y. and Mullen, R. A., 2000a, *Microwave and Optical Technol. Lett.*, **27**, 1.

Lee, S. W., Kuga, Y. and Savrun, E., 2000b, *IEEE AP-S Symposium*, Salt Lake City, Utah

Le Guilly, M., 2004, *Development of Ionic Polymer Actuator Arrays*, Ph.D. dissertation, University of Washington.

Le Guilly, M., Uchida, M. and Taya, M., 2002, In *Proc. SPIE Symp. on Electroactive Polymer Actuators and Devices*, San Diego, March 17–21, 2002, **4685**, pp. 76–84.

Le Guilly, M., Xu, C., Cheng, V., Taya, M., Opperman, L. and Kuga, Y., 2003, In *Proc. SPIE Symp. on Electroactive Polymers and Devices*, ed. Y. Bar-Cohen, San Diego, CA, March 3–6, 2003, **5051**, pp. 362–371.

Levinstein, M. E., Shur, M. S. and Efros, A. L., 1976, *Sov. Phys., JETP*, **42**, 1120–1124.

Li, J. F., Takagi, K., Ono, M., Pan, W., Watanabe, R., Almajid, A. and Taya, M., 2003, *J. Amer. Ceram. Soc.*, **86**(7), 1094–1098.

Li, J. Y., 2000, *J. Mech. Phys. Solids*, **48**, 529–552.

Li, J. Y. and Dunn, M. L., 1998, *Phil. Mag.*, **A77** (5), 1341–1350.

Li, J. Y., Dunn, M. L. and Ledbetter, H. M., 1999, *J. Appl. Phys.*, **86** (8), 4626–4634.

Li, L. and Morris. J. E., 1997, *IEEE Trans. Comp. Packaging, Manufac. Tech.*, A, **20**(1), 3–8.

Liang, Y., Kato, H. and Taya, M., 2000, In *Proc. Plasticity '00: 8th Intl. Symp. on Plasticity and Current Applications, Plastic and Viscoplastic Response of Materials and Metal Forming*, pp. 193–195.

Liang, Y., Kato, H., Taya, M. and Mori., T., 2001, *Scripta Materia*, **45**(5), 569–574.

Liang, Y., Taya, M. and Kuga, Y., 2003, In *Proc. SPIE Symp. on Aerospace Applications*, ed. E. White, San Diego, CA, March 2–6, 2003, **5054**, pp. 45–52.

Lim, S. H., Kim, S. R., Kang, S. Y., Park, J. K., Nam, J. T. and Son, D., 1999, *J. Magnetism and Magnetic Mater.*, **191**, 113–121.

Litman, A. M. and Fowler, N. E., 1981, *36th Ann. Tech. Conf. Soc. Plas. Ind.*, **20-E**, 1–4.

Lobb, C. J. and Frank, D. J., 1979, *J. Phys. C*, **12**, L827–L830.

Lorenz, L., 1880, *Wiedemaunsche Aunaleu*, 11:70.

Lorentz, H. A., 1909, *The Theory of Electrons and Its Applications to the Phenomena of Light and Radiant Heat*, Leipzig: B. G. Teubner (reprint, 1952, New York: Dover).

Madou, M. J., 2002, *Fundamentals of Microfabrication*, 2nd edn, CRC Press, LLC.

Makino, A., Hatannai, T., Inoue, A. and Masumoto, T., 1997, *Mater. Sci. Eng.*, **A226–A228**, 594.

Makino, A., Inoue, A. and Masumoto, T., 1995, *Mater. Trans. JIM*, **36**, 924.

Makino, A., Suzuki, K., Inoue, A. and Masumoto, T., 1994, *Mater. Sci. Eng.*, **A179/A180**, 127.

Malliaris, A. and Turner, D. T., 1971, *J. Appl. Phys.*, **42**, 614–618.

Mani, S. S., Fleming, J. G., Sniegowski, J. J., *et al.* 2000, *Met. Res. Soc. Symp.*, **605**, 135–140.

Mansuripur, M and Li, L., 1997, *Optics and Photonic News*, **8**(5), 50–55.

Mao, M. Y., Wang, T. P., Xie, J. F. and Wang, W. Y., 1995, In *Proc. IEEE Micro. Electro. Mechanical Systems,* Amsterdam, Netherlands, pp. 392–393.

Matsumoto, M. and Miyata, Y., 1997, *IEEE Trans. Magnetics*, **33** (6), 4459–4464.
  1999a, *IEEE Trans. Dielectrics and Electrical Insulation*, **6** (1), 27–34.
  1999b, *NTT R & D*, **48** (3), 343–348.
  2002, *J. Apply. Phys.*, **91** (21), 9635–9637.

Maxwell, J. C., 1904, *Treatise on Electricity and Magnetism, Vol. 1*, 3rd edn, Oxford: Clarendon Press, p. 440.

McKnight, G. P. and Carman, G. P., 2001, In *Proc. SPIE Symp. on Smart Materials and Structures; Active Materials; Behavior and Mechanics*, ed. C. S. Lynch, Newport Beach CA, March 2001, **4333**, pp. 178–183.

McLachlan, D. S., Blaszkiewicz, M. and Newnham, R. E., 1990, *J. Am. Ceram. Soc.*, **73** (8), 2187–2203.

McLaughlin, R., 1990, *Intl. J. Eng. Sci.*, **15**, 237–244.

Mello, L. D. and Kubota, L. T., 2002, *Food Chemistry*, **77**, 237–256.

Mikata, Y., 2000, *Intl. J. Eng. Sci*, **38**, 605–641.
  2001, *Intl. J. Eng. Sci.*, **38**, 7045–7063.

Moffett, M. B., Clark, A. E., Wun-Fogle, M., Jinberg, J., Teter, J. P. and McLaughlin, E. A., 1991, *J. Acoust. Soc. Amer.*, **89** (3), 1448–1455.

Mori, T. and Tanaka, K., 1973, *Acta Metall.*, **21**, 571–574.

Morita, M., Ochiai, K., Kamohara, H., Arima, I., Aisaka, T. and Horie, H., 1985, In *Recent Advances in Composites in USA and Japan*, eds. J. R. Vinson and M. Taya, ASTM STP 864, pp. 401–409.

Mossotti, O. F., 1850, *Memorie di Matematica edi Fisica della Societa Italiana delle Scienze Residente in Modena*, Pt. 2, 24:49–74.

Mukherjee, A. K., Bird, J. E. and Dorn, J. E., 1969, *Trans. ASM*, **62**, 155–179.

Mura, T., 1987, *Micromechanics of Defects in Solids*, 2nd edn, Dordrecht: Martinus Nijhoff Publishing.

Murray, S. J., Marioni, M., Allen, S. M., O'Handley, R. C. and Lagrasso, T. A., 2000, *Appl. Phys. Lett.*, **77**, 886–888.

Murray, S. J., O'Handley, R. C. and Allen, S. M., 2001, *J. Appl. Phys.*, **89**, 1295–1301.

Musick, M. D., Keating, C. D., Keefe, M. H. and Natan, M. J., 1997, *Chem. Mater.*, **9**, 1499–1501.

Narita, K., 1990, *Percolation Model and Mechanical Behavior of Short Fiber Polymer Composites,* M.Sc Thesis, Tohoku University, Japan.

Nathanson, H. C., Newell, W. E., Davis J. R., Jr. and Wickstrom, W. E., 1967, *IEEE Trans., Electron Devices*, **ED4**, 117.

Nemat-Nasser, S., 2004, *A Treatise on Finite Deformations of Heterogeneous Inelastic Solids*, Cambridge: Cambridge University Press.

Nemat-Nasser, S. and Taya, M., 1980, *Intl. J. Solids Structs.*, **16**, 483–494.

Nersessian, N. and Carman, G. P., 2001, In *Proc. SPIE on Smart Materials and Structures; Active Materials; Behavior and Mechanics*, ed. C. S. Lynch, Newport Beach, CA, March 2001, **4333**, pp. 166–177.

Newnham, R. E., Skinner, D. P. and Cross, L. E., 1978, *Mater. Res. Bull.*, **13**, 525–536.

Nicolson, A. M. and Ross, G. E., 1970, *IEEE Trans. Instrum. Meas.*, **IM-19**, 377–382.

Nielsen, L. E., 1974, *Ind. Eng. Chem. Fundam.*, **13**, 17–20.

O'Handley, R. C., 1998, *J. Appl. Phys.*, **83**, 3263–3270.

O'Handley, R. C., Allen, S. M., Paul, D. I., Henry, C. P., Marioni, M., Bono, D., 2003, *In Proc. SPIE Smart Structures and Materials*, **5053**, 200–206.

Onizuka, K., 1975, *J. Phys. Soc. Japan*, **39**, 527–535.

Osborn, J. A., 1945, *Phy. Rev.*, **67** (11–12), 351–357.

Otto, A., 1968, *Z. Physik*, **216**, 398–410.

Pagano, N. J., 1969, *J. Comp. Mater.*, **3**, 398–411.

Park, J. J. and Taya, M., 2003,In *Proc. ICCM-14*, San Diego, July 15, 2003.

Pendry, J. B., Holden, A. J., Robbins, D. J. and Stewart, W. J., 1998, *IEEE Trans. Microwave Theory Tech.*, **47**, 4785.

Pendry, J. B., Holden, A. J., Stewart, W. J. and Youngs, I., 1996, *Phys. Rev Lett.*, **76**, 4773.

Petrou, P. S., Mosei, I. and Jobst, G., 2002, *Biosensors and Bioelectronics*, **17**, 859–865.

Pike, G. E., 1978, *Electrical Transport and Optical Properties of Inhomogeneous Media*, eds. J. C. Garland and D. B. Tanner, New York: Amer. Inst. Phys., p. 366.

Pike, G. E. and Seager, C. H., 1974, *Phys. Rev. B.*, **10**, 1421–1434.

Polla, D. L., Erdman, A. G., Robbins, W. P., *et al.*, 2000, *Ann. Rev. Biomed. Eng.*, **2**, 551–576.

Popovic, S., Tamagawa, H., Taya, M., and Xu, C., In *Proc. SPIE, Electroactive Polymers*, March 2001, Newport Beach CA, **4329**, pp. 238–247.

Post, D., 1987, "Moiré interferametry", Chap. 7 of *Handbook on Expermiental Mechanics*, 2nd edn, ed. A. S. Kobayashi.

Post, D. and Wood, J., 1993, *Experimental Mechanics*, **29** (3), 18–20.

Post, D., Han, B. and Ifju, P., 1994, Experimental Analysis for Mechanics and Materials, New York: Springer-Verlag.

Pozar, D., 1998, *Microwave Engineering*, 2nd edn, New York: John Wiley & Sons Inc.

Pramanik, P. K., Khastgir, D., De, S. K. and Saha, T. N., 1990, *J. Mater. Sci.*, **25**, 3848.

Prandtl, L. 1924, In *Proc. 1st Int. Cong, Appl. Mech.*, Delft, eds. C. B. Biezeno and J. M. Burgers, pp. 43–54.

Reuss, A., 1929, *Zeitschrift fur angewandte Mathematik und Mechanik*, **9**, 49–58.

Reynolds, P. J., Klein, W. and Stanley, H. E., 1977, *J. Phys. C*, **10**, L167–L172.

Reynolds, P. J., Stanley, H. E. and Klein, W., 1978, *J. Phys. A*, **11**, L199–L207. 1980, *Phys. Rev. B*, **21**, 1223–1245.

Rich, R. L. and Myszka, D. G., 2002, *J. Molecular Recognition*, **15**, 352–376.

Richards, M. R., 1996, *Process Development for IrAl Coated SiC–C Functionally Graded Materials for Oxidation Protection of Graphite*, Ph.D. dissertation, University of Washington.

Sahini, M., 1994, *Applications of Percolation Theory*, London: Taylor & Francis.

Sakamoto, T., Fukuda, T., Kakeshita, T., Takeuchi, T. and Kishio, K., 2003, *J. Appl. Phys.*, **93**, 8647–8649.

Sandlund, S., Fahlander, M., Cedell, T., Clark, A. E., Resorff, J. B. and Fogle, M. W., 1994, *J. Appl. Phys.*, **75** (10), 5656–5658.

Sasagawa, K., Nakamura, N., Saka, M. and Abé, H., 1998, *J. Electronic Packaging*, Trans. ASME, **120**, 360–366.

Sasagawa, K., Hasegawa, M., Saka, M. and Abé, H., 2000, *Theor. Appl. Fract. Mech.*, **33**, 67–72.

2002a, *J. Appl. Phys.*, **91** (4), 1882–1890.

2002b, *J. Appl. Phys.*, **91** (11), 9005–9014.

Sasagawa, K., Naito, K., Saka, M. and Abé, H., 1999a, *J. Appl. Physics*, **86** (11), 6043–6051

1999b, *Advances in Electronic Packaging*, ASME EEP, **26**–1, 233–238.

Schedel-Niedrig, T., Sotobayashi, H., Ortega-Villamil, A. and Bradshaw, A. M., 1991, *Surface Sci.*, **247** (2–3), 83–89.

Shante, V. K. S. and Kirkpatrick, S., 1971, *Adv. Phys.*, **20**, 325–357.

Shibata, T., Kitamoto, Y., Unmo, K. and Makino, E., 2000, *J. MEMS*, **9** (1), 47–51.

Shiozawa, N., Isaka, K. and Ohta, T., 1995, *J. Electronics and Manufacturing*, **5** (1), 33–37.

Smith, D. R., Padilla, W. J., Vier, D. C., Nemat-Nasser, S. C. and Schultz, S., 2000, *Phys. Rev. Lett.*, **84** (1), 4184–4187.

Smith, W. A., 1989, In *Proc. IEEE Symp. on Ultrasonics*, October 3–6, 1989, 755–766.

1990, In *Proc. IEEE Symp on Ultrasonics*, December 4–7,1990, **2**, 757–761.

1993, *IEEE Trans. Ultrasonics, Ferroelectrics and Frequency Control*, **40**, 41.

Smith, W. A., and Auld, B. A., 1991, *IEEE Trans. Ultrasonics, Ferroelectrics and Frequency Control*, **38** (1), 40.

Smith, W. A., Shaulov, A. A. and Auld, B. A., 1985, In *Proc. IEEE Symp. on Ultrasonics*, pp. 642–647.

Sniegowski, J. J., 2001, *Interpack '01*, July 8–13, Hawaii. Keynote Talk: *Materials*

Sniegowski, J. J and de Boer, M. P, 2000, *Ann. Rev. Mater. Sci.*, **30**, 299–333.

Sohmura, T., Oshima, R. and Fujita, F. E., 1980, *Scripta Metall.*, **14**, 855–856.

Sozinov, A. Likhachev, A. A. and Ullakko, K., 2001, In *Proc. SPIE on Smart Structures and Materials*, **4333**, 189–196.

Stanley, H. E., 1971, *Introduction to Phase Transitions and Critical Phenomena*, Oxford: Clarendon Press.

Stauffer, D., 1979, *Phys. Rep.*, **54**, 1–74.

1981, In *Lecture Notes in Phys. No. 149: Disordered Systems and Localization*, ed. C. Castellani *et al.*, Berlin: Springer-Verlag, pp. 9–25.

1985, *Introduction to Percolation Theory*, London and Philadelphia: Taylor & Francis.

Stauffer, D. and Aharony, A., 1991, *Introduction to Percolation Theory*, London: Taylor and Francis.

Stoner, E. C., 1945, *Phil. Mag., ser. 7*, **36**, 803–821.

Straley, J. P., 1977, *Phys. Rev. B.*, **15**, 5733–5737.

1978, In *A.I.P. Conf. Proc. No. 40: Electrical Transport and Optical Properties of Inhomogeneous Media*, ed. J. C. Garland and D. B.Tanner, New York: American Institute of Physics, pp. 118–126.

Sullivan, J. P, Friedman, T. A., de Boer, M. P., *et al.*, 2001, *Mater. Res. Soc. Symp. Proc.*, **657**, EE7.1.1–1.9

Suzuki, K., Kataoka, N., Inoe, A., Makino, A. and Masumoto, T., 1990, *Mater. Trans. JIM*, **31**, 743.

Sykes, M. F. and Essam, J. W.,1964, *Phys. Rev.*, **133**, A310–315.

Sykes, M. F., Gaunt, D. S. and Glen, M., 1976a, *J. Phys.*, **A9**, 97.

1976b, *J. Phys.*, **A9**, 1705–1712.

Takagi, K., Li, J. F., Yokoyama, S., Watanabe, R., Almajid, A. and Taya, M., 2002, *Advanced Mater.*, **3**, 217–224.

Takei, T., Hatta, M. and Taya, M., 1991a, *Mater. Sci. Eng.*, **A131** 133–143.
 1991b, *Mater. Sci. Eng.*, **A131**, 145–152.
Takao, Y. and Taya, M., 1985, *J. Appl. Mech.*, **52**, 806–810.
Takao, Y. Chou, T. W. and Taya, M., 1982, *J. Appl. Mech.*, **49**(3), 536–540.
Taya, M., 1995, *J. Eng. Mater. Tech.*, **117**, 462–469.
Taya, M., Almajid, A., Dunn, M. and Takahashi, F., 2003a, *Sensors and Actuators*, **A107**, 248–260.
Taya, M. and Arsenault, R. J., 1989, *Metal Matrix Composites: Thermo-Mechanical Behavior*, Oxford: Pergamon Press.
Taya, M., Kim, W. J. and Ono, K., 1998, *Mech. Mater.*, **28**, 53–59.
Taya, M. and Mori. T., 1994, *J. Eng. Mater. Tech.*, **116**, 408–413.
Taya, M. and Ueda, N., 1987, *J. Eng. Mater. Tech.*, **109**, 252–256.
Taya, M., Wada, T., Lee, C. C. and Kusaka, M., 2003b, In *Proc. SPIE Symp.* on Applications, ed. E. White, San Diego, CA, March 2–6, 2003, **5054**, pp. 156–164.
Thompson, W. (Lord Kelvin), 1856, *Proc. Roy. Soc. London*, **146**, 649–751.
Tran, D. K., Kobayashi, A. S. and White, K. W., 1999, *Experimental Mechanics*, **29** (1), 20–24.
Tsai, S. W. and Hahn, T., 1980, *Introduction to Composite Materials*, Westport CT: Technomic Publication.
Tsutsui, T., 1960, *Experiments in Applied Physics*, Tokyo: Tokyo University Press.
Uchida, M. and Taya, M., 2001, *Polymer*, **42**, 9281–9285.
Uchino, K., 1998, *Acta Materia*, **46**, 3475.
Uchino, K., Yoshizaki, M., Kasai, K., Yamamura, H., Sakai, N. and Asakura, H., 1987, *Japan. J. Appl. Phys.*, **26**, 1046–1049.
Ueda, N. and Taya, M., 1986, *J. Appl. Phys.*, **60** (1), 459–461.
Ullakko, K., Huang, J. K., Kantner, C., O'Handley, R. C. and Kokorin, V. V., 1996, *Appl. Phys. Lett.*, **69**, 1966–1968.
Vaidya, S., Sheng, T. T. and Sinha, A. K., 1980, *Appl. Phys. Lett.*, **36**, 464–466.
Voigt, W., 1889, *Annal. Physik*, **38**, 573–587.
Vyssotsky, V. A., Gordon, S. B., Frisch, H. L. and Hammersley, J. M., 1961, *Phys. Rev.*, **123**, 1566–1567.
Wada, T., Liang, Y., Kato, H., Tagawa, T., Taya, M. and Mori, T., 2003a, *Mat. Sci. Eng. A*, **361**, 75–82.
Wada, T., Taya, M., Chen, H. H. and Kusaka, M., 2003b, In *Proc. SPIE Symp. Applications*, ed. E. White, San Diego, CA, March 2–6, 2003, **5054**, pp. 125–134.
Warfel, R. H., 1980, *35th Ann. Tech. Conf. Soc. Plas. Ind.*, **19-E**, 1–6.
Watanabe, I., Takemura, K., Shiozawa, N., Watanabe, O., Kojima, K. and Hirosawa, Y., 1996, *Hitachi-Kasei Tech. Report*, **26**, 13–16.
Watson, B. P. and Leath, P. L., 1974, *Phys. Rev. B.*, **9**, 4893–4896.
Webman, I., Jortner, J. and Cohen, M. H., 1975, *Phys. Rev. B.*, **11**, 2885–2892.
 1976, *Phys. Rev. B.*, **14**, 4737–4740.
 1977, *Phys. Rev. B.*, **15**, 5712–5713.
Webster, P. J., Ziebeck, K. R. A., Town, S. L. and Peak, M. S., 1984, *Phil. Mag. B.*, **49**, 295.
Whitesides, G. M., Methias, J. P. and Seto, C. T., 1991, *Science*, **254**, 1312–1319.
Wu, C. C., Khau, M. and May, W., 1996, *J. Amer. Ceram. Soc.*, **79**, 809.
Yamada, K. and Kamiya, N., 1995, *Mater. Res. Soc. Symp. Proc.*, **365**, 113.
Yamamoto, T., Taya, M., Sutou, Y., Liang, Y., Wada, T. and Sorensen, L., 2004, *Acta Materia.*, **52**, 5083–5091.
Yoshizawa, Y., Oguma, S. and Yamauchi, K., 1988, *J. Appl. Phys.*, **64**, 6044.

Yuge, Y, 1977, *J. Stat. Phys.*, **16** (4), 339–348.

Yuge, Y. and Onizuka, K., 1978, *J. Phys. C*, **11**, L763–L765.

Zhang, Z., Chen, Q., Knoll, W., Foerch, R., Holcomb, R. and Roitman, D., 2003, *Macromolecules*, **26**, 7689–7694.

Zhu, X. and Meng, Z., 1995, *Sensors and Actuators*, **A48**, 169–176.

Ziaie, B., Baldi, A., Lei, M., Gu, Y. and Siegel, R. A., 2004, *Adv. Drug Delivery Rev.*, **56**, 145–172.

Zweben, C. Z., 1995, "The future of advanced composite electronic packaging", Chap. 6 of *Materials for Electronic Packaging*, ed. D. D. Chung, Oxford: Butterworth–Heinemann, pp. 127–171.

Author index

Subject index

357